# More Than Just a Textbook

## Internet Resources

**Step 1**  Connect to **Math Online** glencoe.com

**Step 2**  Connect to online resources by using **QuickPass** codes. You can connect directly to the chapter you want.

**IM7031c1** — Enter this code with the appropriate chapter number.

### For Students

Connect to **StudentWorks Plus Online** which contains all of the following online assets. You don't need to take your textbook home every night.

- Personal Tutor
- Chapter Readiness Quizzes
- Multilingual eGlossary
- Concepts in Motion
- Chapter Test Practice
- Test Practice

### For Teachers

Connect to professional development content at glencoe.com and **Advance Tracker** at AdvanceTracker.com

### For Parents

Connect to glencoe.com for access to **StudentWorks Plus Online** and all the resources for students and teachers that are listed above.

Glencoe McGraw-Hill

NSF

# IMPACT
## Mathematics

Glencoe

COURSE
1

New York, New York   Columbus, Ohio   Chicago, Illinois   Woodland Hills, California

## About the Cover

Artifacts of a bowling-like sport can be traced to 3200 B.C. Today, 100 million, or $10^8$, people in over 90 countries enjoy bowling, or ten pins, as a recreational activity.

Bowling is played by rolling a ball, or bowl, down a narrow track. The goal is to knock down ten pins, which are placed in a triangular arrangement. A game consists of ten frames. A bowler is allowed two balls per frame, with an opportunity for two bonus balls at the end of the game. Applying the sport's rules, a perfect score of 300 is achieved when a player knocks down all ten pins with each roll.

Typically, three games are played in competitive situations. Suppose a player's average, or mean, score is 240. If the player rolled games of 230 and 235, the player would need to roll a game of 255 to achieve the mean score.

These materials include work supported in part by the National Science Foundation under Grant No. ESI-9726403 to MARS (Mathematics Assessment Resource Service). Any opinions, findings, and conclusions or recommendations expressed in this material are those of the authors and do not necessarily reflect the views of the funding agencies. For more information on MARS, visit http://www.nottingham.ac.uk/education/MARS.

*The McGraw-Hill Companies*

## Macmillan/McGraw-Hill
## Glencoe

The algebra content for *IMPACT Mathematics* was adapted from the series *Access to Algebra*, by Neville Grace, Jayne Johnston, Barry Kissane, Ian Lowe, and Sue Willis. Permission to adapt this material was obtained from the publisher, Curriculum Corporation of Level 5, 2 Lonsdale Street, Melbourne, Australia.

Send all inquiries to:
Glencoe/McGraw-Hill
8787 Orion Place
Columbus, OH 43240-4027

ISBN: 978-0-07-888703-1
MHID: 0-07-888703-8

Printed in the United States of America.

3 4 5 6 7 8 9 10 QDB 17 16 15 14 13 12 11 10

# Contents in Brief

**Focal Points and Connections**
See pages vi and vii for key.

# Principal Investigator

**Faye Nisonoff Ruopp**
Brandeis University
Waltham, Massachusetts

# Consultants and Developers

## Consultants

**Frances Basich Whitney**
Project Director, Mathematics K–12
Santa Cruz County Office of Education
Santa Cruz, California

**Robyn Silbey**
Mathematics Content Coach
Montgomery County Public Schools
Gaithersburg, Maryland

**Dr. Selina Vásquez Mireles**
Associate Professor of Mathematics
Texas State University–San Marcos
San Marcos, Texas

**Teri Willard**
Assistant Professor
Central Washington University
Ellensburg, Washington

## Special thanks to:

**Peter Braunfeld**
Professor of Mathematics Emeritus
University of Illinois

**Sherry L. Meier**
Assistant Professor of Mathematics
Illinois State University

**Judith Roitman**
Professor of Mathematics
University of Kansas

## Developers

**Senior Project Director**
Cynthia J. Orrell

**Senior Curriculum Developers**
Michele Manes, Sydney Foster, Daniel Lynn Watt, Ricky Carter, Joan Lukas, Kristen Herbert

**Curriculum Developers**
Haim Eshach, Phil Lewis, Melanie Palma, Peter Braunfeld, Amy Gluckman, Paula Pace

**Special Contributors**
Elizabeth D. Bjork, E. Paul Goldenberg

# Project Reviewers

Glencoe and Education Development Center would like to thank the curriculum specialists, teachers, and schools who participated in the review and testing of the first edition of *IMPACT Mathematics*. The results of their efforts were the foundation for this second edition. In addition, we appreciate all of the feedback from the curriculum specialists and teachers who participated in review and testing of this edition.

**Debra Allred**
Math Teacher
Wiley Middle School
Leander, Texas

**Tricia S. Biesmann**
Retired Teacher
Sisters Middle School
Sisters, Oregon

**Kathryn Blizzard Ballin**
Secondary Math Supervisor
Newark Public Schools
Newark, New Jersey

**Linda A. Bohny**
District Supervisor of Mathematics
Mahwah Township School District
Mahwah, New Jersey

**Julia A. Butler**
Teacher of Mathematics
Richfield Public School Academy
Flint, Michigan

**April Chauvette**
Secondary Mathematics Facilitator
Leander ISD
Leander, Texas

**Amy L. Chazaretta**
Math Teacher/Math Department Chair
Wayside Middle School, EM-S ISD
Fort Worth, Texas

**Franco A. DiPasqua**
Director of K–12 Mathematics
West Seneca Central
West Seneca, New York

**Mark J. Forzley**
Junior High School Math Teacher
Westmont Junior High School
Westmont, Illinois

**Virginia G. Harrell**
Education Consultant
Brandon, Florida

**Lynn Hurt**
Director
Wayne County Schools
Wayne, West Virginia

**Andrea D. Kent**
7th Grade Math & Pre-Algebra
Dodge Middle School, TUSD
Tucson, Arizona

**Russ Lush**
6th Grade Teacher & Math Dept. Chair
New Augusta–North
Indianapolis, Indiana

**Katherine V. Martinez De Marchena**
Director of Education 7–12
Bloomfield Public Schools
Bloomfield, New Jersey

**Marcy Myers**
Math Facilitator
Southwest Middle School
Charlotte, North Carolina

**Joyce B. McClain**
Middle School Mathematics Consultant
Hillsborough County Schools
Tampa, Florida

**Suzanne D. Obuchowski**
Math Teacher
Proctor School
Topsfield, Massachusetts

**Michele K. Older**
Mathematics Instructor
Edward A. Fulton Jr. High
O'Fallon, Illinois

**Jill Plattner**
Math Program Developer (Retired)
Bend La Pine School District
Bend, Oregon

**E. Elaine Rafferty**
Retired Math Coordinator
Summerville, South Carolina

**Karen L. Reed**
Math Teacher—Pre-AP
Chisholm Trail Intermediate
Fort Worth, Texas

**Robyn L. Rice**
Math Department Chair
Maricopa Wells Middle School
Maricopa, Arizona

**Brian Stiles**
Math Teacher
Glen Crest Middle School
Glen Ellyn, Illinois

**Nimisha Tejani, M.Ed.**
Mathematics Teacher
Kino Jr. High
Mesa, Arizona

**Stefanie Turnage**
Middle School Mathematics
Grand Blanc Academy
Grand Blanc, Michigan

**Kimberly Walters**
Math Teacher
Collinsville Middle School
Collinsville, Illinois

**Susan Wesson**
Math Teacher/Consultant
Pilot Butte Middle School
Bend, Oregon

**Tonya Lynnae Williams**
Teacher
Edison Preparatory School
Tulsa, Oklahoma

**Kim C. Wrightenberry**
Math Teacher
Cane Creek Middle School
Asheville, North Carolina

**Focal Points**

The Curriculum Focal Points identify key mathematical ideas for this grade. They are not discrete topics or a checklist to be mastered; rather, they provide a framework for the majority of instruction at a particular grade level and the foundation for future mathematics study. The complete document may be viewed at www.nctm.org/focalpoints.

**KEY**

**G6-FP1**
Grade 6 Focal Point 1

**G6-FP2**
Grade 6 Focal Point 2

**G6-FP3**
Grade 6 Focal Point 3

**G6-FP4C**
Grade 6 Focal Point 4
Connection

**G6-FP5C**
Grade 6 Focal Point 5
Connection

**G6-FP6C**
Grade 6 Focal Point 6
Connection

**G6-FP1**    Number and Operations: **Developing an understanding of and fluency with multiplication and division of fractions and decimals**

Students use the meanings of fractions, multiplication and division, and the inverse relationship between multiplication and division to make sense of procedures for multiplying and dividing fractions and explain why they work. They use the relationship between decimals and fractions, as well as the relationship between finite decimals and whole numbers (i.e., a finite decimal multiplied by an appropriate power of 10 is a whole number), to understand and explain the procedures for multiplying and dividing decimals. Students use common procedures to multiply and divide fractions and decimals efficiently and accurately. They multiply and divide fractions and decimals to solve problems, including multistep problems and problems involving measurement.

**G6-FP2**    Number and Operations: **Connecting ratio and rate to multiplication and division**

Students use simple reasoning about multiplication and division to solve ratio and rate problems (e.g., "If 5 items cost $3.75 and all items are the same price, then I can find the cost of 12 items by first dividing $3.75 by 5 to find out how much one item costs and then multiplying the cost of a single item by 12"). By viewing equivalent ratios and rates as deriving from, and extending, pairs of rows (or columns) in the multiplication table, and by analyzing simple drawings that indicate the relative sizes of quantities, students extend whole number multiplication and division to ratios and rates. Thus, they expand the repertoire of problems that they can solve by using multiplication and division, and they build on their understanding of fractions to understand ratios. Students solve a wide variety of problems involving ratios and rates.

**G6-FP3**    Algebra: **Writing, interpreting, and using mathematical expressions and equations**

Students write mathematical expressions and equations that correspond to given situations, they evaluate expressions, and they use expressions and formulas to solve problems. They understand that variables represent numbers whose exact values are not yet specified, and they use variables appropriately. Students understand that expressions in different forms can be equivalent, and they can rewrite an expression to represent a quantity in a different way (e.g., to make it more compact or to feature different information). Students know that the solutions of an equation are the values of the variables that make the equation true. They solve simple one-step equations by using number sense, properties of operations, and the idea of maintaining equality on both sides of an equation. They construct and analyze tables (e.g., to show quantities that are in equivalent ratios), and they use equations to describe simple relationships (such as $3x = y$) shown in a table.

## Connections to the Focal Points

**G6-FP4C**  **Number and Operations:** Students' work in dividing fractions shows them that they can express the result of dividing two whole numbers as a fraction (viewed as parts of a whole). Students then extend their work in grade 5 with division of whole numbers to give mixed number and decimal solutions to division problems with whole numbers. They recognize that ratio tables not only derive from rows in the multiplication table but also connect with equivalent fractions. Students distinguish multiplicative comparisons from additive comparisons.

**G6-FP5C**  **Algebra:** Students use the commutative, associative, and distributive properties to show that two expressions are equivalent. They also illustrate properties of operations by showing that two expressions are equivalent in a given context (e.g., determining the area in two different ways for a rectangle whose dimensions are $x + 3$ by 5). Sequences, including those that arise in the context of finding possible rules for patterns of figures or stacks of objects, provide opportunities for students to develop formulas.

**G6-FP6C**  **Measurement and Geometry:** Problems that involve areas and volumes, calling on students to find areas or volumes from lengths or to find lengths from volumes or areas and lengths, are especially appropriate. These problems extend the students' work in grade 5 on area and volume and provide a context for applying new work with equations.

Reprinted with permission from *Curriculum Focal Points for Prekindergarten through Grade 8 Mathematics: A Quest for Coherence,* copyright 2006, by the National Council of Teachers of Mathematics. All rights reserved.

# Table of Contents

**Focal Points and Connections**
See pages vi and vii for key.

**G6-FP1**

**G6-FP6C**

COURSE 1

**②** **Fractions and Decimals**................................ **56**

Focal Points
and Connections
See pages vi and vii
for key.

**G6-FP1**

**G6-FP4C**

**Focal Points and Connections**
See pages vi and vii for key.

**G6-FP3**

**G6-FP5C**

Focal Points and Connections
See pages vi and vii for key.

G6-FP2

Focal Points
and Connections
See pages vi and vii
for key.

**G6-FP1**

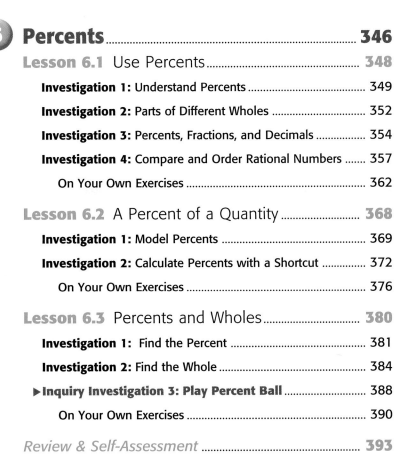

Focal Points
and Connections
See pages vi and vii
for key.

G6-FP1

G6-FP6C

# ⑩ Data and Probability ................................ 576

Focal Points
and Connections
See pages vi and vii
for key.

G6-FP2

G6-FP4C

# Polygons, Angles, and Circles

## Real-Life Math

**Geometry in Sports** Geometric figures are used in many aspects of everyday life, including sporting events. The shapes of soccer pitches, volleyball courts, and baseball diamonds are quadrilaterals. Spherical tennis balls are packaged in cylindrical cans. Skateboarding aerials and figure skating jumps are often named according to the number of revolutions performed by the athlete.

**Think About It** List as many other sports-related geometry examples as you can.

### Contents in Brief

**Math Online**
Take the **Chapter Readiness Quiz** at glencoe.com.

# Dear Family,

Mathematics has been called the "science of patterns." Recognizing and describing patterns and using patterns to make predictions are important mathematical skills.

The class will begin by looking for patterns in geometry.

## Key Concept—Polygons and Circles

Polygons are flat, two-dimensional, geometric figures that have the following characteristics.

- They are made of line segments.
- Each segment touches exactly two other segments, one at each endpoint.

The class will classify polygons by the number of their sides. A few polygons are shown below.

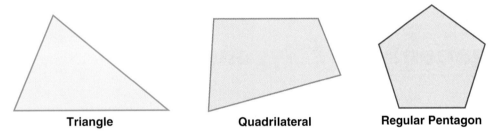

| **Triangle** | **Quadrilateral** | **Regular Pentagon** |

The class will explore circles. Even though circles are two-dimensional figures, they are not made of line segments. So, circles are not classified as polygons.

## Chapter Vocabulary

| | |
|---|---|
| **angle** | **perpendicular** |
| **circumference** | **polygon** |
| **concave polygon** | **radius** |
| **diameter** | **regular polygon** |
| **line symmetry** | **right angle** |
| **perimeter** | **vertex** |

## Home Activities

- Go on a family walk. Identify different-shaped objects around your home, in a neighborhood park, or along a city street.
- Look at a building or house and discuss the figures that you see.

# Patterns in Geometry

In this lesson, you will work with two-dimensional geometric figures. You will classify polygons and find angle measures.

**Explore**

How many squares are in this design? (Hint: The answer is more than 16.)

## Investigation 1 Polygons

**Vocabulary**

**polygon**

**vertex**

**Polygons** are flat, two-dimensional geometric figures that have these characteristics.

- They are made of line segments.
- Each segment touches exactly two other segments, one at each of its endpoints.

These shapes are polygons.

These shapes are not polygons.

  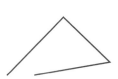 

**Think & Discuss**

Look at the shapes above that are not polygons. Explain why each of these shapes does not fit the definition of a polygon.

Polygons can be classified according to the number of sides they have. You have probably heard many of these names.

| Name | Sides | Examples |
|---|---|---|
| Triangle | 3 | |
| Quadrilateral | 4 | |
| Pentagon | 5 | |
| Hexagon | 6 | |
| Heptagon | 7 | |
| Octagon | 8 | |
| Nonagon | 9 | |
| Decagon | 10 | |

Most polygons with more than ten sides have no special name. A polygon with 11 sides is described as an *11-gon*, a polygon with 12 sides is a *12-gon*, and so on. Each of the polygons below is a 17-gon.

Each corner of a polygon, where two sides meet, is called a **vertex**. The plural of vertex is *vertices*. Labeling vertices with capital letters makes it easy to refer to a polygon by name.

## ─Example

This figure can be seen as two triangles and one quadrilateral.

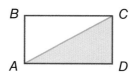

To name one of the polygons in the figure, list its vertices in order as you move around it in either direction. One name for the white triangle is △*ABC*. Other names are possible, including △*BCA* and △*ACB*. One name for the green triangle is △*ADC*.

The quadrilateral in the figure could be named quadrilateral *ABCD*, or *BCDA*, or *DCBA*, or *DABC*. All of these names list the vertices in order as you move around the quadrilateral. The name *ACBD* is *not* correct.

**Math Link**

A *diagonal* is a segment that connects two vertices of a polygon but is not a side of the polygon. In quadrilateral *ABCD,* the diagonal is $\overline{AC}$.

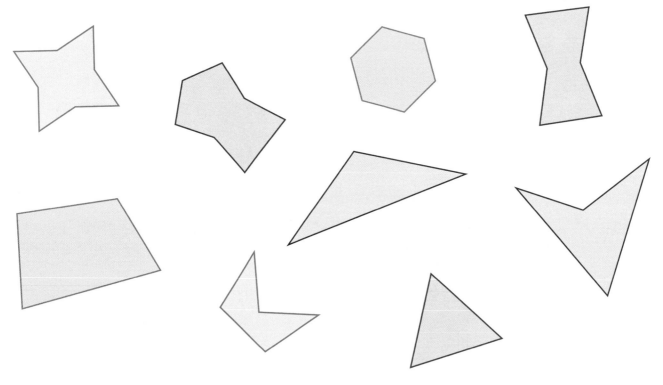

## *Real-World Link*

Polygons, such as triangles and octagons, are used for traffic signs.

. . . . . . . . . . . . . . . . . . . . . .

## ☑️ *Develop & Understand: A*

You will now search for polygons in given figures. Each figure has a total score that is calculated by adding the following.

- 3 points for each triangle
- 4 points for each quadrilateral
- 5 points for each pentagon
- 6 points for each hexagon

As you work, try to discover a systematic way to find and list all the polygons in a figure. Be careful to give only one name for each polygon.

Record your work in a table like this one, which has already been started for Exercise 1.

| Polygon | Names | Score |
|---|---|---|
| Triangle | *ABC, ADC* | 6 |
| Quadrilateral | | |
| Pentagon | | |
| Hexagon | | |
| | Total Score | |

1.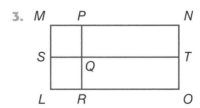

2.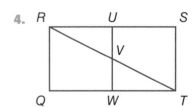

3. M  P  N / S  T / Q / L  R  O

4. R  U  S / V / Q  W  T

5. Now create your own figure that is worth at least 30 points. Label the vertices. List each of the triangles, quadrilaterals, pentagons, and hexagons in your figure.

1. Draw two polygons. Also draw two shapes that are not polygons. Explain why the shapes that are not polygons do not fit the definition of a polygon.

2. In Exercises 1–5, you had to find ways to list all the polygons in a figure without repeating any. Describe one strategy you used.

# Investigation 2 Angles

## Vocabulary

angle

## Materials

• paper polygons or pattern blocks

You probably already have a good idea about what an angle is. You may think about an angle as a rotation, or a turn, about a point. Examples include an arm bending at the elbow or hinged boards snapping shut at the start of a movie scene.

You may also think about an angle as two sides that meet at a point, like the hands of a clock or the vanes of a windmill.

Or you may think of an angle as a wedge, like a piece of cheese or a slice of pizza.

In mathematics, an **angle** is defined as two rays with the same endpoint. A ray is straight, like a line. It has an endpoint where it starts, and it goes forever in the other direction.

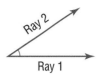

Angles can be measured in *degrees*. Below are some angles with which you may be familiar.

- The angle at the vertex of a square measures 90°. You can think of a 90° angle as a rotation $\frac{1}{4}$ of the way around a circle.

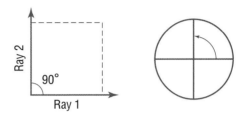

- Two rays pointing in opposite directions form a 180° angle. A 180° angle is a rotation $\frac{1}{2}$ of the way around a circle.

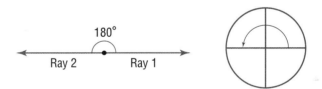

- A 360° angle is a rotation around a complete circle. In a 360° angle, the rays point in the same direction.

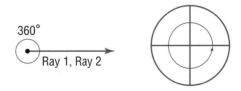

You can use 90°, 180°, and 360° angles to help estimate the measures of other angles. For example, the angle below is about a third of a 90° angle, so it has a measure of about 30°.

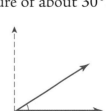

## Think & Discuss

Copies of the polygon at right can be arranged to form a star.

What is the measure of the angle that is marked in the star? How do you know?

## Develop & Understand: A

1. You will be given several copies of each polygon below. Your job is to determine the angle measures for each polygon.

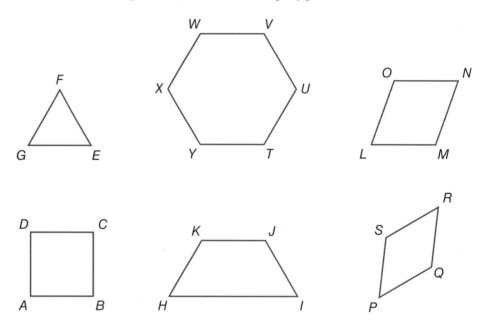

To find the measures of the angles, you can use 90°, 180°, and 360° angles as guides, and you can compare the angles of the polygons with one another.

Your answers should be a record of each vertex, *A–Y*, and the measure of the angle at that vertex. For many of the polygons, two or more of the angles are identical. So, you only have to find the measure of one of them.

You will now use the angles you found in Exercise 1 to help estimate the measures of other angles.

## Real-World Link

Think about the corners of index cards when estimating angle measures. They form approximate 90° angles.

### ✓ Develop & Understand: B

Estimate the measure of each angle. To help make your estimates, you can compare the angles to 90°, 180°, and 360° angles and to the angles of the polygons in Exercise 1. For each angle, explain how you made your estimate.

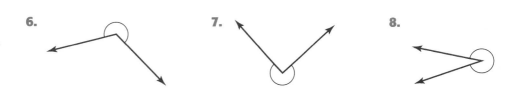

2.     3.     4.     5.

6.     7.     8.

### Share & Summarize

1. Describe how you can estimate the measure of an angle.

2. Moria said the angles below have the same measure. Hannah said Angle 2 is larger than Angle 1. Who is correct? Explain.

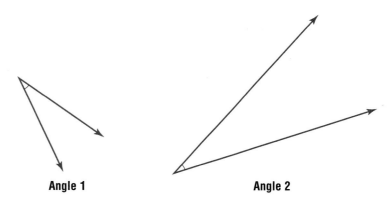

Angle 1        Angle 2

3. Explain the difference between a $\frac{1}{4}$-rotation and a $\frac{1}{2}$-rotation.

# Investigation ③ Classify Polygons

## Vocabulary

concave polygon

line symmetry

regular polygon

## Materials

• set of polygons and category labels

• large Venn diagram

Polygons can be divided into groups according to certain properties.

**Concave polygons** look like they are "collapsed" or have a "dent" on one or more sides. Any polygon with an angle measuring more than 180° is concave. The polygons below are concave.

 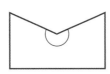  

The polygons below are not concave. Such polygons are sometimes called *convex polygons*.

**Regular polygons** have sides that are all the same length and angles that are all the same size. The polygons below are regular.

The polygons below are not regular. Such polygons are sometimes referred to as *irregular*.

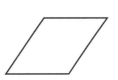   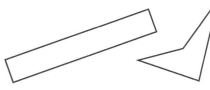

A polygon has **line symmetry**, or *reflection symmetry*, when you can fold it in half along a line and the two halves match exactly. The "folding line" is called the *line of symmetry*.

**Real-World Link**

The United Nations building located in New York City is an example of line symmetry in modern-day architecture.

The polygons below have line symmetry. The lines of symmetry are shown as dashed lines. Notice that three of the polygons have more than one line of symmetry.

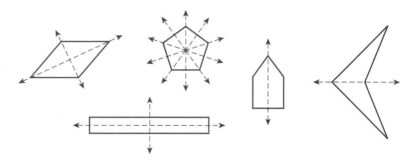

These polygons do not have line symmetry.

**Think & Discuss**

Consider the polygons below.

This diagram shows how these four polygons can be grouped into the categories *concave* and *not concave*.

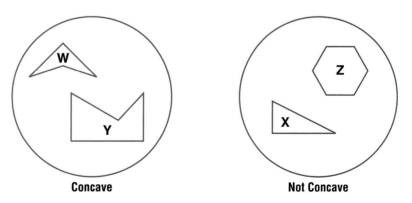

Now make a diagram to show how the four polygons can be grouped into the categories *line symmetry* and *not concave*. Use a circle to represent each category.

## ✅ *Develop & Understand: A*

You will now play a polygon-classification game with your group. Your group will need a set of polygons, category labels, and a large Venn diagram.

Here are the polygons used for the game.

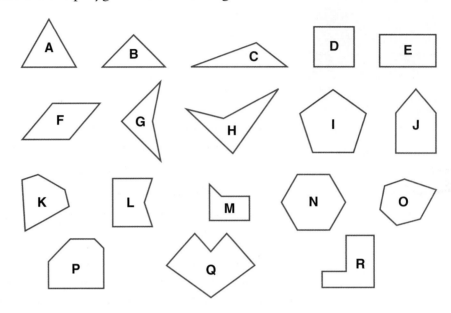

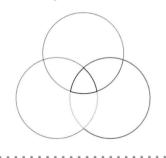

**Math Link**

A Venn diagram uses circles to represent relationships among sets of objects.

Here are the category labels.

| Regular | Concave | Triangle |
|---|---|---|
| Not Regular | Not Concave | Not Triangle |
| Quadrilateral | Pentagon | Hexagon |
| Not Quadrilateral | Not Pentagon | Not Hexagon |
| Line Symmetry | No Line Symmetry | |

1. As a warm-up for the game, put one of the labels *Regular*, *Concave*, and *Triangle* next to each of the circles on the diagram. Work with your group to place each of the polygons in the correct region of the diagram.

   Record your work. Sketch the three-circle diagram, label each circle, and record the polygons you placed in each region of the diagram. Record just the letters. You do not need to draw the polygons.

**2.** Now you are ready to play the game. Choose one member of your group to be the leader. Use the following rules.

- The leader selects three category cards and looks at them *without showing them to the other group members.*
- The leader uses the cards to label the regions, placing one card *face down* next to each circle.
- The other group members take turns selecting a polygon. The leader places the polygon in the correct region of the diagram.
- After a player's shape has been placed in the diagram, he or she may guess what the labels are. The first player to guess all three labels correctly wins.

At the end of each game, work with your group to place the remaining shapes. Then copy the final diagram. Take turns being the leader until each member of the group has had a chance.

**3.** Work with your group to create a diagram in which no polygons are placed in an overlapping region, that is, no polygon belongs to more than one category.

**4.** Work with your group to create a diagram in which all of the polygons are placed either in the overlapping regions or outside the circles, that is, no polygon belongs to just one category.

## Share & Summarize

**1.** Determine what the labels on this diagram must be. Use the category labels on page 14.

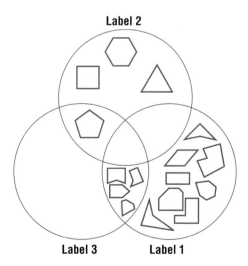

**2.** Explain why there are no polygons in the overlap of the Label 1 circle and the Label 2 circle.

**3.** Explain why there are no polygons in the Label 3 circle that are not also in one of the other circles.

**Target Market**

Men    18–34

Exercise 3–5
times per
week

Ideal Customer:
1.62 Million

Possible Customer:
6.8 Million

### *Real-World Link*

Venn diagrams are named after John Venn (1834–1923) of England, who made them popular. Venn diagrams are used in business to create visual models.

# Inquiry

## Investigation  Triangle Sides

### Materials

- linkage strips and fasteners

In many ways, triangles are the simplest polygons. They are the polygons with the fewest sides. Any polygon can be split into triangles. For this reason, learning about triangles can help you understand other polygons as well.

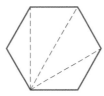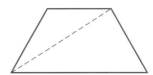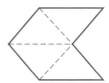

In this investigation, you will build triangles from linkage strips. The triangles will look like the one below. The sides of this triangle are 2, 3, and 4 units long. Notice that a "unit" is the space between two holes.

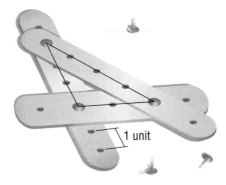

Do you think any three segments can be joined to make a triangle? You will investigate this question.

### Build the Triangles

1. Copy the table on the next page. Do the following steps for each row.
   - Try to build a triangle with the given side lengths.
   - In the "Triangle?" column, enter "yes" if you could make a triangle and "no" if you could not.
   - If you could make a triangle, try to make a *different* triangle from the same side lengths. (Hint: For two triangles to be different, they must have different shapes.) In the "Different Triangle?" column, enter "yes" if you could make another triangle and "no" if you could not.

| Side 1 | Side 2 | Side 3 | Triangle? | Different Triangle? |
|--------|--------|--------|-----------|---------------------|
| 4 units | 4 units | 4 units | | |
| 5 units | 4 units | 3 units | | |
| 4 units | 4 units | 2 units | | |
| 4 units | 4 units | 1 unit | | |
| 4 units | 3 units | 1 unit | | |
| 4 units | 2 units | 2 units | | |
| 3 units | 5 units | 6 units | | |
| 3 units | 3 units | 1 unit | | |
| 3 units | 2 units | 2 units | | |
| 3 units | 2 units | 1 unit | | |
| 3 units | 1 unit | 1 unit | | |

## Analyze the Results

**2.** Do you think you could make a triangle with segments 4, 4, and 10 units long? Explain your answer.

**3.** Do you think you could make a triangle with segments 10, 15, and 16 units long? Explain your answer.

## What Did You Learn?

**4.** Describe a rule you can use to determine whether three given segments will make a triangle. Test your rule on a few cases different from those in the table until you are convinced it is correct.

**5.** Do you think you can make more than one triangle with the same set of side lengths? Explain.

**Practice & Apply**

1. How many triangles are in this figure? Do not count just the smallest triangles.

2. Look at the figure in Exercise 1.

   **a.** Copy the figure. Label each vertex with a capital letter.

   **b.** In your figure, find at least one of each of the following polygons.

   • quadrilateral

   • pentagon

   • hexagon

   Use your vertex labels to name each shape.

   **c.** Find the polygon with the maximum number of sides in your figure. Use the vertex labels to name the shape.

3. List all the polygons in the figure below. Compute the figure's score using the following point values.

   • 3 points for each triangle

   • 4 points for each quadrilateral

   • 5 points for each pentagon

   • 6 points for each hexagon

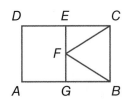

**Record your work in a table like the one below.**

| Polygon | Names | Score |
|---|---|---|
| Triangle | | |
| Quadrilateral | | |
| Pentagon | | |
| Hexagon | | |
| | Total Score | |

**In Exercises 4–7, several identical angles have the same vertex. Find the measure of the marked angle. Explain how you found it.**

**4.**

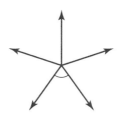

**5.**

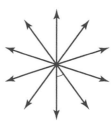

**6.**

**7.**

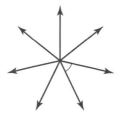

**Real-World Link**

During a Ferris wheel ride, the wheel makes several complete rotations.

**8.** A 180° angle is sometimes called a *straight angle*. Explain why that name makes sense.

**9.** You know that a 360° rotation is one complete rotation around a circle. Find the degree measures for each of these rotations.

   **a.** half a rotation

   **b.** two complete rotations

   **c.** $1\frac{1}{2}$ rotations

   **d.** three complete rotations

   **e.** $2\frac{1}{4}$ rotations

   **f.** five and one-half rotations

**10.** Draw two angles that each measure more than 90°. Explain how you know they measure more than 90°.

**11.** Draw two angles that each measure less than 90°. Explain how you know they measure less than 90°.

**12.** The diagram shows the result of one round of the game of polygon classification described on page 14.

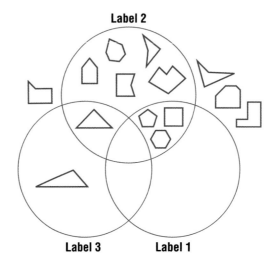

**Label 2**

**Label 3**    **Label 1**

**a.** Figure out what the labels must be. Use the category labels from the polygon-classification game.

**b.** Where would you place each of these shapes?

**In Exercises 13–15, draw a polygon that fits the given description, if possible. If it is not possible, say so.**

**13.** a regular polygon with four sides

**14.** a concave polygon with a line of symmetry

**15.** a triangle with just one line of symmetry

**Connect & Extend**

**16.** A *diagonal* of a polygon is a segment that connects two vertices but is not a side of the polygon. In each polygon below, the dashed segment is one of the diagonals.

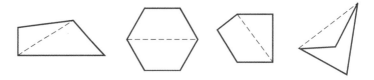

The number of diagonals you can draw from a vertex of a polygon depends on the number of vertices the polygon has.

**Math Link**
Triangles are the only polygons that are rigid. If you use linkage strips to build a polygon with more than three sides, you can push on the sides or vertices to create an infinite number of different shapes.

**a.** Copy each of these regular polygons. On each polygon, choose a vertex. Draw every possible diagonal from that vertex.

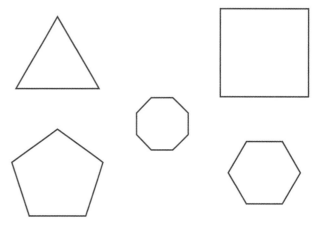

**b.** Copy and complete the table.

| Polygon | Vertices | Diagonals from a Vertex |
|---|---|---|
| | 3 | |
| Quadrilateral | | |
| | 5 | |
| Hexagon | | |
| Heptagon | 7 | |
| Octagon | | |

**c.** Describe a rule that connects a polygon's number of vertices to the number of diagonals that can be drawn from each vertex.

**d.** Explain how you know your rule will work for polygons with any number of vertices.

**e. Challenge** Describe a rule for predicting the *total number of diagonals* you can draw if you know the number of vertices in a polygon. Explain how you found your rule. Add a column to your table to help you organize your thinking.

**17.** Look for polygons in your home or school. Describe at least three different polygons that you find. Tell where you found them.

| Total Diagonals |
|---|
| 0 |
| 2 |
| |
| |
| |
| |

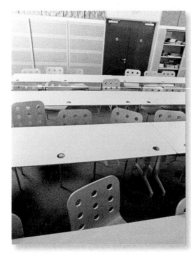

**18.** Find three angles in your home or school with measures equal to 90°, three with measures less than 90°, and three with measures greater than 90°. Describe where you found each angle.

**19.** Order the angles below from smallest to largest.

**20.** **Statistics** In a survey for the school yearbook, students were asked to name their favorite class. Conor made a circle graph to display the results. He forgot to label the wedges.

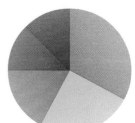

**a.** Of the students surveyed, $\frac{1}{3}$ liked math best. Which color wedge represents these students? What is the angle measure of that wedge?

**b.** Conor remembers that he used light blue to represent students who like science best. What fraction of the students surveyed chose science as their favorite subject?

**c.** Drama and English tied with $\frac{1}{8}$ of the students choosing each. Which wedges represent drama and English? What is the angle measure of each wedge?

In Exercises 21–23, describe a rule for creating each shape based on the preceding shape. Then draw the next two shapes.

**21.**   **22.**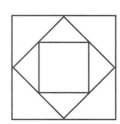

**23.**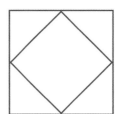

**24.** Circle diagrams, like those you used to classify polygons, are sometimes used to solve logic puzzles like this one.

*Camp Maple Leaf offers two sports, soccer and swimming. Of 30 campers, 24 play soccer, 20 swim, and 4 play no sport at all. How many campers both swim and play soccer?*

The diagram below includes a circle for each sport. The 4 outside the circles represents the four campers that do not play either sport. Use the diagram to help you solve the logic puzzle.

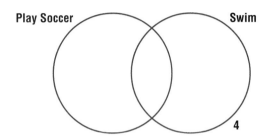

**25. In Your Own Words** Explain what each of the following words means. Give at least two facts related to each word.

• polygon       • angle       • triangle

**Mixed Review**

For Exercises 26–28, find each sum or difference without using a calculator.

**26.** $5,853 - 788$       **27.** $1,054 + 1,492$       **28.** $47,745 - 2,943$

**29.** Write *thirty-two thousand, five hundred sixty-three* in standard form.

**30.** Write *fourteen million, three hundred two thousand, two* in standard form.

For Exercises 31–36, write each number in words.

**31.** 324       **32.** 614       **33.** 1,025

**34.** 4,601       **35.** 10,809       **36.** 12,640

For Exercises 37–42, find each product or quotient without using a calculator.

**37.** $15 \cdot 10$       **38.** $24 \cdot 3$       **39.** $51 \cdot 4$

**40.** $72 \div 9$       **41.** $56 \div 7$       **42.** $480 \div 80$

# Angles

In Lesson 1.1, you investigated angles. You learned that an angle is defined as two rays with a common endpoint called the *vertex*.

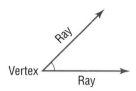

Angles are measured in degrees. In Lesson 1.1, you used 90°, 180°, and 360° angles as benchmarks to help estimate the measures of other angles.

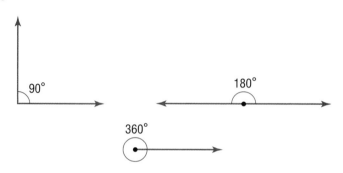

## Math Link

You can think of an angle as a rotation. A 360° angle is a rotation around a complete circle. A 180° angle is a rotation $\frac{1}{2}$ of the way around a circle. A 90° angle is a rotation $\frac{1}{4}$ of the way around a circle.

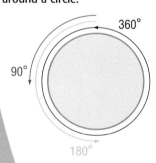

### Think & Discuss

Each diagram is constructed from angles of the same size. Estimate the measure of each marked angle. Explain how you found it.

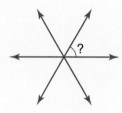

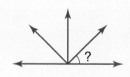

# Investigation  Measure Angles

## Vocabulary

**acute angle**

**obtuse angle**

**perpendicular**

**protractor**

**reference line**

**right angle**

## Materials

- protractor
- copies of the angles
- ruler

A **protractor** is a tool for measuring angles. A protractor has two sets of degree labels around the edge of a half circle or is sometimes a full circle. The line that goes through 0° is called the **reference line**.

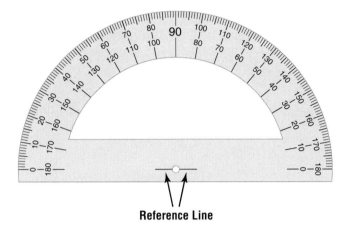

Reference Line

To measure an angle, follow these steps.

- Place the bottom center of the protractor at the vertex of the angle.
- Line up the reference line with one ray of the angle. Make sure the other ray can be seen through the protractor. You may need to extend this ray so that you can see where it meets the tick marks along the edge of the protractor.
- Read the angle measurement.

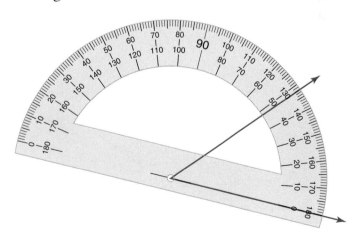

***Real-World Link***

When in-line skating, 360s, or full turns, and 540s, or one and a half turns, are two of the more difficult stunts.

The angle below measures about 48°. Or is it 132°? How do you know which number to use?

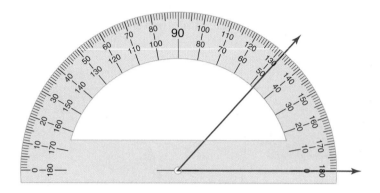

Is the measure of the angle below a little more than 90° or a little less than 90°? How do you know?

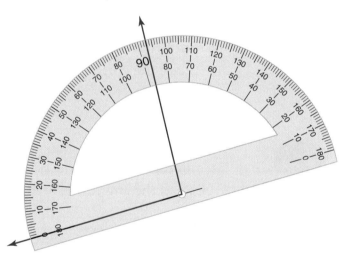

Measure these two angles. How do the measures compare?

Find the measure of the angle below. Describe the method that you used.

**Real-World Link**

A *mariner's astrolabe* was a navigational instrument of the 15th and 16th centuries. Sailors used it to measure the angle of elevation of the sun or other star.

You have seen that when you measure an angle with a protractor, you must determine which of two measurements is correct. One way to decide is to compare the angle to a 90° angle, which is an important benchmark. Angles are sometimes classified by how their measures compare to 90°.

**Acute angles** measure less than 90°.

**Obtuse angles** measure more than 90° and less than 180°.

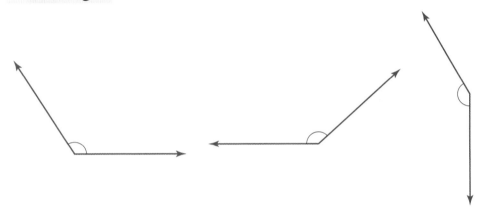

**Right angles** measure exactly 90°. Right angles are often marked with a small square at the vertex.

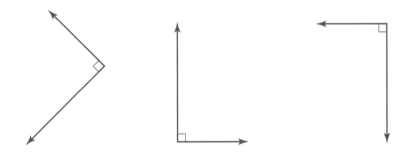

Two lines or segments that form a right angle are said to be **perpendicular**.

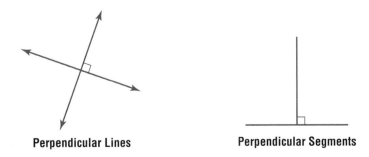

**Perpendicular Lines**          **Perpendicular Segments**

**Tell whether each angle is acute or obtuse. Then find its measure.**

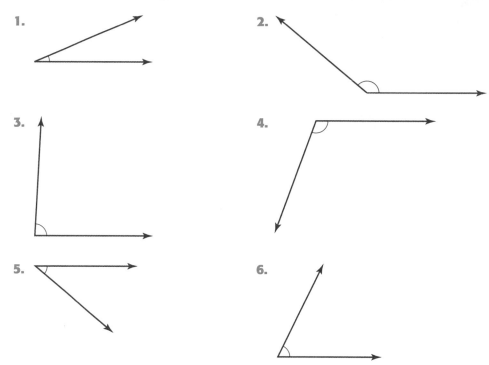

1.

2.

3.

4.

5.

6.

You have seen that a protractor is a useful tool for measuring angles. You can also use a protractor to draw angles with given measures.

## Example

To create a 25° angle, start by drawing a line segment. This segment will be one side of the angle.

Line up the reference line of the protractor with the segment. Place the center of the protractor at one endpoint of the segment, which will be the angle's vertex.

Draw a mark next to the 25° label on the protractor. Be sure to choose the correct 25° label.

Remove the protractor. Draw a segment from the vertex through the mark.

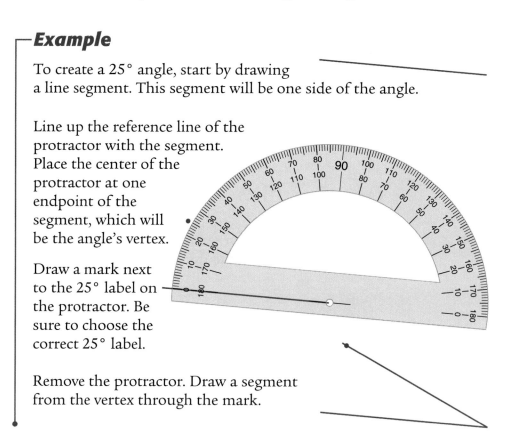

## Math Link

Notice the small arcs inside the angles for Exercises 1–6 on page 28. These are called *angle marks* and should be used when drawing angles.

## Develop & Understand: B

7. Draw a 160° angle. Include a curved angle mark to show which angle is 160°.

8. Draw a 210° angle. Include an angle mark to show which angle is 210°.

9. Draw two perpendicular segments.

10. Draw a triangle in which one angle measures 50° and the other angles have measures greater than 50°. Label each angle with its measure.

11. Draw a triangle with one obtuse angle. Label each angle with its measure.

12. Draw a triangle with two 60° angles.

13. Measure the sides of the triangle you drew in Exercise 12. What do you notice?

14. Draw a square. Make sure all the sides are the same length and all the angles measure 90°.

15. Draw a polygon with any number of sides and one angle that has a measure greater than 180°. Mark that angle.

## Share & Summarize

1. When you measure an angle with a protractor, how do you know which of the two possible numbers to choose?

2. The protractor on page 28 has a scale up to 180°. Describe how you would use such a protractor to draw an angle with a measure greater than 180°. Give an example if it helps you to explain your thinking.

**Real-World Link**

A *theodolite* is an instrument used in navigation, meteorology, and surveying to measure angles.

# Investigation ② Angle Relationships

## Vocabulary

intersecting lines

vertical angles

## Materials

• protractor

• ruler

You can refer to the angles in a drawing more easily if you label them with numbers or letters.

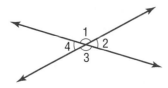

In the drawing above, the measure of Angle 1 is 135°. You can write this in symbols as $m\angle 1 = 135°$. The "m" stands for "measure," and $\angle$ is the symbol for "angle."

## ✅ Develop & Understand: A

Two lines that cross each other, like the lines in the drawing above, are called **intersecting lines**.

1. Measure angles 1, 2, 3, and 4 in the drawing above.

2. Which angles have the same measure?

3. Use a ruler to draw another pair of intersecting lines. Measure each of the four angles formed. Label each with its measure.

4. Draw one more pair of intersecting lines. Label each angle with its measure.

5. What patterns do you see relating the measure of the angles formed by two intersecting lines?

When two lines intersect, two angles that are not directly next to each other are called **vertical angles**. In the drawing below, $\angle a$ and $\angle c$ are vertical angles, and $\angle b$ and $\angle d$ are vertical angles.

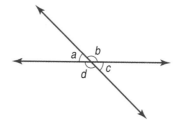

6. What do you think might be true about the measures of $\angle a$, $\angle b$, $\angle c$, and $\angle d$?

Below, Conor explains the relationship he discovered.

## Think & Discuss

In the cartoon, Conor showed that $m\angle 1 = m\angle 3$. Explain why $m\angle 2 = m\angle 4$.

The *interior angles* of a polygon are the angles inside the polygon. In this quadrilateral, the interior angles are marked.

## Develop & Understand: B

The sum of the measures of the interior angles of any triangle is 180°. If you need a convincing argument for this, cut out any triangle, tear off the three vertices, and line the vertices up. What do you notice about the measure of the angles they form? In the following exercises, you will look for similar rules about the angle sums of other polygons.

7. Use a ruler and a pencil to draw a triangle. Measure each interior angle. Then find the sum of the three angles.

8. Draw a quadrilateral. Measure each interior angle. Then find the sum of the four angles.

9. Now draw a pentagon. Measure each interior angle. Find the sum of the five angles.

10. Finally, draw a hexagon. Measure each interior angle. Find the sum of the six angles.

For Exercises 7–10, you and your classmates probably all drew different polygons. Compare the angle sums you found with the sums found by your classmates. What patterns do you see?

Describe a rule that you could use to predict the sum of the interior-angle measures of a polygon when you know only the number of angles.

Use your rule to predict the interior-angle sums for each concave polygon below. Check your predictions by measuring the angles. Be sure to measure the interior angles.

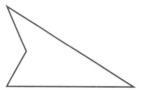 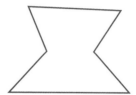

### Math Link

A *concave polygon* has at least one interior angle with measure greater than 180°. Concave polygons look "dented."

By now, you have probably concluded that the sum of the angle measures in a polygon depends only on the number of angles or the number of sides. You may have also discovered a rule for predicting the angle sum of any polygon when you know the number of angles.

Hannah and Jahmal wondered whether they could use what they know about the angle sum for triangles to think about the angle sums for other polygons.

In the following exercises, you will investigate whether Hannah's strategy applies to other polygons. You will also see how her strategy leads to a rule for calculating angle sums.

## ✅ *Develop & Understand: C*

**11.** First consider pentagons.

   **a.** Draw two pentagons. Make one of the pentagons concave. Divide each pentagon into triangles by drawing diagonals from one of the vertices.

   **b.** Into how many triangles did you divide each pentagon?

   **c.** Use your answer to Part b to find the sum of the interior angles in a pentagon.

**12.** Now consider hexagons.

   **a.** Draw two hexagons. Make one of the hexagons concave. Divide each hexagon into triangles by drawing diagonals from one of the vertices.

   **b.** Into how many triangles did you divide each hexagon?

   **c.** Use your answer to Part b to find the sum of the interior angles in a hexagon.

**13.** Suppose a quadrilateral has three 90° angles.

   **a.** What is the measure of the fourth angle? How do you know?

   **b.** What kind of quadrilateral is it?

**14.** Now think about octagons, which are 8-sided polygons.

   **a.** Without making a drawing, predict how many triangles you would divide an octagon into if you drew all the diagonals from one of the vertices. Explain how you made your prediction.

   **b.** Draw an octagon. Check your prediction.

   **c.** Use your answer to find the interior-angle sum for an octagon.

**15.** Suppose you drew a 15-sided polygon and divided it into triangles by drawing diagonals from one of the vertices.

   **a.** How many triangles would you make?

   **b.** Use your answer to find the interior-angle sum for a 15-sided polygon.

### Math Link
A 15-sided polygon is called a *pentadecagon*.

**16.** Suppose you know the number of angles a polygon has.

    **a.** How would you find the number of triangles you could make if you divided the polygon into triangles by drawing diagonals from one of the vertices?

    **b.** How would you find the sum of the angle measures?

## *Share* & *Summarize*

**1.** Martin said the sum of the angle measures for a quadrilateral must be 720° because a quadrilateral can be split into four triangles by drawing both diagonals.

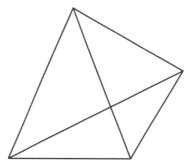

Explain what is wrong with Martin's argument.

**2.** What is the sum of the angle measures of a nonagon, a 9-sided polygon?

**Practice & Apply**     **Find the measure of each angle.**

**1.**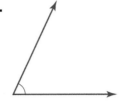

**2.**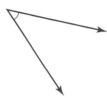

**3.**

**4.**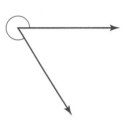

**Without measuring, find the missing angle measures.**

**5.**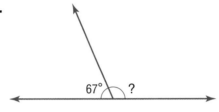

**6.**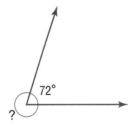

**Draw an angle with the given measure.**

**7.** 17°          **8.** 75°          **9.** 164°          **10.** 290°

**Draw the figure described. Label every angle in the figure with its measure.**

**11.** a quadrilateral with two 60° angles

**12.** a pentagon with two 90° angles

**13.** a quadrilateral with one 200° angle

**Without measuring, find the measure of each lettered angle.**

**14.**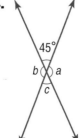

**15.**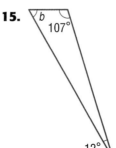

**16.**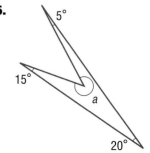

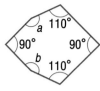

**17.** In this polygon, ∠a and ∠b have the same measure. What is it?

**18.** What is the measure of each angle of a regular pentagon?

**19.** What is the measure of each angle of a regular hexagon?

**Connect & Extend**

**20.** The drawings below show angles formed by a soccer player and the goalposts. The greater the angle, the better chance the player has of scoring a goal. For example, the player has a better chance of scoring from Position A than from Position B.

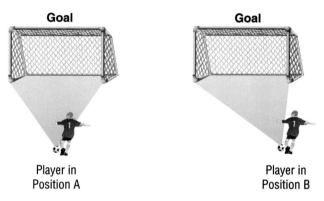

Player in Position A

Player in Position B

In Parts a and b, it may help to trace the diagrams and draw and measure angles.

**a.** Seven soccer players are practicing their kicks. They are lined up as shown in front of the goalposts. Which player has the best, or greatest, kicking angle?

**b.** Now the players are lined up as shown. Which player has the best kicking angle?

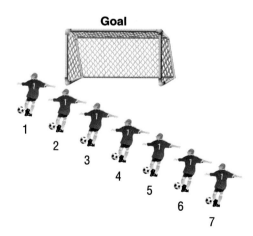

. . . . . . . . . . . . . . . . . . . .
**Math Link**

In a regular polygon, all sides are the same length, and all angles the same measure.
. . . . . . . . . . . . . . . . . . . .

**21.** The *diameter* of a circle is a segment that passes through the center of the circle and has both its endpoints on the circle. The four triangles below have all three vertices on a circle and the diameter as one side.

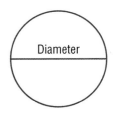

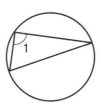 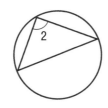 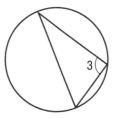 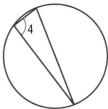

Measure each numbered angle. What do the measures have in common?

**22.** You discovered a rule about the sums of the interior angles of polygons. Polygons also have *exterior* angles, which can be found by extending their sides. In the drawings below, the exterior angles are marked.

**In Parts a–d, find the measure of each exterior angle. Find the sum of the measures. Then describe any pattern you find in the measures.**

**a.**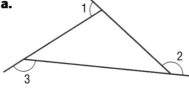

**b.**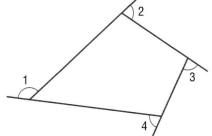

**c.**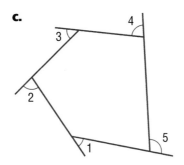

**d.**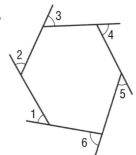

**23.** The angle at which a pool ball hits the side of a table has the same measure as the angle at which it bounces off the side. This is shown in the drawing below. The marked angles have the same measure, and the arrow shows the ball's path.

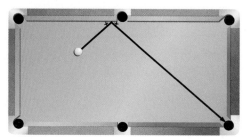

In Parts a–c, trace the drawing. Then use your protractor to find the path the ball will take when it bounces off the side. Tell whether the ball will go into a pocket or hit another side. Draw just one bounce.

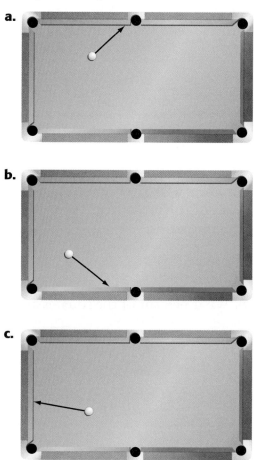

a.

b.

c.

**d. Challenge** Trace this drawing. Draw a path for which the ball will bounce off a side and land in the lower-right pocket.

← Land
Here

**24. In Your Own Words** Describe how you can find the interior-angle sum of any polygon without measuring any angles. Then explain how you know that your method works.

*Mixed Review*

**For Exercises 25–30, find each sum or difference without using a calculator.**

**25.** 73.97 − 12.43          **26.** 4.642 − 2.1          **27.** 37.13 − 16.4

**28.** 194.5 + 73.94          **29.** 54.32 + 45.68          **30.** 73.7654 − 5

**31.** Lucida drew the following grid.

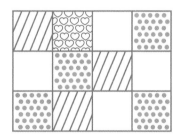

**a.** What fraction of the squares contain dots?

**b.** What percent of the squares are striped?

**c.** What fraction of the squares have hearts?

**d.** Describe how Lucita could fill in the blank squares to create a grid in which 50% of the squares contain dots, $\frac{1}{4}$ have hearts, and 25% have stripes.

**e.** Describe how Lucita could fill in the blank squares to create a grid in which $\frac{2}{3}$ of the squares have the same pattern.

# Measure Around

The **perimeter** of a two-dimensional shape is the distance around the shape. The perimeter of the shape to the right is 10.8 cm.

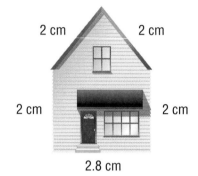

2 cm     2 cm

2 cm     2 cm

2.8 cm

## Vocabulary

**perimeter**

### Think & Discuss

Describe as many methods as you can for measuring the perimeter of the floor of your classroom.

Which of your methods do you think will give the most accurate measurement?

Which of your methods do you think is the most practical?

## Investigation ① Perimeter

### Vocabulary

**formula**

### Materials

• metric ruler, ruler, string
• copies of the auditorium floor

To find the perimeter of a polygon, add the lengths of its sides.

### ✔ Develop & Understand: A

This is the floor plan of the second floor of Millbury Middle School. On the drawing, each centimeter equals 2 meters.

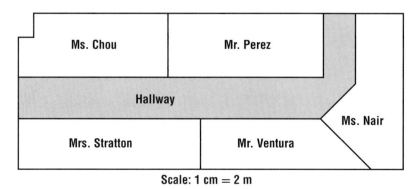

Scale: 1 cm = 2 m

1. Without measuring, tell whose classroom you think has the greatest perimeter. Explain why you think so.

2. Look at the floor plan for Ms. Nair's room.

   a. What type of polygon is the floor of Ms. Nair's room?

   b. Find the perimeter of Ms. Nair's floor plan to the nearest tenth of a centimeter. Then calculate the perimeter of the actual floor in meters.

3. Find the perimeter of Mrs. Stratton's floor plan to the nearest tenth of a centimeter. Then calculate the perimeter of the actual floor in meters.

4. Look at the floor plan for Mr. Perez's room.

   a. Is Mr. Perez's floor a rectangle? How do you know?

   b. Describe how to find the perimeter of Mr. Perez's floor plan by making only two measurements.

   c. Measure the perimeter of Mr. Perez's floor plan to the nearest tenth of a centimeter. Then calculate the perimeter of the actual floor in meters.

5. To find the perimeter of Ms. Chou's floor plan, Althea made the measurements labeled below. She claims these are the only measurements she needs to make.

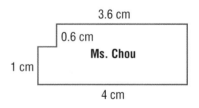

   a. Is Althea correct? If so, explain how to find the perimeter of Ms. Chou's room using only these measurements. If not, tell what other measurements you would need.

   b. Find the perimeter of Ms. Chou's floor plan to the nearest tenth of a centimeter. Then calculate the perimeter of the actual floor in meters.

6. Which teacher's classroom floor has the greatest perimeter? What is the greatest classroom perimeter?

***Real-World Link***
Many colleges and universities offer classes over the Internet. Students who are unable to travel to a college campus can earn college credits.

In the previous exercises, you probably realized you could find the perimeter of a rectangle without measuring every side. This is because the opposite sides of a rectangle are the same length. If you measure the length and the width of a rectangle, you can find the perimeter using either of two rules.

*Add the length and the width. Double the result.*

*Double the length and double the width. Add the results.*

If you use $P$ to represent the perimeter and $L$ and $W$ to represent the length and width, you can write these rules in symbols.

Geometric rules expressed using symbols, like those above, are often called **formulas**.

| Perimeter of a Rectangle |
|---|
| $P = 2 \cdot (L + W)$ $\qquad\qquad$ $P = 2L + 2W$ |
| In these formulas, $P$ represents the perimeter and $L$ and $W$ represent the length and width. |

## ✅ Develop & Understand: B

7. Use one of the perimeter formulas to find the perimeter of a rectangle with length 5.7 meters and width 2.9 meters.

8. The floor of a rectangular room has a perimeter of 42 feet. What are three possibilities for the dimensions of the floor?

9. A square floor has a perimeter of 32.4 meters. How long are the sides of the floor?

10. Write a formula for the perimeter of a square, using $P$ to represent the perimeter and $s$ to represent the length of a side. Explain why your formula works.

This floor plan is of the auditorium at Marshville Middle School.

Since part of the floor is curved, it is difficult to find the perimeter using just a ruler. You could use a measuring tape or a piece of string to find the length of the curved part. Another method is to use a polygon to *approximate* the shape of the floor.

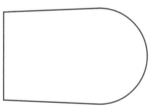

Scale: 1 cm = 5 m

## Example

Luke drew a pentagon to approximate the shape of the floor.

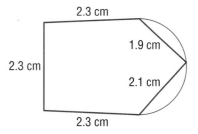

**Scale: 1 cm = 5 m**

Then he found the pentagon's perimeter.

$$2.3 + 2.3 + 2.3 + 1.9 + 2.1 = 10.9 \text{ cm}$$

## ✓ Develop & Understand: C

**11.** You can get a closer approximation than Luke's by using a polygon with more sides.

**a.** Try using a hexagon, a polygon with six sides, to approximate the shape of the floor plan. What perimeter estimate do you get using a hexagon?

**b.** Is the actual perimeter greater than or less than your estimate? Explain.

**c.** Now try a heptagon, a polygon with seven sides, to approximate the shape of the floor plan. What perimeter estimate do you get using a heptagon?

**d.** Is the actual perimeter greater than or less than your estimate? Explain.

**e.** Using a polygon with more than seven sides, make another estimate. What is your estimate?

**12.** Wrap a piece of string around the floor plan. Try to keep the string as close to the sides of the floor plan as possible. Then mark the string to indicate the length of the perimeter. Measure the string's length up to the mark. What is your perimeter estimate?

**13.** Which of your estimates do you think is most accurate? Explain.

Write a paragraph discussing what you know about finding the perimeter of two-dimensional shapes. Include the following.

- polygons and nonpolygons
- ruler measurements and string measurements
- formulas

# Investigation 2 Circumference

## Vocabulary

chord

circumference

diameter

radius

## Materials

- 5 objects with circular faces (for example, a soup can, a coffee can, a roll of tape, a plate, and a quarter)
- string or measuring tape
- ruler
- scissors

In the last investigation, you found perimeters of polygons. You estimated perimeters of a shape with curved sides. In this investigation, you will focus on circles.

The perimeter of a circle is called its **circumference**. Although you can estimate the circumference of a circle by using string or by approximating with polygons, there is a formula for finding the exact circumference. Before you begin thinking about circumference, you need to learn some useful words for describing circles.

A **chord** is a segment connecting two points on a circle. The **diameter** is a chord that passes through the center of the circle. *Diameter* also refers to the distance across a circle through its center. The **radius** is a segment from the center to a point on the circle. *Radius* also refers to the distance from the center to a point on the circle. The plural of *radius* is *radii*.

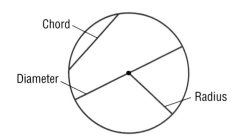

### Think & Discuss

Are all the chords of a circle the same length? If not, which are the longest?

Are all the diameters of a circle the same length? Are all the radii the same length?

Describe the relationship between the radius of a circle $r$ and its diameter $d$.

This quote from the novel *Contact* by Carl Sagan mentions a relationship between the circumference and diameter of any circle.

> *In whatever galaxy you happen to find yourself,*
> *you take the circumference, divide it by its diameter,*
> *measure closely enough, and uncover a miracle.*

In the following exercises, you will examine the relationship that Sagan is describing.

## ✅ Develop & Understand: A

**For Exercises 1–5, your group will need five objects with circular faces, for example, a soup can, a plate, or a quarter.**

1. Follow these steps for each object.
   - Use string or a measuring tape to approximate the circumference of the object.
   - Trace the circular face of the object. Cut out the tracing. Fold it in half to form a crease along the diameter of the circle. Measure the diameter.

   Record your measurements in a table like this one.

| Object | Circumference, C | Diameter, d |
|--------|------------------|-------------|
|        |                  |             |
|        |                  |             |
|        |                  |             |
|        |                  |             |
|        |                  |             |

2. Do you see a relationship between the circumference and the diameter of each circle? If so, describe it.

3. The quotation from *Contact* mentions dividing the circumference by the diameter. Add a column to your table showing the quotient $C \div d$ for each object. Describe any patterns you see.

4. Share your group's $C \div d$ results with the class.

5. Does the $C \div d$ value depend on the size of the circle? Explain.

No matter what size a circle is, the circumference divided by the diameter is always the same value. You probably discovered that this quotient is a little more than 3. The exact value is a decimal number whose digits never end or repeat. This value has been given the special name "pi" and is represented by the Greek letter $\pi$.

**Math Link**

Decimal numbers that never end or repeat are called *irrational numbers*. Whole numbers, fractions, and the decimals with which you have worked to this point are called *rational numbers*.

The symbol $\pi$ is used to represent the ratio $\frac{C}{d}$, where $C$ is the circumference of a circle and $d$ is the diameter. The ratio $\frac{C}{d}$ can be written as $C \div d$.

Since the digits of $\pi$ never end or repeat, it is impossible to write its exact numeric value. The number 3.14 is often used as an approximation of $\pi$. You can press the $\boxed{\pi}$ key on your calculator to get a closer approximation.

Use the division equation $\pi = C \div d$ to write the related multiplication equation $C = \pi \cdot d$. This is the formula for computing the circumference $C$ of a circle when you know its diameter $d$.

| **Circumference of a Circle** |
| --- |
| $C = \pi \cdot d$ |
| In this formula, $C$ is the circumference and $d$ is the diameter. Since the diameter of a circle is twice the radius $r$, you can also write the formula in the following ways. |
| $C = \pi \cdot 2 \cdot r$        $C = 2 \cdot \pi \cdot r$ |

Since the radius of this circle is 2.5 cm, the diameter is 5 cm.

$$C = \pi \cdot d$$
$$= \pi \cdot 5$$

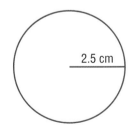

2.5 cm

The exact circumference of the circle is $5 \cdot \pi$ cm. Although you cannot write the circumference as an exact numeric value, you can use the $\boxed{\pi}$ key on your calculator to find an approximation.

$$C = \pi \cdot 5\text{cm} \approx 15.71 \text{ cm}$$

The symbol $\approx$ means "is approximately equal to."

## ✅ *Develop & Understand: B*

For Exercises 6 and 7, write your answer in terms of π.

**6.** Find the circumference of a circle with diameter 9 centimeters.

**7.** Find the circumference of this circle where a radius has been drawn.

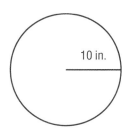

10 in.

**Real-World Link**

Earth is not perfectly round. The southern end bulges a bit, creating a slight pear shape. However, data collected by satellite indicate that Earth is gradually rounding out itself.

For Exercises 8 and 9, write your answer as a decimal rounded to the nearest hundredth. Use your calculator's [ π ] key to approximate π. If your calculator does not have a [ π ] key, use 3.14 as an approximation for π.

**8.** A circular pool has a circumference of about 16 meters. What is the pool's diameter?

**9.** The radius of Earth at the equator is about 4,000 miles.

**a.** Suppose you could wrap a string around Earth's equator. How long would the string have to be to reach all the way around? Assume the equator is a perfect circle.

**b.** Now suppose you could raise the string one mile above Earth's surface. How much string would you have to add to your piece from Part a to go all the way around?

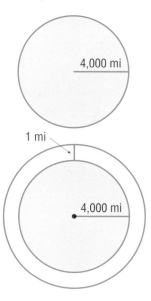

4,000 mi

1 mi

4,000 mi

## **Share & Summarize**

Explain what π is in your own words. Be sure to discuss the following.

- how it is related to circles
- its approximate value

**Practice & Apply**

In Exercises 1–3, use this diagram of a baseball field.

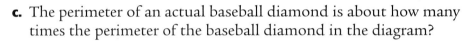

1. Consider the baseball diamond in this diagram.

   a. Find the perimeter of the diamond to the nearest $\frac{1}{4}$ inch.

   b. An actual baseball diamond is a square with sides 90 feet long. What is the perimeter of an actual baseball diamond?

   c. The perimeter of an actual baseball diamond is about how many times the perimeter of the baseball diamond in the diagram?

2. Rosita approximated the perimeter of the infield using a quadrilateral. She found a perimeter of about $6\frac{3}{4}$ inches.

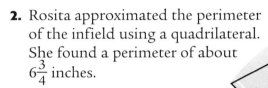

   a. Trace the shape of the infield. Use a polygon with more than four sides to find a better approximation of the infield's perimeter. Make all measurements to the nearest $\frac{1}{8}$ of an inch.

   b. How does your approximation compare to Rosita's?

3. Suppose the manager tells a player to run five laps around the entire baseball field, including the outfield. The player stays as close to the outer edge as possible.

   a. Measure the perimeter of the field in the diagram at the top of this page to the nearest $\frac{1}{8}$ of an inch.

**Math Link**

1 mile = 5,280 feet

   b. Suppose one inch on the diagram represents approximately 100 feet on the actual field. About how many miles will the player run in his five laps around the field?

**Real-World Link**

Founded in 1800 in Washington, D.C., the Library of Congress is one of the greatest national libraries. In addition to 15,000,000 books, it houses impressive collections of manuscripts, music, prints, and maps.

· · · · · · · · · · · · · · · ·

**4.** This is the floor plan of the Harperstown Library. What is the perimeter of the floor?

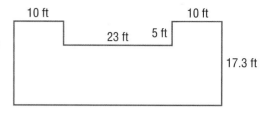

**5.** Give the dimensions of five rectangles that have a perimeter of 50 feet.

**6.** Find the circumference of a circle with diameter 7 meters. Write your answer in terms of $\pi$.

**7.** Find the circumference of a circle with radius 4.25 inches. Write your answer in terms of $\pi$ and as a decimal rounded to the nearest hundredth.

**8.** The circumference of a tire is 150 inches. What is the tire's radius? Use the $\boxed{\pi}$ key on your calculator or 3.14 to approximate $\pi$. Round your answer to the nearest hundredth.

**9. Challenge** The radius of the wheel on Jahmal's bike is 2 feet.

   **a.** If he rides 18.9 feet, how many full turns will the wheel make?

   **b.** If the wheel on Jahmal's bike turned 115 times, how many feet did Jahmal ride? About how many miles is this?

   **c.** If Jahmal rides 20 miles, how many times will his wheel turn?

**Connect & Extend**

**10.** Two shapes are *nested* when one is completely inside the other.

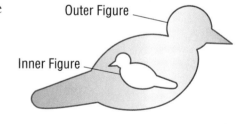

Outer Figure

Inner Figure

   **a.** Draw two nested shapes so that the outer shape has a greater perimeter than the inner shape. Give the perimeters of both shapes.

   **b.** Draw two nested shapes so that the inner shape has a greater perimeter than the outer shape. Give the perimeters of both shapes.

   **c.** Draw two nested shapes so that the outer shape has the same perimeter as the inner shape. Give the perimeters of both shapes.

   **d.** Look at your shapes from Parts a–c. In each case, which shape has more space inside, the inner shape or the outer shape? How do you know?

**11.** Many artists incorporate mathematics into their artwork. The artwork at the right is a tessellation. A *tessellation* is a design made of identical shapes that fit together without gaps or overlaps.

One way to make a shape that will tessellate is to cut a rectangle into two pieces and slide one piece to the other side.

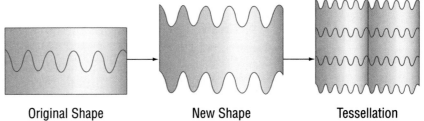

Original Shape    New Shape    Tessellation

**a.** Find the perimeter of the original shape in the artwork above.

**b.** Trace the new shape. Estimate its perimeter by using a polygon approximation or a piece of string.

**c.** When the new shape is formed from the original, the space inside the shape, the *area*, stays the same. However, the perimeter changes. Explain why this happens.

**12.** This is a diagram of the outer lane of the track at Albright Middle School. The lane is made of two straight segments and two semicircles, or half circles. Suppose a student runs one lap around the track in this lane. How many yards will she run?

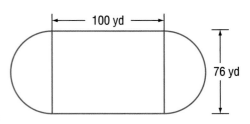

|← 100 yd →|

76 yd

**13.** Caroline wrapped a piece of string around the circumference of a circle with a diameter of 23 inches. She cut the string to the length of the circumference and then formed a rectangle with the string. Give the approximate dimensions of three rectangles she could make.

**14.** A circle with radius 6.5 inches is cut into four wedges and rearranged to form another shape.

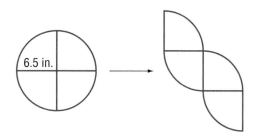

Does the perimeter change? How do you know? If it does change, by how much does it increase or decrease?

**15. In Your Own Words** Describe what perimeter is and how to find it for various shapes. Give an example of a situation in which finding a shape's perimeter would be useful.

*Mixed Review*

For Exercises 16–21, find each sum or difference.

**16.** $\frac{3}{7} + \frac{1}{7}$
**17.** $\frac{1}{4} + \frac{1}{4} + \frac{1}{4}$
**18.** $\frac{13}{32} + \frac{13}{32} + \frac{6}{32}$

**19.** $\frac{9}{5} - \frac{6}{5}$
**20.** $\frac{12}{15} - \frac{1}{15} - \frac{1}{15}$
**21.** $\frac{5}{7} - \frac{2}{7} - \frac{3}{7}$

*Math Link*

A *line of symmetry,* or a *reflection line,* divides a figure into mirror-image halves. If you fold a figure on a line of symmetry, the two halves match exactly.

**Earth Science** The symbols in Exercises 22–24 are used in *meteorology,* the study of weather. Copy each symbol, and draw all its lines of symmetry.

**22.** violent rain showers          **23.** ice pellets          **24.** hurricane

**Give the next four terms in each sequence.**

**25.** 64, 32, 16, 8, ...          **26.** 4, 6, 5, 7, 6, 8, 7, ...

# Review & Self-Assessment

## Chapter Summary

In this chapter, you focused on patterns in geometry. You learned to identify, name, and classify polygons. You worked with angles and studied some important properties about the side lengths and angle measures of triangles.

Next, you explored ideas about geometry and measurement. You started by working with angles. You measured angles and drew angles with given measures. You looked at relationships among the angles formed by intersecting lines. You explored the relationship between a polygon's number of angles or sides and the sum of its interior angles.

You then found the perimeters of polygons by adding side lengths. You estimated the perimeters of curved objects by using string and by approximating with polygons. You also learned that the ratio of the circumference of any circle to its diameter is equal to $\pi$.

## Strategies and Applications

The questions in this section will help you review and apply the important ideas and strategies developed in this chapter.

### Identifying, naming, and classifying polygons

**Tell whether each figure is a polygon. If it is not, explain why.**

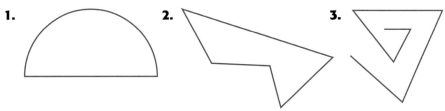

1.   2.   3.

**Draw a polygon that fits the given description, if possible. If it is not possible, say so.**

**4.** a concave hexagon with line symmetry

**5.** a regular quadrilateral without line symmetry

**6.** a concave pentagon with no line symmetry

## Vocabulary

- acute angle
- angle
- chord
- circumference
- concave polygon
- diameter
- formula
- intersecting lines
- line symmetry
- obtuse angle
- perimeter
- perpendicular
- polygon
- protractor
- radius
- reference line
- regular polygon
- right angle
- vertex
- vertical angles

### Understanding and applying properties of triangles

**7.** Explain how you can tell whether three segments can be joined to form a triangle. Give the lengths of three segments that can form a triangle and the lengths of three segments that cannot form a triangle.

**8.** If you know the measures of two angles of a triangle, how can you find the measure of the third angle? Explain why your method works.

### Measuring angles and drawing angles with given measures

**9.** Victor measured these angles with a protractor. He said both angles have measure of 130°.

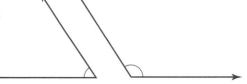

    **a.** How do you know that Victor is incorrect?

    **b.** What mistake do you think Victor made?

    **c.** What advice would you give to help him measure angles correctly?

**10.** Draw an angle with measure 320°. Explain the steps you followed.

## Demonstrating Skills

**Find the measure of each angle.**

**11.**

**12.**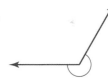

**13.**

**14.**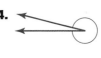

**Draw an angle with the given measure.**

**15.** 72°

**16.** 160°

**17.** 210°

**18.** 295°

**19.** Find the measures of Angles 1, 2, and 3.

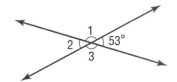

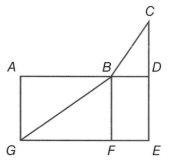

**In Exercises 20–22, refer to this figure.**

**20.** Name all the triangles in the figure.

**21.** Name all the quadrilaterals in the figure.

**22.** Name all the pentagons in the figure.

**In Exercises 23–28, tell which of these terms describe each polygon. List all terms that apply.**

| triangle | pentagon | concave | line symmetry |
| quadrilateral | hexagon | regular | |

**23.**

**24.**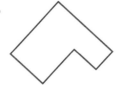

**25.**

**26.**

**27.**

**28.**

**Estimate the measure of each angle.**

**29.**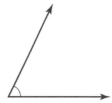

**30.**

**31.**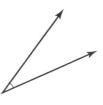

**Tell whether the given measures could be the angle measures of a triangle.**

**32.** 45°, 45°, 45°　　**33.** 80°, 40°, 80°　　**34.** 54°, 66°, 60°

**35.** Explain how you can determine if a polygon is concave.

**36.** Draw two angles that each measure more than 180°. Explain how you know they measure more than 180°.

**37.** A quadrilateral has three angle measures of 45°, 60°, and 100°. What is the missing angle measure?

**38.** A rectangle has a length of 4.2 meters and a width of 3.6 meters. What is its perimeter?

**For Exercises 39 and 40, write each answer in terms of π and as a decimal rounded to the nearest hundredth.**

**39.** Find the circumference of a circle with a diameter of 5 inches.

**40.** Find the circumference of this circle.

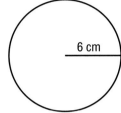

## Test-Taking Practice

### SHORT RESPONSE

**1** Lisa is putting a fence around a circular garden. If the radius of the garden is 4 feet, how many feet of fencing will Lisa need around the garden?

**Show your work.**

**Answer** _____

### MULTIPLE CHOICE

**2** Which of the following could be the side lengths of a triangle?

**A** 3, 5, 8

**B** 5, 17, 9

**C** 13, 4, 6

**D** 14, 6, 10

**3** Which term best describes the polygon shown?

**F** octagon

**G** hexagon

**H** pentagon

**J** regular

**4** A triangle has two angle measures of 23° and 46°. What is the measure of the third angle of the triangle?

**A** 21°

**B** 69°

**C** 111°

**D** 157°

**5** A rectangle has a length of 3.4 cm and a width of 1.2 cm. What is its perimeter?

**F** 4.08 cm

**G** 4.6 cm

**H** 9.2 cm

**J** 20.16 cm

# Fractions and Decimals

## Real-Life Math

**Market Place Values** *Buy low, sell high!* You may have heard this piece of wisdom about stock investing. Stocks allow people to own parts of companies, from fast-food chains to software developers to retail stores. Stockowners hope the value of their stock will rise over time, allowing them to sell their stocks at a higher price than they paid for them.

**Think About It** Stock-market reports use decimals to describe how a stock is doing. How might you show *a gain of $3* using a decimal? How about *a gain of $1.25*?

### Contents in Brief

**Math Online**

Take the **Chapter Readiness Quiz** at glencoe.com.

# Dear Family,

Chapter 2 extends mathematical ideas to look at patterns in fractions and decimals, numbers you use everyday.

## Key Concept—Fractions and Decimals

There are many number patterns found in fractions and decimals. Here is a pattern shown by fractions with the same denominator. Try to predict the numbers in the next column.

| Fraction | $\frac{1}{5}$ | $\frac{2}{5}$ | $\frac{3}{5}$ | $\frac{4}{5}$ | $\frac{5}{5}$ | $\frac{6}{5}$ | $\frac{7}{5}$ | ? |
|---|---|---|---|---|---|---|---|---|
| Decimal | 0.2 | 0.4 | 0.6 | 0.8 | 1.0 | 1.2 | 1.4 | ? |

Knowing these patterns and the decimal equivalents of common fractions will make it easier to calculate with fractions and decimals.

## Chapter Vocabulary

decimal-fraction

equivalent fractions

improper fraction

lowest terms

mixed number

repeating decimal

terminating decimal

## Home Activities

Fractions and decimals are everywhere.

- Ask your student to note the many ways fractions and decimals, in addition to money, are used in his or her day-to-day life.
- Discuss recipes as a meal is made. Compare the fractions used in different ingredients.
- When running an errand, have your student describe distances with fractions of miles.

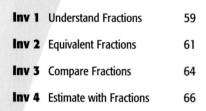

# Patterns in Fractions

You probably know quite a bit about fractions already. A fraction can be used to describe part of a whole or to name a number between two whole numbers.

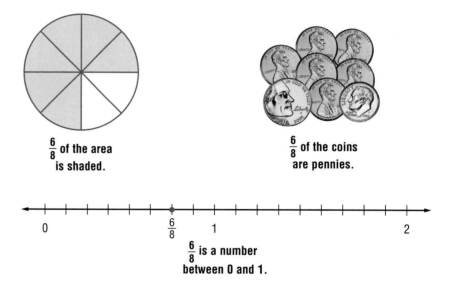

$\frac{6}{8}$ of the area is shaded.

$\frac{6}{8}$ of the coins are pennies.

$\frac{6}{8}$ is a number between 0 and 1.

> **Explore** · · · · · · · · · · · · · · · · · · · · · · · · · · · · · · · · · · · · ·
>
> In each representation above, what does the 8 in $\frac{6}{8}$ represent? What does the 6 represent?
>
> Trace the circle, and shade $\frac{3}{4}$ of its area. How does the area that you shaded compare to the above shaded area? Explain why your answer makes sense.
>
> Trace the number line. Indicate where $\frac{12}{16}$ is located. How does the location of $\frac{12}{16}$ compare with the location of $\frac{6}{8}$? Explain why your answer makes sense.

In this lesson, you will see how factor and multiple relationships can help you think about and work with fractions.

# Investigation ① Understand Fractions

**Vocabulary**

improper fraction

mixed number

In this investigation, you will review some ideas about fractions.

## ✓ Develop & Understand: A

The students in Mr. Jacobs' art class are sitting in four groups. Mr. Jacobs gives each group some bricks of clay to equally share among its members. All the bricks are the same size.

- Group 1 has 5 students and receives 4 bricks.
- Group 2 has 6 students and receives 4 bricks.
- Group 3 has 12 students and receives 9 bricks.
- Group 4 has 5 students and receives 6 bricks.

1. For each group, determine what fraction of a brick each student will get. Explain how you found your answers.

2. Did Mr. Jacobs pass out the clay fairly? Explain your answer.

The fraction of clay each member of Group 4 received was greater than 1. The next example shows how Miguel, Luke, and Hannah thought about dividing six bricks of clay among five students.

## ─Example

First, I gave each student one brick. I divided the extra brick into fifths & gave $\frac{1}{5}$ to each student.

So, each student got $1\frac{1}{5}$ bricks.

I drew 6 bricks and divided each into fifths. Each student got one of the fifths from each brick for a total of $\frac{6}{5}$.

I solved the division problem $6 \div 5$ and found that each student receives $1\frac{1}{5}$ bricks.

The example on page 59 shows two ways of expressing a fraction greater than 1. Luke's answer, $\frac{6}{5}$, is an **improper fraction**. This is a fraction in which the numerator is greater than the denominator. Miguel and Hannah's answer, $1\frac{1}{5}$, is a **mixed number**. A mixed number is a whole number and a fraction.

## ✓ Develop & Understand: B

In Exercises 3–6, give your answer as a mixed number and as a fraction.

3. If 12 bricks of clay are divided among 5 students, what portion of a brick will each student receive?

4. Mr. Davis' geese laid 18 eggs. What fraction of a dozen is this?

5. Each grid below has 100 squares. What fraction of a grid is the entire shaded portion?

6. What number is indicated by the point?

**Real-World Link**

The oldest known piece of pottery was made in China around 7900 B.C. The potter's wheel was invented in China around 3100 B.C.

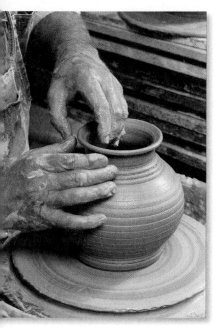

## Share & Summarize

Write a word problem, like those in Exercises 3 and 4 above, that leads to a fraction greater than 1. Show two ways of expressing the fraction.

# Investigation  Equivalent Fractions

## Vocabulary

**equivalent fractions**

**lowest terms**

In this investigation, you will see how different fractions can represent the same part of a whole and how different fractions can represent the same number.

### ✓ Develop & Understand: A

Casey, Collin, Manuel, Regina, and Jovan baked fruit bars in their Family and Consumer Science class. Each student cut his or her bar into a different number of equal pieces.

**Casey**

**Collin**

**Manuel**

**Regina**

**Jovan**

1. Casey wants to trade some of her lemon bar for an equal portion of Regina's raspberry bar. She could trade $\frac{1}{3}$ of her bar for $\frac{2}{6}$ of Regina's bar.

   What other fair trades could they make? List all of the possibilities. Give your answers as fractions of fruit bars.

2. Collin wants to trade some of his apple bar for an equal portion of Jovan's peach bar. Describe all of the fair trades they could make.

3. Describe all of the fair trades Manuel and Jovan could make.

4. Which pairs of students can trade only whole fruit bars?

5. List some fractions of a fruit bar that are fair trades for $\frac{1}{2}$ of a bar.

In Exercises 1–5, $\frac{1}{5}$ of Manuel's fruit bar is the same as $\frac{2}{10}$ of Jovan's fruit bar. Fractions such as $\frac{1}{5}$ and $\frac{2}{10}$ *describe the same portion of a whole or name the same number.* Such fractions are called **equivalent fractions**.

You can find a fraction equivalent to a given fraction by multiplying or dividing the numerator and denominator by the same number. Althea worked out an example to convince herself that dividing by the same number gives an equivalent fraction.

### Think & Discuss

Use an argument similar to Althea's to convince yourself that multiplying the numerator and denominator of a fraction by a number gives an equivalent fraction.

A fraction is in **lowest terms** if its numerator and denominator are relatively prime. For example, the fractions $\frac{2}{3}$, $\frac{12}{18}$, and $\frac{20}{30}$ are all equivalent. However, only $\frac{2}{3}$ is in lowest terms because the only common factor of 2 and 3 is 1.

## ✅ Develop & Understand: B

**Math Link**

Two numbers are prime if their only common factor is 1.

**Real-World Link**

The sizes of wrenches and drill bits are often given as fractions.

6. Fractions can be grouped into "families" of equivalent fractions. A fraction family is named for the member that is in lowest terms. The following is part of the "$\frac{3}{4}$ fraction family."

$$\frac{3}{4} \quad \frac{6}{8} \quad \frac{9}{12} \quad \frac{12}{16} \quad \frac{15}{20} \quad \frac{18}{24} \quad \frac{21}{28} \quad \frac{24}{32} \quad \frac{27}{36} \quad \frac{30}{40}$$

a. What do the numerators of these fractions have in common? What do the denominators have in common?

b. How do you know that all of the fractions in this family are equivalent?

c. Find at least three more fractions in this family. Explain how you found them.

d. Is $\frac{164}{216}$ in this fraction family? Explain how you know.

7. Now consider the $\frac{3}{8}$ fraction family.

a. List four members of this family with numerators greater than 15.

b. List four members of this family with numerators less than 15.

8. The fractions below belong to the same family.

$$\frac{15}{21} \quad \frac{20}{28} \quad \frac{25}{35}$$

a. Find four more fractions in this family, two with denominators less than 21 and two with denominators greater than 35.

b. What is the name of this fraction family?

9. How can you determine to which family a fraction belongs?

10. Are $\frac{6}{10}$ and $\frac{16}{20}$ in the same fraction family? Explain how you know.

## Share & Summarize

1. In Exercises 1–5, how did you determine which trades could be made? Give an example to help explain your answer.

2. Explain how you can find fractions equivalent to a given fraction. Demonstrate your method by choosing a fraction and finding four fractions equivalent to it.

3. Describe a method for determining whether two given fractions are equivalent.

# Investigation 3 Compare Fractions

In Investigation 2, you explored families of equivalent fractions. You saw how you could find fractions equivalent to a given fraction by multiplying or dividing the numerator and denominator by the same number. You will now use what you learned to compare fractions.

## ✓ Develop & Understand: A

1. The following are some members of the $\frac{1}{7}$ and the $\frac{2}{11}$ fraction families.

### The $\frac{1}{7}$ Fraction Family

$$\frac{1}{7} \quad \frac{2}{14} \quad \frac{3}{21} \quad \frac{4}{28} \quad \frac{5}{35} \quad \frac{6}{42} \quad \frac{7}{49} \quad \frac{8}{56} \quad \frac{9}{63} \quad \frac{10}{70} \quad \frac{11}{77} \quad \frac{12}{84}$$

### The $\frac{2}{11}$ Fraction Family

$$\frac{2}{11} \quad \frac{4}{22} \quad \frac{6}{33} \quad \frac{8}{44} \quad \frac{10}{55} \quad \frac{12}{66} \quad \frac{14}{77} \quad \frac{16}{88} \quad \frac{18}{99} \quad \frac{20}{110}$$

a. Recall that all the fractions in the $\frac{1}{2}$ fraction family equal $\frac{1}{2}$. Choose a pair of fractions, one from each family above, that you could use to easily compare $\frac{1}{7}$ and $\frac{2}{11}$.

b. Which fraction is greater, $\frac{1}{7}$ or $\frac{2}{11}$? Explain how you know.

2. Consider the fractions $\frac{3}{4}$ and $\frac{7}{12}$.

a. List some members of their fraction families.

b. List two pairs of fractions you could use to compare $\frac{3}{4}$ and $\frac{7}{12}$.

c. Which fraction is greater, $\frac{3}{4}$ or $\frac{7}{12}$?

You can compare two fractions by finding members of their fraction families with a *common denominator* or with a *common numerator*.

The diagram on the left shows how fractions with a common denominator of 10 compare. The diagram on the right shows how fractions with a common numerator of 7 compare.

**Common Denominators**

$\frac{2}{10}$
$\frac{3}{10}$
$\frac{4}{10}$
$\frac{5}{10}$
$\frac{6}{10}$
$\frac{7}{10}$

**Common Numerators**

$\frac{7}{5}$
$\frac{7}{6}$
$\frac{7}{7}$
$\frac{7}{8}$
$\frac{7}{9}$
$\frac{7}{10}$

## Think & Discuss

If two fractions have the same denominator but different numerators, how can you tell which of the fractions is greater? Explain why your reasoning works.

If two fractions have the same numerator but different denominators, how can you tell which of the fractions is greater? Explain why your reasoning works.

When you are given two fractions to compare, how can you quickly find equivalent fractions with a common denominator? With a common numerator?

## ✅ Develop & Understand: B

### Math Link

- $>$ means "is greater than"
- $<$ means "is less than"
- These statements have the same meaning.

$$3 < 4 \qquad 4 > 3$$

3. Rewrite $\frac{4}{17}$ and $\frac{3}{10}$ with a common denominator. Tell which fraction is greater.

4. Rewrite $\frac{4}{9}$ and $\frac{8}{15}$ with a common numerator. Tell which fraction is greater.

5. Consider the fractions $\frac{5}{8}$ and $\frac{7}{10}$.

   a. Rewrite the fractions with the least common denominator.

   b. Which fraction is greater, $\frac{5}{8}$ or $\frac{7}{10}$?

   c. What is the relationship between the least common denominator and the multiples of the original denominators, 8 and 10?

**Replace each $\bigcirc$ with $<$, $>$, or $=$ to make a true statement.**

6. $\frac{3}{4} \bigcirc \frac{7}{12}$   7. $\frac{8}{13} \bigcirc \frac{12}{19}$   8. $\frac{48}{120} \bigcirc \frac{12}{39}$

9. $\frac{17}{11} \bigcirc \frac{11}{7}$   10. $\frac{13}{12} \bigcirc \frac{6}{5}$   11. $\frac{19}{36} \bigcirc \frac{10}{24}$

## Share & Summarize

Order these fractions from least to greatest using any method you like.

$$\frac{1}{3} \qquad \frac{2}{9} \qquad \frac{5}{3} \qquad \frac{5}{4} \qquad \frac{7}{9} \qquad \frac{1}{2} \qquad 1\frac{1}{8}$$

In this lesson, you will see how you can use familiar fractions to estimate the values of other fractions.

## Think & Discuss

Keisha wants to make $\frac{1}{3}$ of a recipe of her grandmother's spaghetti sauce. When she divided the amount of each ingredient by 3, she found that she needed $\frac{4}{9}$ cup of olive oil. Keisha has only the measuring cups at right. How can she use these cups to measure *approximately* $\frac{4}{9}$ cup of oil?

Most people are familiar with such fractions as $\frac{1}{4}$, $\frac{1}{3}$, $\frac{1}{2}$, $\frac{2}{3}$, and $\frac{3}{4}$, and have a good sense of their value. For this reason, familiar fractions like these, along with the numbers 0 and 1, are often used as benchmarks. *Benchmarks,* or reference points, can help you approximate the values of other fractions.

When you estimate with benchmark fractions, you should ask yourself questions like these.

- Is the fraction closest to 0, $\frac{1}{2}$, or 1?
- Is it greater than $\frac{1}{2}$ or less than $\frac{1}{2}$?
- Is it greater than $\frac{1}{4}$ or less than $\frac{1}{4}$?
- Is it greater than $\frac{2}{3}$ or less than $\frac{2}{3}$?

## Develop & Understand: A

Heather and Miranda were working in their school's computer lab. They began installing the same software program onto their computers at the same time. Each computer displayed a progress bar indicating how much of the program had been installed. The following shows what the progress bars looked like after one minute.

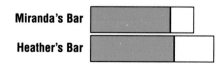

1. After one minute, about what fraction of the program had been installed on Miranda's computer? Explain how you made your estimate.

2. After one minute, about what fraction of the program had been installed on Heather's computer? Explain how you made your estimate.

3. Heather noticed that after one minute the shaded parts of the progress bars were about the same length. She said, "Our progress bars are the same length. Your computer has completed just as much of the installation as my computer." Is she correct? Explain.

4. Miranda said, "Your computer is only $\frac{1}{2}$ as fast as my computer." Is she correct? Explain your reasoning.

Fractions involving real data can be unfamiliar and complicated. Using familiar fractions to approximate actual fractions often makes information easier to understand.

## Think & Discuss

In a sixth-grade gym class, 28 out of the 40 students are girls. The gym teacher said, "Girls make up about $\frac{3}{4}$ of this class." Do you agree with this statement? Explain.

## Develop & Understand: B

5. Rewrite each statement using a more familiar fraction to approximate the actual fraction. Explain how you decided which fraction to use. Tell whether your approximation is a little greater than or a little less than the actual fraction.

   a. I have been in school for $\frac{43}{180}$ of the school year.

   b. Mrs. Stratton's class is $\frac{48}{60}$ of an hour long.

   c. The air distance from Washington, D.C., to Los Angeles is $\frac{2,300}{4,870}$ the air distance from Washington, D.C., to Moscow.

   d. The Volga River in Europe is $\frac{2,290}{3,362}$ the length of the Ob-Irtysh River in Asia.

6. Make up your own situation involving a "complicated" fraction. Tell which benchmark fraction you could use as an approximation.

## Share & Summarize

In what types of situations is it useful to approximate the value of a fraction with a more familiar fraction?

**Real-World Link**
The Volga River, the longest river in Europe, is located in western Russia. It is about 2,300 miles long (3,700 km).

**Practice & Apply**

1. Suppose ten friends share four medium-sized pizzas. Each friend gets the same amount of pizza. What fraction of a pizza does each friend receive?

2. Suppose six people share eight submarine sandwiches. Each person gets the same amount. What fraction of a sandwich does each person receive? Express your answer as a fraction and as a mixed number.

3. Suppose ten people share fifteen cinnamon rolls. Each person gets the same amount. What fraction of a roll does each person receive? Express your answer as a fraction and as a mixed number.

4. What number is indicated by the point? Give your answer as a mixed number and as a fraction.

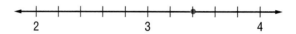

5. A coin roll holds 50 pennies. What fraction of a roll is 237 pennies? Give your answer as a mixed number and as a fraction.

6. Althea baked a plum bar and cut it into ninths. What fair trades can she make with Regina, whose raspberry bar is divided into sixths?

**Althea**

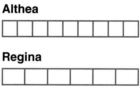

**Regina**

7. Alicia's kiwifruit bar is divided into tenths, and Rob's strawberry bar is divided into fifteenths. List all the fair trades they could make.

**Give two fractions that are equivalent to each given fraction.**

8. $\frac{2}{3}$

9. $\frac{5}{8}$

10. $\frac{1}{5}$

11. The following are four members of the $\frac{5}{9}$ fraction family. List four more.

$$\frac{5}{9} \qquad \frac{10}{18} \qquad \frac{15}{27} \qquad \frac{20}{36}$$

12. All of these fractions are in the same family.

$$\frac{33}{27} \qquad \frac{44}{36} \qquad \frac{66}{54}$$

   a. Find four more fractions in this family.

   b. What is the name of this fraction family?

**Real-World Link**

Although kiwifruit was first grown in China over 700 years ago, it was not widely available in the United States until the 1970s.

**13.** Consider the fraction $\frac{27}{36}$.

   **a.** To what fraction family does $\frac{27}{36}$ belong?

   **b.** List all the members of this fraction family with numerators less than 27.

**14.** Are $\frac{34}{64}$ and $\frac{18}{36}$ in the same fraction family? Explain how you know.

**Rewrite each fraction or mixed number in lowest terms.**

**15.** $\frac{12}{3}$  **16.** $\frac{9}{24}$  **17.** $5\frac{6}{9}$  **18.** $\frac{18}{45}$

**Tell whether the fractions in each pair are equivalent. Explain how you know.**

**19.** $\frac{4}{8}$ and $\frac{15}{30}$  **20.** $\frac{4}{12}$ and $\frac{8}{32}$  **21.** $\frac{50}{60}$ and $\frac{15}{18}$

**Replace each $\bigcirc$ with <, >, or = to make a true statement.**

**22.** $\frac{7}{8} \bigcirc \frac{2}{3}$  **23.** $\frac{5}{9} \bigcirc \frac{3}{5}$  **24.** $\frac{5}{16} \bigcirc \frac{5}{17}$

**25.** $\frac{90}{70} \bigcirc \frac{45}{35}$  **26.** $\frac{1}{2} \bigcirc \frac{7}{11}$  **27.** $\frac{13}{8} \bigcirc 1\frac{2}{3}$

**Order each set of fractions from least to greatest.**

**28.** $\frac{3}{4}, \frac{3}{3}, \frac{3}{8}, \frac{3}{5}, \frac{3}{16}, \frac{3}{7}, \frac{3}{1}$  **29.** $\frac{7}{7}, \frac{3}{4}, \frac{1}{2}, \frac{2}{5}, \frac{2}{3}, \frac{11}{8}, \frac{1}{3}$

**30.** Three computers begin installing the same program at the same time. The following are their progress bars after one minute.

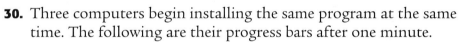

Computer A    Computer B    Computer C

   **a.** About what fraction of the installation has been completed by computer A? By computer B? By computer C?

   **b.** Order the machines from fastest to slowest. Explain how you determined the ordering.

**Tell whether each fraction is closest to 0, $\frac{1}{4}, \frac{1}{3}, \frac{1}{2}, \frac{2}{3}, \frac{3}{4}$, or 1. Explain how you decided.**

**31.** $\frac{5}{18}$  **32.** $\frac{1}{5}$  **33.** $\frac{9}{10}$

**34.** Determine whether $\frac{33}{40}$ is greater than or less than $\frac{21}{50}$ by comparing the fractions to benchmark fractions. Explain your thinking.

**Connect & Extend**

**35. Measurement** Five segments are shown above the yardstick. Give the length of each segment **in feet**. Give your answers as fractions or mixed numbers in lowest terms.

E •————————————————————————————————•

D •————————————————————————•

C •———————————————————————•

B •——————————•

A •————•

```
|''''|''''|''''|''''|''''|''''|''''|''''|''''|''''|''''|''''|''''|''''|''''|''''|''''|''''|
 0   2   4   6   8  10  12  14  16  18  20  22  24  26  28  30  32  34  36
in.
```

36 in. = 1 yd

**36. Number Sense** People often use mixed numbers to compare two quantities or to describe how much something has changed or grown.

**a.** Dion's height is about $1\frac{1}{2}$ times his younger brother Jamil's height. Jamil is about 40 inches tall. How tall is Dion?

**b.** Bobbi spends 40 minutes each night practicing her violin. She said, "That's $1\frac{1}{3}$ times the amount of time I spent last year." How much time did Bobbi practice each night last year?

**c.** The 2007 population of Seattle was about $7\frac{2}{5}$ times the 1900 population. Seattle's 1900 population was about 80,000. Estimate Seattle's population in 2007.

**37.** Imagine that you have baked a delicious mango-papaya bar the same size as the bars baked by the students in Investigation 2 on page 61.

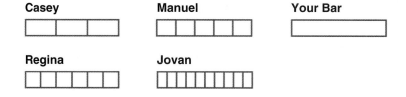

**a.** You want to trade portions of your bar with Jovan and Regina and still have some left for yourself. Tell how many equal-sized pieces you would cut your bar into. Describe the trade you would make with each student. Tell what fraction of your bar you would have left.

**b.** Give the same information for trading with Casey and Regina.

**c.** Give the same information for trading with Casey and Manuel.

**38.** Ying baked a coconut bar and would like to trade with all three students.

**Casey**

| | | |
|---|---|---|

**Collin**

| | | | |
|---|---|---|---|

**Regina**

| | | | | |
|---|---|---|---|---|

    **a.** Into how many equal-sized pieces should Ying divide her bar?

    **b.** List the trades Ying could make. Tell how much of the bar she would have left for herself.

**39. Prove It!** Write a convincing argument to show that $\frac{3}{4}$ of a fruit bar is not a fair trade for $\frac{3}{5}$ of a fruit bar.

**40. Biology** Water makes up about $\frac{2}{3}$ of a person's body weight.

    **a.** A student weighs 90 pounds. Determine how many pounds of the student's weight are attributed to water. Find a fraction equivalent to $\frac{2}{3}$ with a denominator of 90.

    **b.** A student weighs 75 pounds. Determine how many pounds of the student's weight are attributed to water.

**41. Preview** Percent means "out of 100." You can think of a percent as the numerator of a fraction with a denominator of 100. For example, 25% means $\frac{25}{100}$. You can change a fraction to a percent by first finding an equivalent fraction with a denominator of 100.

    **a.** Change the following fractions to percents: $\frac{3}{4}, \frac{1}{5}, \frac{20}{50}, \frac{8}{25}$.

    **b.** In Eva's homeroom, 14 of the 20 students ride the bus to school. What percent of the students take the bus?

    **c.** Of the 500 people in the audience at the school play, 350 bought their tickets in advance. What percent of the audience bought tickets in advance?

**42. Measurement** Between which two-twelfths of a foot will you find each measurement? For example, $\frac{1}{8}$ is between $\frac{1}{12}$ and $\frac{2}{12}$.

    **a.** $\frac{3}{5}$ of a foot

    **b.** $\frac{1}{10}$ of a foot

    **c.** $\frac{5}{8}$ of a foot

**43.** Of the 560 students at Roosevelt Middle School, 240 participate in after-school sports. Of the 720 students at King Middle School, 300 participate in after-school sports.

    **a.** In which school does the greater *number* of students participate in sports?

    **b.** In which school does the greater *fraction* of students participate in sports?

**44. Statistics** In a recent year, about 58,300,000 people lived in the western region of the United States. About 16,200,000 of these people were under 18 years of age. At the same time, about 23,900,000 of the 91,700,000 people living in the southern region were under 18 years of age. Which region had the greater fraction of children and teenagers?

**45.** A survey asked all the sixth graders at Belmont Middle School how much time they spent on homework each week.

    **a.** Malik said, "About half of the students in the sixth grade spend five to six hours each week doing homework." Do you agree with this statement? Explain why or why not.

| Time Spent Doing Homework | Fraction of Class |
|---|---|
| 0 to 1 hours | $\frac{0}{100}$ |
| 1 to 2 hours | $\frac{1}{100}$ |
| 2 to 3 hours | $\frac{12}{100}$ |
| 3 to 4 hours | $\frac{3}{100}$ |
| 4 to 5 hours | $\frac{22}{100}$ |
| 5 to 6 hours | $\frac{57}{100}$ |
| More than 6 hours | $\frac{5}{100}$ |

    **b.** Teresa said, "Hardly anyone spends four to five hours each week on homework." Do you agree with this statement? Explain why or why not.

**46. Challenge** The table shows the populations of the five most heavily populated countries in 2007.

**2007 Population**

| Country | Population |
|---------|-----------|
| China | 1,320 million |
| India | 1,169 million |
| United States | 303 million |
| Indonesia | 231 million |
| Brazil | 187 million |

  **a.** The world population in 2007 was about 6,671 million people. Which of these countries had about $\frac{1}{5}$ of the world population?

  **b.** About what fraction of the total 2007 world population lived in the United States?

**47. In Your Own Words** Explain how factors and multiples can be used to find members of a fraction family.

*Mixed Review*    **Find the measure of each angle.**

**48.**

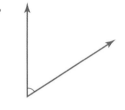

**49.**

**50.**

**51.**

**52.** The formula for the perimeter of a rectangle is $2L + 2W$, where $L$ is the length of the rectangle and $W$ is the width. Find the perimeter of a rectangle with length 6 cm and width 2 cm.

# Patterns in Decimals

You encounter decimals everyday. Prices displayed in stores and statistics in the sports section of the newspaper are often given as decimals. In this lesson, you will review the meaning of decimals. You will also practice working with decimals in a variety of situations.

## Materials

- Spare Change cards
  (1 set per group)
- dollar charts (1 per player)

### Explore

Read the rules of the *Spare Change* game. Play two rounds with your group.

*Spare Change* Game Rules

- Place four *Spare Change* cards face up on the table. Place the rest of the deck face down in a pile.

- To take a turn, a player chooses one of the four cards showing and shades that fraction of a dollar on his or her dollar chart. The player then places the card on the bottom of the deck and replaces it with the top card from the deck.

- Play continues until one player has shaded the entire card or is unable to shade any of the four amounts showing on the face up cards.

- The player who has shaded the amount closest to a dollar at the end of the game is the winner.

Describe some strategies you used while playing the game.

Discuss what you learned about decimals.

# Investigation (1) Understand Decimals

## Materials

- dollar charts

Decimals are equivalent to fractions whose denominators are 10, 100, 1,000, 10,000, and so on. Each decimal place has a name based on the fraction it represents.

### Math Link

The prefix "deci" comes from the Latin word *decem*, meaning "ten."

| Decimal | Equivalent Fraction | In Words |
|---------|---------------------|----------|
| 0.1 | $\frac{1}{10}$ | one tenth |
| 0.01 | $\frac{1}{100}$ | one hundredth |
| 0.001 | $\frac{1}{1,000}$ | one thousandth |
| 0.0001 | $\frac{1}{10,000}$ | one ten-thousandth |

## Example

How is 9.057 different from 9.57?

These decimals look similar, but they represent different numbers. You can see this by looking at the place values of the digits.

| 1000 | 100 | 10 | 1 | 0.1 | 0.01 | 0.001 |
|------|-----|----|----|-----|------|-------|
| thousands | hundreds | tens | ones | tenths | hundredths | thousandths |
| | | | 9. | 0 | 5 | 7 |
| | | | 9. | 5 | 7 | |

9.057 means $9 + \frac{0}{10} + \frac{5}{100} + \frac{7}{1,000}$,

or $9\frac{57}{1,000}$,

or $\frac{9,057}{1,000}$.

9.57 means $9 + \frac{5}{10} + \frac{7}{100}$,

or $9\frac{57}{100}$,

or $\frac{957}{100}$.

The number 9.057 is read "nine and fifty-seven thousandths."
The number 9.57 is read "nine and fifty-seven hundredths."

In Exercises 1–3, you will see how the ideas discussed in Investigation 1 relate to the *Spare Change* game.

## ✓ Develop & Understand: A

1. Consider the values $0.3 and $0.03.

   a. Are these values the same?

   b. Explain your answer to Part a by writing both amounts as fractions.

   c. Illustrate your answer to Part a by shading both amounts on a dollar chart.

2. In her first four turns of the *Spare Change* game, Kristina chose $0.45, $0.1, $0.33, and $0.05.

   a. Complete a dollar chart showing the amount Kristina should have shaded after the first four turns.

   b. What part of a dollar is shaded on Kristina's chart? Express your answer as a fraction and as a decimal.

   c. How much more does Kristina need to have $1.00? Express your answer as a decimal and as a fraction.

3. How could you shade a dollar chart to represent $0.125?

Now you will explore how multiplying or dividing a number by 10, 100, 1,000, and so on changes the position of the decimal point.

### Math Link

In many countries, a *decimal comma* is used instead of a decimal point, and a space is used to separate groups of three digits.

# ✅ *Develop & Understand: B*

4. Copy the table.

| Calculation | | Result |
|---|---|---|
| 81.07 | = 81.07 • 1 | 81.07 |
| 81.07 • 10 | = 81.07 • 10 | |
| 81.07 • 10 • 10 | = 81.07 • 100 | |
| 81.07 • 10 • 10 • 10 | = 81.07 • 1,000 | |
| 81.07 • 10 • 10 • 10 • 10 | = 81.07 • 10,000 | |
| 81.07 • 10 • 10 • 10 • 10 • 10 | = 81.07 • 100,000 | |

   a. Enter the number 81.07 on your calculator. Multiply it by 10. Record the result in the second row of the table.

   b. Find 81.07 • 100 by multiplying your result from Part a by 10. Record the result in the table.

   c. Continue to multiply each result by 10 to find 81.07 • 1,000; 81.07 • 10,000; and 81.07 • 100,000. Record your results.

   d. Describe how the position of the decimal point changed each time you multiplied by 10.

5. In Parts a–c, predict the value of each product without doing any calculations. Check your prediction by using your calculator.

   a. 7.801 • 10,000      b. 0.003 • 100      c. 9,832 • 1,000

   d. When you predicted the results of Parts a–c, how did you determine where to put the decimal point?

6. Think about how the value of a number changes as you move the decimal point to the right.

   a. How does the value of a number change when you move the decimal point one place to the right? Two places to the right? Three places to the right? (Hint: Look at your completed table from Exercise 5, or test a few numbers to see what happens.)

   b. **Challenge** In general, what is the relationship between the number of places a decimal is moved to the right and the change in the value of the number?

7. Tell what number you must multiply the given number by to get 240. Explain how you found your answer.

   a. 2.4      b. 0.24      c. 0.00024

**8.** Copy the table.

| Calculation | | Result |
|---|---|---|
| 81.07 | | 81.07 |
| 81.07 ÷ 10 | $= \frac{1}{10}$ of 81.07 | |
| 81.07 ÷ 10 ÷ 10 | $= \frac{1}{100}$ of 81.07 | |
| 81.07 ÷ 10 ÷ 10 ÷ 10 | $= \frac{1}{1,000}$ of 81.07 | |
| 81.07 ÷ 10 ÷ 10 ÷ 10 ÷ 10 | $= \frac{1}{10,000}$ of 81.07 | |
| 81.07 ÷ 10 ÷ 10 ÷ 10 ÷ 10 ÷ 10 | $= \frac{1}{100,000}$ of 81.07 | |

**a.** Find $\frac{1}{10}$ of 81.07 by entering 81.07 on your calculator and dividing by 10. Record the result in the second row of the table.

**b.** Find $\frac{1}{100}$ of 81.07 by dividing your result from Part a by 10. Record the result in the table.

**c.** Continue to divide each result by 10 to find $\frac{1}{1,000}$ of 81.07, $\frac{1}{10,000}$ of 81.07, and $\frac{1}{100,000}$ of 81.07. Record your results.

**d.** Describe how the position of the decimal point changed each time you divided by 10, that is, each time you found $\frac{1}{10}$.

**9.** In Parts a–c, predict each result without doing any calculations. Check your prediction by using your calculator.

**a.** $\frac{1}{10,000}$ of 14.14   **b.** 34,372 ÷ 100   **c.** $\frac{1}{1,000}$ of 877

**d.** When you predicted the results of Parts a–c, how did you determine where to put the decimal point?

**10.** Think about how the value of a number changes as you move the decimal point to the left.

**a.** How does the value of a number change when you move the decimal point one place to the left? Two places to the left? Three places to the left? (Hint: Look at your completed table from Exercise 8, or test a few numbers to see what happens.)

**b.** **Challenge** In general, what is the relationship between the number of places a decimal is moved to the left and the change in the value of the number?

**11.** Tell what number you must divide the given number by to get 1.8. Explain how you found your answer.

    **a.** 18                 **b.** 180               **c.** 18,000

### Share & Summarize

**1.** A shirt is on sale for $16.80. Write $16.80 as a mixed number.

**2.** A big-screen television costs 100 times as much as the shirt. How much does the TV cost?

**3.** A fresh-cooked pretzel costs $\frac{1}{10}$ as much as the shirt. How much is the pretzel?

## Investigation ② Measure with Decimals

### Materials

- tape
- meterstick

**Math Link**

The abbreviation for meter is m.

The abbreviation for centimeter is cm.

The abbreviation for millimeter is mm.

In the metric system, units of measure are based on the number 10. This makes converting from one unit to another as easy as moving a decimal point.

The basic unit of length in the metric system is the meter. Each meter can be divided into 100 centimeters.

Each centimeter can be divided into 10 millimeters. (Note: Ruler is not to scale.)

### Think & Discuss

Fill in the blanks. Give your answers as both decimals and fractions.

1 cm = _____ m       1 mm = _____ cm       1 mm = _____ m

5 cm = _____ m       15 mm = _____ cm       15 mm = _____ m

## ✅ *Develop & Understand: A*

Convert each measurement to meters. Write your answers as fractions and as decimals.

**1.** 35 cm      **2.** 9 mm      **3.** 23 mm

**4.** Give the lengths of segments A and B **in meters**. Express your answers as fractions and as decimals.

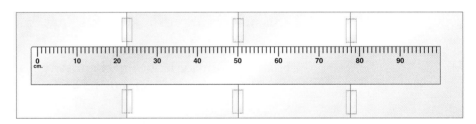

For Exercises 5–6, tape four sheets of paper together lengthwise. Tape a meterstick on top of the paper as shown, so you can draw objects above and below the meterstick.

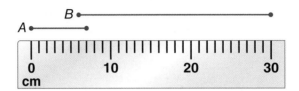

**5.** Collect objects whose lengths you can measure, such as pencils, books, staplers, and screwdrivers.

   **a.** Place the objects end to end along your meterstick until the combined length is as close to one meter as possible. Sketch the objects *above* the meterstick at their actual lengths.

   **b.** Find the length of each object to the nearest millimeter. Label the sketch of each object with its length in millimeters, centimeters, and meters.

> **Math Link**
>
> Since 1983, the meter has been defined as the distance light travels in a vacuum in $\frac{1}{299{,}792{,}458}$ of a second.

6. In this exercise, you will try to find the combination of the following objects with a length as close to one meter as possible.

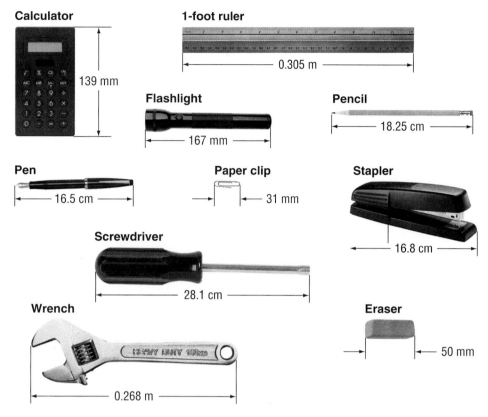

Calculator — 139 mm

1-foot ruler — 0.305 m

Flashlight — 167 mm

Pencil — 18.25 cm

Pen — 16.5 cm

Paper clip — 31 mm

Stapler — 16.8 cm

Screwdriver — 28.1 cm

Wrench — 0.268 m

Eraser — 50 mm

a. Choose one of the objects. Below your meterstick, begin at 0 and sketch the object at its actual length.

b. Choose a second object. Starting at the right end of the previous drawing, sketch the second object at its actual length.

c. Continue to choose objects and sketch them until the total length is as close to one meter as possible.

d. How much of a meter is left? Express your answer as a decimal and as a fraction.

e. Which object is longest? Which object is shortest? Explain how you found your answers.

## Share & Summarize

Suppose you are given a measurement in centimeters.

1. How would you move the decimal point to change the measurement to meters? Explain why this technique works.

2. How would you move the decimal point to change the measurement to millimeters? Explain why this technique works.

# *Inquiry*
# Investigation ③ Compare and Order Decimals

You will now play a game that will give you practice finding decimals between other decimals.

## Guess My Number

### *Guess My Number* Game Rules

- Player 1 thinks of a number between 0 and 10 with no more than four decimal places and writes it down so that Player 2 cannot see it.
- Player 2 asks "yes" or "no" questions to try to figure out the number. Player 2 writes down each question and answer on a record sheet.

| Question | Answer | What I Know about the Number |
|---|---|---|
| Is the number greater than 6? | No | It is less than or equal to 6. |
| Is the number less than 3? | No | It is between 3 and 6 including 3 and 6. |
|  |  |  |

- Play continues until Player 2 guesses the number. Player 1 receives one point for each question Player 2 asked.
- The winner is the player with the most points after four rounds.

## Try It Out

Play four rounds with your partner, switching roles for each round. Then, with your partner, look closely at your record sheet for the game.

1. What strategies did you find helpful when you asked questions?

2. Once you knew which whole numbers the answer was between, did your strategy change? Explain.

3. Who was able to ask the least questions to find the numbers? What pattern, if any, did you notice?

**4.** Jessica and Kali are playing *Guess My Number.* Kali is the Asker. Here is her record sheet.

| Question | Answer |
|---|---|
| Is it greater than 1? | Yes |
| Is it between 4 and 10? | Yes |
| Is it between 5 and 10? | Yes |
| Is it greater than 8? | No |
| Is it between 7 and 8? | Yes |

**a.** What do you think of Kali's questions? What, if anything, do you think she should have done differently?

**b.** What question do you think Kali should ask next?

## What Did You Learn?

**5.** Suppose you have asked several questions and you know the number is between 4.71 and 4.72. List at least four possibilities for the number.

**6.** What is the greatest decimal that can be made in this game? What is the least decimal that can be made in this game?

## On Your Own Exercises

**Lesson 2.2**

***Practice & Apply***

1. In his first two turns in the *Spare Change* game, Maurice chose $0.03 and $0.8.

   **a.** Complete a dollar chart showing the amount Maurice should have shaded after his first two turns.

   **b.** What part of a dollar is shaded on Maurice's chart? Express your answer as a fraction and as a decimal.

   **c.** How much more does Maurice need to have $1.00? Express your answer as a fraction and as a decimal.

2. Ms. Picó added cards with three decimal places to the *Spare Change* game deck. In her first three turns, Una chose $0.77, $0.1, and $0.115.

   **a.** Complete a dollar chart showing the amount Una should have shaded after her first three turns.

   **b.** What part of a dollar is shaded on Una's chart? Express your answer as a fraction and as a decimal.

**Write each decimal as a mixed number.**

**3.** 1.99  **4.** 7.016  **5.** 100.5

**Find each product without using a calculator.**

**6.** $100 \cdot 0.0436$  **7.** $100,000 \cdot 754.01$  **8.** $1,000 \cdot 98.9$

**Find each quantity without using a calculator.**

**9.** $\frac{1}{10}$ of 645  **10.** $7.7 \div 1,000$  **11.** $\frac{1}{10,000}$ of 55.66

**Measurement** In Exercises 12–14, convert each measurement to meters. Write your answers as both fractions and decimals.

**12.** 50 cm  **13.** 50 mm  **14.** 700 mm

**15.** Give the length of the baseball bat in centimeters and in meters.

**Give the nearest tenths of a centimeter that each given measurement is between, such as 3.66 is between 3.6 and 3.7.**

**16.** 5.75 cm  **17.** 0.25 cm  **18.** 1.01 cm

**Give the nearest hundredths of a meter that each given measurement is between, such as 2.865 is between 2.86 and 2.87.**

**19.** 0.555 m  **20.** 1.759 m  **21.** 0.0511 m

**Order each set of numbers from least to greatest.**

**22.** 7.31, 7.4, 7.110, 7.3, 7.04, 7.149

**23.** 21.5, 20.50, 22.500, 20.719, 21.66, 21.01, 20.99

**24.** Participants in the school gymnastics meet are scored on a scale from 1 to 10 with 10 being the highest score. To the right are the scores for the first event. Ryan has not yet had his turn.

| Student | Score |
|---------|-------|
| Kent | 9.4 |
| Elijah | 8.9 |
| Santiago | 9.25 |
| Matthew | 8.85 |
| Ernesto | 9.9 |
| Ryan | |
| Tyler | 9.1 |
| Craig | 8.0 |
| Alvin | 8.7 |
| Pierce | 9.2 |

**a.** List the students from highest score to lowest score.

**b.** Ryan is hoping to get third place in this event. List five possible scores that would put him in third place.

*Connect & Extend*

**25. Economics** The FoodStuff market is running the following specials.

• Bananas: $0.99 per pound

• Swiss cheese: $3.00 per pound

• Rolls: $0.25 each

**a.** Jenna paid $9.90 for bananas. How many pounds did she buy?

**b.** Allie bought $\frac{1}{10}$ of a pound of Swiss cheese. How much did the cheese cost?

**c.** Ms. Washington is organizing the school picnic. How many rolls can she purchase with $250?

**d.** Ms. Washington decides to buy some rolls, cheese, and bananas for $250. What combination of food could she buy without going over her budget?

**26. Economics** A grocery store flyer advertises bananas for 0.15¢ each. Does this make sense? Explain.

**27. In Your Own Words** Explain the purpose of the decimal point in decimal numbers.

**28.** Today is Tony's 10th birthday. His parents have decided to start giving him a monthly allowance, but they each suggest a different plan.

• Tony's mother wants to give him $0.01 each month this year, $0.10 each month next year, $1.00 each month the third year, and so on, multiplying the monthly amount by 10 each year until Tony's 16th birthday.

**Real-World Link**

A nanoguitar is about the size of a human being's white blood cell.

• Tony's father wants to give him $10 each month this year, $20 each month next year, $30 each month the next year, and so on, adding $10 to the monthly amount each year until Tony's 16th birthday.

His parents told Tony he could decide which plan to use. Which plan do you think he should choose? Explain your reasoning.

29. **Science** *Nanotechnology* is a branch of science that focuses on building very small objects from molecules. These tiny objects are measured with units such as microns and nanometers.

   • 1 micron = 1 millionth of a meter
   • 1 nanometer = 1 billionth of a meter

   **a.** The photo to the left is a nanoguitar. Although this guitar is only 10 microns long, it actually works. However, the sound it produces cannot be heard by the human ear. Express the length of the nanoguitar in meters. Give your answer as a decimal and as a fraction.

   **b.** Two human hairs, side by side, would be about 0.001 meter wide. What fraction of this width is the length of the nanoguitar?

   **c.** Microchips inside the processors of computers can have widths as small as 350 nanometers. Express this width in meters. Give your answer as a fraction and as a decimal.

   **d.** A paper clip is about 0.035 meter long. What fraction of the length of a paper clip is the width of a microchip?

30. How much greater than 5.417 meters is 5.42 meters?

31. If a person is 2 meters 12 centimeters tall, we can say that he is 2.12 meters tall. If a person is 5 feet 5 inches tall, can we say that she is 5.5 feet tall? Why or why not?

**Economics** The table below gives the value of foreign currencies in U.S. dollars in 2007.

| Currency | Value in U.S. Dollars |
|----------|----------------------|
| Australian dollar | 0.8989 |
| British pound | 2.0486 |
| Canadian dollar | 1.0326 |
| Chinese renminbi | 0.1333 |
| Danish krone | 0.1913 |
| Russian rouble | 0.0402 |
| Mexican new peso | 0.0923 |
| Singapore dollar | 0.6837 |

BUREAU DE CHANGE

| RATES PER £1.00 | WE SELL | WE BUY |
|-----------------|---------|--------|
| U.S.A. | 1.52 | 1.60 |
| BELGIUM | 47.87 | 51.47 |
| CANADA | 2.035 | 2.24 |
| DENMARK | 8.75 | 9.65 |
| FRANCE | 7.80 | 8.40 |
| GERMANY | 2.30 | 2.53 |
| IRELAND | 0.96 | 1.05 |
| ITALY | 2350 | 2548 |
| JAPAN | 165 | 181 |
| NETHERLANDS | 2.60 | 2.80 |
| PORTUGAL | 239 | 256 |
| SPAIN | 192.00 | 208.73 |
| SWEDEN | 10.20 | 11.10 |
| SWITZERLAND | 1.86 | 2.02 |
| MIN'CHARGE £ | 3.00 | 3.00 |
| COM'RATE % | 2.00 | 5.00 |

**32.** If you exchanged one Canadian dollar for U.S. currency, how much money would you receive? Assume that values are rounded to the nearest penny.

**33.** Of those listed in the table, which currency is worth the most in U.S. dollars?

**34.** Of those listed in the table, which currency is worth the least in U.S. dollars?

**35.** Which currency listed in the table is worth closest to one U.S. dollar? How much more or less than one U.S. dollar is this currency worth?

**36.** How many Russian roubles could you exchange for one dime?

**Mixed Review**

**Geometry** Draw a polygon matching each description, if possible. If it is not possible, say so.

**37.** a concave pentagon

**38.** a triangle with exactly two lines of symmetry

**39.** a quadrilateral that is not regular and that has two lines of symmetry

**Geometry** Find the perimeter of each figure.

**40.**

2.19 cm

3.65 cm

**41.**

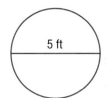

5 ft

**42.**

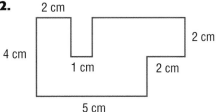

2 cm

4 cm

1 cm

2 cm

2 cm

5 cm

**43.**

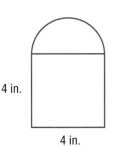

4 in.

4 in.

# LESSON 2.3

# Fraction and Decimal Equivalents

Fractions and decimals are two ways of expressing quantities that are not whole numbers. You already know how to write a decimal in fraction form by thinking about the place values of its digits. In this lesson, you will find decimals that estimate the values of fractions. You will also learn to write fractions in decimal form.

## Think & Discuss

Use both a fraction and a decimal to describe the approximate location of each point.

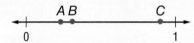

What methods did you use to make your estimates?

# Investigation 1 Estimate Equivalents

## Materials

• copy of the number-line diagram

Number lines can help you understand the relationship between decimals and fractions. The diagram on page 89 shows ten number lines. The first is labeled with decimals, and the others are labeled with fractions. The fractions on each number line have the same denominator.

## ✓ Develop & Understand: A

1. Describe at least two patterns you notice in the diagram on page 89.

2. Consider decimal values greater than 0.5.

   a. Choose any decimal greater than 0.5 and less than 1.

   b. Find all the fractions in the diagram that appear to be equivalent to the decimal that you chose. State if there are no equivalent fractions.

   c. Find two fractions in the diagram that are a little less than your decimal and two fractions that are a little greater. Try to find fractions as close to your decimal as possible.

# Decimal and Fraction Number Lines

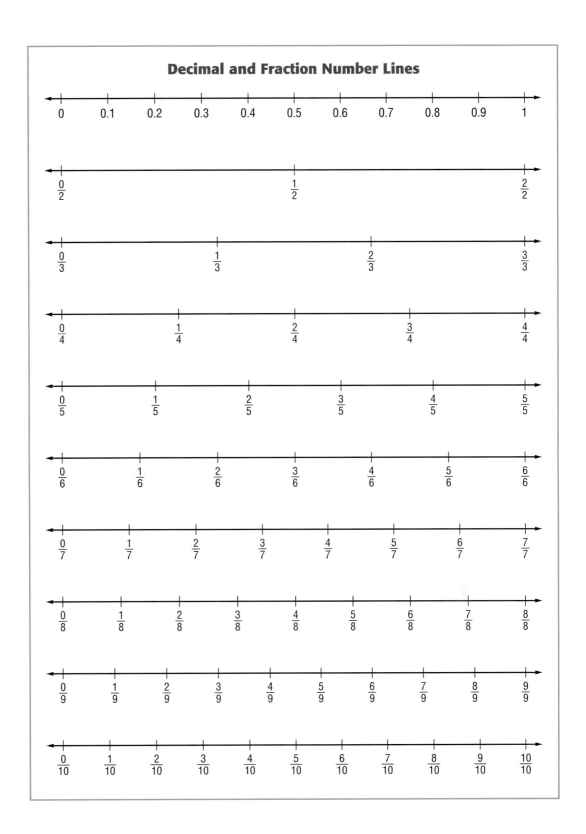

3. Refer to the diagram on page 89. Consider the fractions that are not equivalent to $\frac{1}{2}$.

   a. Choose any fraction in the diagram that is not equivalent to $\frac{1}{2}$.

   b. Name the fraction on each of the other fraction number lines that is closest to the fraction that you chose.

   c. If possible, find an exact decimal value for your fraction to the hundredths place. If this is not possible, tell which two decimals your fraction is between, to the hundredths place.

   d. Repeat Parts a–c for two more fractions. Each fraction you choose should have a different denominator.

4. Decimals that are familiar make good benchmarks for estimating the values of other decimals and fractions.

   a. Which decimals do you think would be useful as benchmarks? Explain why you chose those numbers

   b. For each fraction labeled in the diagram, find the benchmark decimal from Part a closest to that fraction. Organize your answers in any way you like.

In Exercises 5–10, you will estimate fractions and decimals greater than 1.

## ✓ Develop & Understand: B

For Exercises 5–7, use both a mixed number and a decimal to describe the approximate location of each point.

5.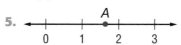

6.

7. 
0  1  2  3  4  5  6  7  8  C

8. Consider the mixed number $3\frac{1}{3}$.

   a. Which of the following decimals is closest to $3\frac{1}{3}$? Explain how you decided.

   $$3.1 \qquad 3.2 \qquad 3.3 \qquad 3.5$$

   b. Find a decimal that is even closer to $3\frac{1}{3}$. Explain how you found your answer.

9. Consider the mixed number $81\frac{5}{6}$.

   a. Which of the following decimals is closest to $81\frac{5}{6}$? Explain how you decided.

   $$81.5 \qquad 81.6 \qquad 81.8 \qquad 81.9$$

   b. Find a decimal that is even closer to $81\frac{5}{6}$. Explain how you found your answer.

10. Use your estimation skills to order these numbers from least to greatest.

   $$\frac{5}{6} \qquad 0.7 \qquad \frac{2}{5} \qquad 3.1 \qquad 0.5 \qquad 3\frac{1}{3} \qquad 3.6 \qquad 5.2$$

## Share & Summarize

Below is part of a student's homework assignment. Using your estimation skills, find the answers that are definitely incorrect. Explain how you know.

1. $\frac{2}{3} = 0.9$  2. $3\frac{7}{10} = 3.7$  3. $\frac{1}{5} = 2.0$  4. $2\frac{3}{4} = 2.75$

# Investigation 2 Change Decimals to Fractions

## Vocabulary

**terminating decimal**

## Materials

• 100-grid

In this investigation, you will explore the process for writing fractions as decimals. You will also look at problems that involve ordering numbers in both decimal and fraction form as you consider the merits of each.

## ✅ Develop & Understand: A

1. In each of the following, represent the shaded portion as a decimal. Then write the decimal as a fraction. How does place value help you do this?

   a.

   b.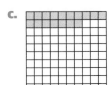

   c.

A decimal that terminates, stops, or ends is called a **terminating decimal**. Some examples of terminating decimals are 0.25, 1.38, 4.5.

2. Copy or draw blank grids. Shade in the following decimals.

   a. 0.3

   b. 0.62

   c. 1.07

   d. 0.065

3. Write each of the following decimals as a fraction in simplest form.

   a. 0.34

   b. 0.453

   c. 1.3

   d. 0.2875

4. Look at the decimals you converted to fractions in Exercise 3.

   a. Write the decimals in order from least to greatest.

   b. Would it be easier to order these numbers as decimals or as fractions? Explain your thinking.

## ✓ Develop & Understand: B

5. Name a decimal between 0.99 and 1. Explain the strategy you used.

6. Name the decimal which is exactly halfway between 0.99 and 1.

7. Name a decimal between 2.34 and 2.341. Explain the strategy you used.

8. Name the decimal that is exactly halfway between 2.34 and 2.341.

9. How many fractions are there between 2.34 and 2.341? Explain your thinking.

**Here is a game that you can play with a partner. It will sharpen your skills with decimals and fractions.**

   • Label you and your partner A and B. Player A will go first.

   • You will be given two decimals for each game.

Round 1: Player A states a decimal that is between the two given decimals.

Player B also states a decimal (different from Players A's choice) that is between the two given decimals. Players A and B check each other to see if he or she has given correct decimals.

Round 2: Player B states a decimal between the two decimals that were provided by Players A and B in the first round. Player A does the same. Players A and B check each other to see if he or she has given correct decimals.

Round 3: The selection continues for three more rounds. Play five rounds total. Players should alternate who goes first in each round.

Players score 1 point if they:
- correctly identify a decimal between the two decimals in any given round
- find a mistake that the other player made

Here are decimals you can use for playing the game.

Game 1: 1.5 and 1.6

Game 2: 0.71 and 0.72

Repeat the same game, but use fractions this time. Here are fractions you can use for playing the game.

Game 1: $\frac{3}{5}$ and $\frac{4}{5}$

Game 2: $\frac{5}{11}$ and $\frac{6}{11}$

10. Which game do you think is easier? Explain.

## Share & Summarize

1. Explain the process for converting decimals to fractions.

2. When comparing numbers, is it easier to compare them as fractions or as decimals? Explain your reasoning.

## Vocabulary

repeating decimal

In Investigation 1, you found decimals approximations for given fractions. Now you will find exact decimal values for fractions.

### Think & Discuss

Find a decimal equivalent to each fraction.

$$\frac{47}{1,000} \qquad \frac{59}{10}$$

$$\frac{7}{20} \qquad \frac{3}{5}$$

Here is how Jing and Marcus found a decimal equivalent to $\frac{7}{20}$.

### Example

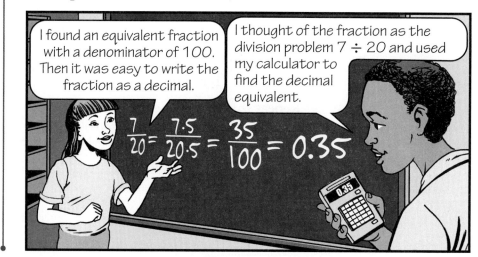

### ✓ Develop & Understand: A

**Find a decimal equivalent to each given fraction.**

1. $\frac{16}{25}$

2. $\frac{5}{8}$

3. $\frac{320}{200}$

4. $\frac{8}{125}$

**5.** Without using a calculator, try to find a decimal equivalent to $\frac{2}{3}$ by dividing. What happens?

When you divided to find a decimal equivalent to $\frac{2}{3}$, you got 0.6666 ..., where the 6s repeat forever. Decimals with a pattern of digits that repeat without stopping are called **repeating decimals**. Repeating decimals are usually written with a bar over the repeating digits.

- $0.\overline{6}$ means 0.66666 ...
- $3.1\overline{24}$ means 3.1242424 ...

All fractions whose numerators and denominators are whole numbers have decimal equivalents that end, like 0.25, or that repeat forever, like $0.4\overline{6}$.

Calculators are useful for determining whether a fraction is equivalent to a repeating decimal. However, because the number of digits a calculator can display is limited, you sometimes cannot be certain.

## Think & Discuss

Use your calculator to find a decimal approximation for each fraction below. Write your answers exactly as they appear in the calculator's display.

$$\frac{6}{9} \qquad \frac{6}{13} \qquad \frac{6}{15}$$

$$\frac{6}{17} \qquad \frac{6}{21} \qquad \frac{666}{1,000}$$

Which of the fractions above definitely *are* equivalent to repeating decimals?

Which of the fractions definitely *are not* equivalent to repeating decimals?

Which of the fractions above do you *think* might be repeating decimals? How could you find out for sure?

6. Copy the chart below. Fill in the decimal equivalents for each fraction in the chart. Some of the cells have been filled for you. Start by writing the decimal equivalents you know. Then use your calculator to find the others.

| $\frac{0}{1}$ | $\frac{0}{2}$ | $\frac{0}{3}$ | $\frac{0}{4}$ | $\frac{0}{5}$ | $\frac{0}{6}$ | $\frac{0}{7}$ | $\frac{0}{8}$ | $\frac{0}{9}$ | $\frac{0}{10}$ |
|---|---|---|---|---|---|---|---|---|---|
| $\frac{1}{1}$ | $\frac{1}{2}$ | $\frac{1}{3}$ | $\frac{1}{4}$ <br> 0.25 | $\frac{1}{5}$ | $\frac{1}{6}$ | $\frac{1}{7}$ <br> $0.\overline{142857}$ | $\frac{1}{8}$ | $\frac{1}{9}$ | $\frac{1}{10}$ |
| $\frac{2}{1}$ | $\frac{2}{2}$ | $\frac{2}{3}$ | $\frac{2}{4}$ | $\frac{2}{5}$ | $\frac{2}{6}$ | $\frac{2}{7}$ <br> $0.\overline{285714}$ | $\frac{2}{8}$ | $\frac{2}{9}$ | $\frac{2}{10}$ |
| $\frac{3}{1}$ | $\frac{3}{2}$ | $\frac{3}{3}$ | $\frac{3}{4}$ | $\frac{3}{5}$ | $\frac{3}{6}$ | $\frac{3}{7}$ <br> $0.\overline{428571}$ | $\frac{3}{8}$ | $\frac{3}{9}$ <br> $0.\overline{3}$ | $\frac{3}{10}$ |
| $\frac{4}{1}$ | $\frac{4}{2}$ | $\frac{4}{3}$ | $\frac{4}{4}$ | $\frac{4}{5}$ | $\frac{4}{6}$ | $\frac{4}{7}$ <br> $0.\overline{571428}$ | $\frac{4}{8}$ | $\frac{4}{9}$ | $\frac{4}{10}$ |
| $\frac{5}{1}$ | $\frac{5}{2}$ | $\frac{5}{3}$ | $\frac{5}{4}$ | $\frac{5}{5}$ | $\frac{5}{6}$ | $\frac{5}{7}$ <br> $0.\overline{714285}$ | $\frac{5}{8}$ | $\frac{5}{9}$ | $\frac{5}{10}$ |
| $\frac{6}{1}$ | $\frac{6}{2}$ <br> 3 | $\frac{6}{3}$ | $\frac{6}{4}$ | $\frac{6}{5}$ | $\frac{6}{6}$ | $\frac{6}{7}$ <br> $0.\overline{857142}$ | $\frac{6}{8}$ | $\frac{6}{9}$ | $\frac{6}{10}$ |
| $\frac{7}{1}$ | $\frac{7}{2}$ | $\frac{7}{3}$ | $\frac{7}{4}$ | $\frac{7}{5}$ | $\frac{7}{6}$ | $\frac{7}{7}$ | $\frac{7}{8}$ | $\frac{7}{9}$ | $\frac{7}{10}$ |
| $\frac{8}{1}$ | $\frac{8}{2}$ | $\frac{8}{3}$ | $\frac{8}{4}$ | $\frac{8}{5}$ | $\frac{8}{6}$ | $\frac{8}{7}$ | $\frac{8}{8}$ | $\frac{8}{9}$ | $\frac{8}{10}$ |
| $\frac{9}{1}$ | $\frac{9}{2}$ | $\frac{9}{3}$ | $\frac{9}{4}$ | $\frac{9}{5}$ | $\frac{9}{6}$ | $\frac{9}{7}$ | $\frac{9}{8}$ | $\frac{9}{9}$ | $\frac{9}{10}$ |
| $\frac{10}{1}$ | $\frac{10}{2}$ | $\frac{10}{3}$ | $\frac{10}{4}$ | $\frac{10}{5}$ | $\frac{10}{6}$ | $\frac{10}{7}$ | $\frac{10}{8}$ | $\frac{10}{9}$ | $\frac{10}{10}$ |

7. Which columns contain fractions equivalent to repeating decimals?

8. Describe at least two patterns that you see in the completed chart.

Save your chart for the On Your Own Exercises and for Investigation 4.

1. Describe a method for finding a decimal equivalent to a given fraction.

2. Describe a repeating decimal. Give an example of a fraction whose decimal equivalent is a repeating decimal.

# Investigation 4 Patterns in Fractions and Decimals

## Materials

- completed fraction and decimal equivalents chart

In Exercises 1–14, you will look for patterns in your chart of fraction and decimal equivalents from Investigation 3.

### ✓ Develop & Understand: A

For Exercises 1–2, find all of the fractions in the chart that are equivalent to each given decimal.

1. 1.25

2. $0.\overline{6}$

3. Color all of the cells with fractions equivalent to $\frac{1}{2}$. What pattern do you notice? Why does this happen?

4. Does the chart have a column showing fractions with a denominator of 0? Why or why not?

5. Look at the column containing fractions with denominator 10.

   a. Describe the pattern in the decimals in this column. Explain why this pattern occurs.

   b. Write $5\frac{7}{10}$ as a decimal.

   c. Write 68.3 as a mixed number.

6. Look at the column containing fractions with denominator 2.

   a. How do the decimal values change as you move down the column? Why?

   b. Use the pattern from Part a to find the decimal equivalent of $\frac{11}{2}$, the number that would be next in the chart if a row was added.

7. Look at the column containing fractions with denominator 4.

   a. How do the decimal values change as you move down the column? Why?

   b. Use the pattern from Part a to find the decimal equivalent of $\frac{11}{4}$, the number that would be next in the chart if a row was added.

8. Look again at the fractions with denominator 2. The decimals in this column are 0, 0.5, 1, 1.5, 2, 2.5, 3, 3.5, 4, 4.5, 5. Notice that the "decimal parts" alternate between 0 (no decimal part) and 0.5.

   a. Look for a similar pattern in the column for fractions with denominator 4. Describe the pattern.

   b. Look for a similar pattern in the column for fractions with denominator 6. Describe the pattern.

   c. Look at a few other columns. Do similar patterns hold?

   d. **Challenge** Explain why these patterns occur.

**Use the chart to help find the decimal equivalent for each fraction or mixed number.**

9. $10\frac{1}{2}$

10. $32\frac{7}{9}$

11. $62\frac{4}{5}$

12. $23\frac{5}{6}$

**Use the chart to help find a fraction equivalent to each decimal.**

13. $14.125$

14. $4.\overline{6}$

# ✅ *Develop & Understand: B*

In Exercises 15–16, you will use what you have learned about comparing and converting between fractions and decimals to build a fraction tower.

## Building a Fraction Tower

- Choose a fraction less than 1 whose numerator and denominator are whole numbers between 1 and 9. Write both the fraction and its decimal equivalent on the bottom level of the tower.

- Choose another fraction whose numerator and denominator are between 1 and 9 and that is *less than* the fraction in the bottom level. Write the fraction and its decimal equivalent on the next level of the tower.

- Continue this process of choosing fractions and adding levels until you are unable to make a fraction with a value less than the fraction in the top level.

**Real-World Link**

The tallest building in the world, as measured to the top of the roof, is the Burj Dubai in Dubai, United Arab Emirates, at 1,885 feet.

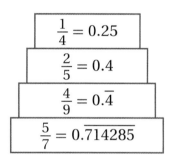

$$\frac{1}{4} = 0.25$$
$$\frac{2}{5} = 0.4$$
$$\frac{4}{9} = 0.\overline{4}$$
$$\frac{5}{7} = 0.\overline{714285}$$

**15.** Work with your partner to build several fraction towers. Record each tower that you build. Try to build the highest tower possible.

**16.** What strategies did you use when building your towers?

## Share & Summarize

1. Choose a pattern in the chart of fraction and decimal equivalents. It can be a pattern you discussed in class or a new pattern you have discovered. Describe the pattern, and explain why it occurs.

2. Write a letter to a student who is just learning how to build fraction towers. Explain strategies that he or she might use to build a high tower.

# On Your Own Exercises

**Lesson 2.3**

**Practice & Apply**

For Exercises 1–3, use the number-line diagram on page 89.

**1.** Name the three fractions in the diagram that are closest to 0.8.

**2.** Tell which two decimals $\frac{6}{7}$ is between, to the hundredths place.

**3.** Tell which two decimals $\frac{7}{8}$ is between, to the hundredths place.

**4.** Use a mixed number and a decimal to approximate the location of each point.

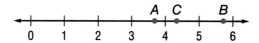

**5.** Consider the fraction $\frac{17}{8}$.

**a.** Which of these decimals is closest to $\frac{17}{8}$? Explain how you decided.

   2.1    2.2    2.8    2.9    17.1    17.8

**b.** Give a decimal that would be even closer to $\frac{17}{8}$. Explain how you found your answer.

**Write the decimal and fraction from the model. Simplify each fraction.**

**6.**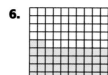

**7.** ☐☐☐☐☐☐☐☐☐☐

**8.** Which of the following is equivalent to $1.2\overline{34}$?

   1.23412341234 …    1.23434343434 …    1.23444444 …

**9.** Which of the following is equivalent to 2.393939 …?

   $2.3\overline{9}$    $2.\overline{39}$

**Find a decimal equivalent for each fraction or mixed number.**

**10.** $\frac{16}{5}$

**11.** $\frac{15}{11}$

**12.** $\frac{70}{250}$

**13.** $\frac{33}{24}$

**14.** $\frac{14}{3}$

**15.** $\frac{376}{20,000}$

**16.** $5\frac{1}{16}$

**17.** $\frac{9}{12}$

**18.** In the chart you completed for page 96, color all of the cells with fractions equivalent to $\frac{1}{3}$.

    **a.** What pattern do you notice? Why does this happen?

    **b.** Why doesn't every column have a colored cell?

**Use the chart that you completed for page 96 to help find a decimal equivalent to each fraction or mixed number.**

**19.** $\frac{12}{5}$      **20.** $32\frac{7}{10}$      **21.** $65\frac{2}{3}$      **22.** $3\frac{1}{7}$

**Use the chart you completed for page 96 to help find a fraction or mixed number equivalent to each decimal.**

**23.** 4.125      **24.** 32.5      **25.** 4.75      **26.** $8.\overline{1}$

**27.** Refer to the chart that you completed for page 96. Look at the column containing fractions with denominator 5.

    **a.** How do the decimal values change as you move down the column? Why?

    **b.** Use the pattern in Part a to find the decimal equivalent of $\frac{11}{5}$, the number that would be next in the chart if a row was added.

In Exercises 28 and 29, the start of a fraction tower is given. If you want to build the highest tower possible, what number should you choose for the next level? Explain your answer. Give your number in both fraction and decimal form.

**28.**

| $\frac{6}{9} = 0.\overline{6}$ |
| $\frac{7}{9} = 0.\overline{7}$ |
| $\frac{8}{9} = 0.\overline{8}$ |

**29.**

| $\frac{3}{9} = 0.\overline{3}$ |
| $\frac{4}{5} = 0.8$ |

**Real-World Link**

Devils Tower National Monument in northeast Wyoming rises 1,267 feet above the nearby Belle Fourche River.

**Represent each number as a fraction or mixed number. Simplify each fraction.**

**30.** Twenty-eight and six-tenths

**31.** Nine and two-hundredths

**32.** One hundred fifteen and twelve-thousandths

**33.** Order these numbers from least to greatest.

$$2.3 \qquad \frac{11}{5} \qquad 2\frac{3}{7} \qquad 2.05$$

**34. In Your Own Words** Explain how to find a fraction between any two given fractions. Give an example to illustrate your method.

**Connect & Extend**

**35.** Copy the number line. Mark and label the point corresponding to each number.

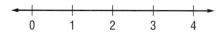

Point A: 2.4    Point B: $3\frac{8}{9}$    Point C: 0.67    Point D: 1.2

**36.** Look back at the number-line diagram on page 89. Imagine creating two more number lines to represent 11ths and 12ths.

**a.** Is $\frac{1}{11}$ less than or greater than 0.1? How do you know?

**b.** Is $\frac{1}{12}$ less than or greater than $\frac{1}{11}$? How do you know?

**c.** Is $\frac{10}{11}$ less than or greater than 0.9? How do you know?

**d.** Is $\frac{11}{12}$ less than or greater than $\frac{10}{11}$? How do you know?

**37.** Look back at the number-line diagram on page 89. Notice that 0.5 is between $\frac{1}{3}$ and $\frac{2}{3}$.

**a.** Between which two fifths is 0.5?

**b.** Between which two sevenths is 0.5?

**c.** Between which two ninths is 0.5?

Now think beyond the diagram.

**d.** Between which two 15ths is 0.5?

**e.** Between which two 49ths is 0.5?

**In each pair, tell which fraction is closer to 0.5.**

**38.** $\frac{2}{5}$ or $\frac{3}{7}$

**39.** $\frac{4}{9}$ or $\frac{11}{23}$

**40.** $\frac{2}{5}$ or $\frac{4}{5}$

**41.** $\frac{4}{9}$ or $\frac{6}{9}$

**42.** $\frac{3}{17}$ or $\frac{16}{17}$

**43.** $\frac{19}{44}$ or $\frac{27}{44}$

**44.** Use prime factorization to name two fractions that are repeating decimals, and two fractions that are terminating decimals. Explain how you know.

**45.** Refer to your completed chart from page 96.

    **a.** Look at the column containing fractions with denominator 9. What pattern do you see as you move down this column?

    **b.** Find the decimal equivalents of $\frac{1}{99}, \frac{2}{99}, \frac{3}{99}$, and so on, up to $\frac{9}{99}$. What pattern do you see?

    **c.** Find the decimal equivalents of $\frac{1}{999}, \frac{2}{999}, \frac{3}{999}$, and so on, up to $\frac{9}{999}$. What pattern do you see?

    **d.** Predict the decimal equivalents for $\frac{1}{9,999}, \frac{5}{9,999}$, and $\frac{8}{9,999}$. Use your calculator to check your prediction.

**46.** Consider fractions with denominator 11.

    **a.** Find decimal equivalents for $\frac{1}{11}, \frac{2}{11}, \frac{3}{11}, \frac{4}{11}$, and $\frac{5}{11}$. What pattern do you see?

    **b.** Use the pattern you discovered to predict the decimal equivalents of $\frac{7}{11}$ and $\frac{9}{11}$.

**47.** Refer to your completed chart from page 96. Compare the *row* containing fractions with *numerator* 10 to the row containing fractions with numerator 1. The decimals for fractions with numerator 10 are the same as those for fractions with numerator 1, except the decimal point is moved one place to the right. Explain why.

**48.** In your completed chart from page 96, color the cells containing fractions equivalent to 1. Use a different color.

    **a.** Describe the pattern you see.

    **b.** The numbers below the line of 1s are greater than 1. The numbers above the line of 1s are less than 1. Why?

In Exercises 49–52, the rules for building a fraction tower have been changed. Tell what fraction you would choose as your starting number in order to build the tallest possible tower.

**49.** Use whole numbers between 1 and 20 for numerators and denominators. As before, fractions must be less than 1.

**50.** Use whole numbers between 1 and 9 for numerators and denominators. There is no limit on the value of the fraction.

**51.** Use whole numbers between 1 and 9 for numerators and denominators. The fractions must be less than $\frac{1}{2}$.

**52.** Suppose the rules for building a fraction tower are changed so you can choose only numbers between 1 and 4 for numerators and denominators. Fractions still must be less than 1. Draw a tower with the maximum number of levels. Explain why it is the maximum.

**Mixed Review**

**Geometry** Estimate the measure of each angle.

**53.**

**54.**

**55.**

**56.**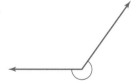

Use the fact that $13 \cdot 217 = 2{,}821$ to find each product *without* using a calculator.

**57.** $1.3 \cdot 217$

**58.** $13 \cdot 2.17$

**59.** $0.013 \cdot 21.7$

**60.** $0.13 \cdot 0.217$

**61.** $1{,}300 \cdot 2{,}170$

**62.** $13 \cdot 0.0217$

# Review & Self-Assessment

## Chapter Summary

In this chapter, you used what you learned about factors and multiples to find *equivalent fractions* and to compare fractions. You also saw how you could use *benchmark fractions* like $\frac{1}{4}$, $\frac{1}{2}$, and $\frac{2}{3}$ to estimate the values of more complicated fractions.

You reviewed the meaning of decimals, and you investigated how multiplying or dividing by 10, 100, 1,000, 10,000, and so on affects the position of the decimal point. Then you compared and ordered decimals and found a number between two given decimals.

You saw how you could write a fraction as a decimal. You discovered that sometimes the decimal representation of a fraction is a *repeating decimal* with a pattern of digits that repeats forever.

## Vocabulary

equivalent fractions

lowest terms

mixed number

repeating decimal

## Strategies and Applications

The questions in this section will help you review and apply the important ideas and strategies developed in this chapter.

### Finding equivalent fractions and comparing fractions

1. Find two fractions equivalent to $\frac{4}{6}$. Use diagrams or another method to explain why the fractions are equivalent to $\frac{4}{6}$.

2. Explain what it means for a fraction to be in *lowest terms*. Then describe a method for writing a given fraction in lowest terms.

3. Describe two methods for comparing fractions. Then use one of the methods to determine whether $\frac{7}{10}$ is greater than, less than, or equal to $\frac{8}{11}$.

### Understanding and comparing decimals

4. Describe a rule for comparing two decimals. Demonstrate your rule by comparing 307.63 with 308.63 and by comparing 3.786 with 3.779.

5. How does the value of a number change when you move the decimal point three places to the right? Illustrate your answer with an example.

6. How does the value change when you move the decimal point two places to the left? Illustrate your answer with an example.

## Converting decimals to fractions and fractions to decimals

**7.** Explain how you would write a decimal in fraction form. Illustrate by writing 0.97 and 0.003 as fractions.

**8.** Explain how you would find a decimal equivalent to a given fraction. Give an example.

**9.** Explain what a *repeating decimal* is. Give an example of a fraction that is equivalent to a repeating decimal.

**List three fractions equivalent to each given fraction.**

**10.** $\frac{6}{7}$        **11.** $\frac{13}{39}$        **12.** $\frac{32}{720}$

**Replace each $\bigcirc$ with $<$, $>$, or $=$ to make a true statement.**

**13.** $\frac{5}{16} \bigcirc \frac{7}{24}$        **14.** $\frac{14}{22} \bigcirc \frac{35}{55}$        **15.** $\frac{9}{49} \bigcirc \frac{3}{14}$

**Order each set of decimals from least to greatest.**

**16.** 0.7541, 1.754, 0.754, 0.75411, 0.7641

**17.** 251.889, 249.9, 251.9, 251.8888, 252.000001

**Compute each result mentally.**

**18.** $0.00012 \cdot 1{,}000$      **19.** $\frac{1}{10{,}000}$ of 344      **20.** $100 \cdot 77.5$

**Find a number between each given pair of numbers.**

**21.** 11.66 and 11.67

**22.** 0.0001 and 0.001

**23.** 3.04676 and 3.04677

**Write each fraction or mixed number in decimal form.**

**24.** $\frac{7}{8}$                 **25.** $\frac{8}{6}$

**26.** $\frac{11}{15}$               **27.** $5\frac{17}{20}$

**Represent each number as a fraction or mixed number. Simplify each fraction.**

**28.** forty-two hundredths

**29.** six and eight tenths

**Give your answer as a mixed number and a fraction.**

**30.** If 14 slices of pizza are divided among four people, what portion of pizza will each person receive?

**31.** Joan worked for 36 hours last week. What portion of a day is this?

**32.** Find two fractions equivalent to $\frac{3}{9}$. Use diagrams or another method to explain why the fractions are equivalent to $\frac{3}{9}$.

**Replace each $\bigcirc$ with <, >, or = to make a true statement.**

**33.** $\frac{9}{36} \bigcirc \frac{10}{40}$

**34.** $\frac{17}{4} \bigcirc \frac{9}{12}$

**35.** $\frac{28}{52} \bigcirc \frac{12}{50}$

**36.** $\frac{72}{27} \bigcirc \frac{8}{3}$

**Write each fraction or mixed number in decimal form.**

**37.** $\frac{3}{5}$

**38.** $\frac{4}{12}$

**39.** $\frac{70}{15}$

**40.** $\frac{52}{8}$

# Test-Taking Practice

**SHORT RESPONSE**

**1** Determine whether $\frac{5}{7}$ is greater than, less than, or equal to $\frac{7}{9}$.

**Show your work.**

Answer _____

**MULTIPLE CHOICE**

**2** Which set of fractions are equivalent?

**A** $\frac{6}{24}, \frac{3}{8}, \frac{12}{36}, \frac{5}{16}$

**B** $\frac{20}{32}, \frac{5}{9}, \frac{15}{24}, \frac{10}{18}$

**C** $\frac{4}{5}, \frac{8}{10}, \frac{15}{20}, \frac{11}{15}$

**D** $\frac{18}{21}, \frac{6}{7}, \frac{12}{14}, \frac{30}{35}$

**3** Which list shows the decimals in order from the least to the greatest?

**F** 0.362 0.252 0.237 0.31 0.27

**G** 0.27 0.31 0.237 0.252 0.362

**H** 0.237 0.252 0.27 0.31 0.362

**J** 0.362 0.31 0.27 0.252 0.237

**4** Which of the following is equivalent to 0.0037 · 100?

**A** 0.000037

**B** 0.003700

**C** 0.037

**D** 0.37

**5** Which decimal is equivalent to $\frac{7}{12}$?

**F** $0.58\overline{3}$

**G** $0.\overline{583}$

**H** 0.712

**J** 1.714285

# Patterns, Numbers, and Rules

## Contents in Brief

## Real-Life Math

**A Bee Tree** Although a female honeybee has two parents, a male honeybee has only a mother. The family tree of a male honeybee's ancestors reveals an interesting pattern of numbers.

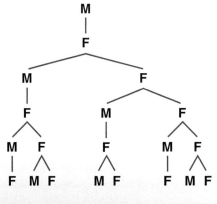

M — 1 male bee

F — 1 parent

M   F — 2 grandparents

F   M   F — 3 great-grandparents

M   F   F   M   F — 5 great-great-grandparents

F   M F   M F   F   M F — 8 great-great-great-grandparents

The numbers of bees in the generations, which are 1, 1, 2, 3, 5, 8, and so on, form a famous list of numbers known as the *Fibonacci sequence*.

**Think About It** Can you discover a pattern in the family tree or the list of numbers that will help you find the next two or three numbers in the Fibonacci sequence?

**Math Online**

Take the **Chapter Readiness Quiz** at glencoe.com.

# Dear Family,

In this chapter, students will study two important mathematical concepts, number sense and patterns. The chapter begins with a discussion of millions, billions, and trillions. The class will ponder questions such as, "How high would a stack of one million quarters reach?"

Next, students will look for patterns in diagrams, in sequences of numbers, and in a triangular array. Your student will use patterns to write mathematical rules. Mathematics has been called the "science of patterns." Recognizing and describing patterns and using patterns to make predictions are important mathematical skills.

## Key Concepts—Patterns and Rules
Patterns can be represented using diagrams, words, numbers, and equations.

**Diagram**

| Words | Numbers | Equation |
|-------|---------|----------|
| A square is added each time. | 4, 8, 12 | Let $s$ represent the number of squares. Let $t$ represent the total number of sides. $t = 4s$ |

## Chapter Vocabulary

exponent        property

input        sequence

order of operations        term

output        variable

## Home Activities
- Look and listen for real-world references that use millions, billions, and trillions. Discuss the size of these numbers.
- Work together to represent a pattern using a diagram, words, numbers, and an equation.
- Challenge your student to a game of *What's My Rule*.
- Apply properties to simplify daily calculations.

# Number Sense

Can you imagine what a million quarters look like? If you stacked a million quarters, how high would they reach? How long is a million seconds? How old is someone who has been alive a billion seconds? How much money is a trillion dollars?

In this lesson, you will explore these questions, and you will learn a new tool for writing large numbers.

### Think & Discuss

Consider the following numbers. Which number is the greatest?

9 trillion      300 million      6.6 billion

This newspaper says the national debt is 9 trillion dollars!

Can you believe that there are 300 million people in the United States?

The world's population is estimated to be 6.6 billion people.

# Investigation  Millions, Billions, and Trillions

## Materials

• a quarter

It is difficult to understand just how large one million, one billion, and one trillion are. It helps to think about these numbers in contexts that you can imagine. The exercises in this investigation will help you get a better sense of the size of a million, a billion, and a trillion.

### Think & Discuss

How many zeros follow the 1 in 1 million?

How many zeros follow the 1 in 1 billion?

How many millions are in 1 billion?

### ✅ Develop & Understand: A

1. How old is someone who has been alive one million seconds?

2. How old is someone who has been alive one billion seconds?

3. Find the diameter of a quarter in inches. Give your answer to the nearest inch.

4. The distance around Earth's equator is about 24,830 miles.

   a. If you lined up one million quarters end to end, how far would they reach? Give your answer in miles.

   b. Would they reach around the equator? If not, how many quarters would you need to reach around the equator?

5. How many quarters do you need to make a stack one inch high?

6. The average distance from Earth to the moon is 238,855 miles.

   a. If you stacked one million quarters, how high would they reach?

   b. Would they reach the moon? If not, how many quarters would you need to reach the moon? Estimate your answer without using a calculator. Explain how you found your answer.

**Math Link**

5,280 feet = 1 mile

**7.** A baseball diamond is a square, 90 feet on each side. Would one million quarters spread out cover a baseball diamond? Explain how you found your answer.

Hint: Think of one million quarters forming a square with 1,000 quarters on each side. Since quarters are round, they do not fit together exactly. Do not worry about the extra space left uncovered.

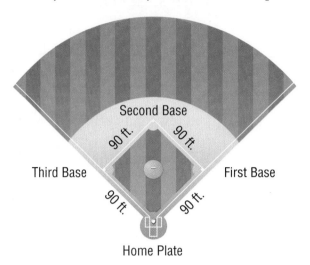

Now you will think about one trillion in a real-world context to help you get a better sense of its size. As with one million and one billion, it can be difficult to understand how large one trillion is.

## Think & Discuss

How many zeros follow the 1 in 1 trillion?

How many billions are in 1 trillion?

How many millions are in 1 trillion?

## ✓ Develop & Understand: B

Consider the following statistics. Use this information for Exercises 8–10.

- In 2006, the real median household income in the United States was about $50,000.
- During the 2006–2007 basketball season, the median salary for a player on the San Antonio Spurs team was $3,000,000.
- In 2007, the United States national debt was estimated at 7 trillion dollars.

**8.** How many years would an American family work to earn the amount of money earned by a San Antonio Spurs player in one year?

9. About how many years would a San Antonio Spurs player work to earn 1 trillion dollars?

10. About how many years would a San Antonio Spurs player work to earn enough money to pay off the 7 trillion dollar national debt?

## Share & Summarize

1. The average distance from Earth to the Sun is about 93,000,000 miles. How many quarters would you need to stack for the pile to reach the sun?

2. Why do you think many people do not understand the difference between a million, a billion, and a trillion?

# Investigation ② Exponents

## Vocabulary

exponent

factor

input

output

In Investigation 1, you worked with millions, billions, and trillions. In this lesson, you will learn an operation that can be used to represent big numbers.

Imagine a machine that takes in a number called an **input**. The machine applies a rule to the input. After the rule is applied, a number comes out of the machine. The number that comes out of the machine is called an **output**.

Suppose the rule is take any input and multiply it by itself. You can use □ to represent the original number and □ · □ to represent the rule. Let 3 be the input number. The rule, input, and resulting output are shown below.

### ✅ *Develop & Understand: A*

For Exercises 1–6, input the following numbers into the □ • □ machine. Find the output for each input number.

**1.** 2    **2.** 5    **3.** 8

**4.** 4    **5.** 7    **6.** 1

A **factor** is a number that divides into another number without a remainder. For example, 5 is a factor of 30 because $30 \div 5$ equals 6.

When a factor is repeated, you can use an *exponent* to write the multiplication in a shorter form. An **exponent** is a small, raised number that tells how many times a factor is multiplied.

**2 is the exponent.**

$3^2$

$3^2$ is read "3 to the second power" or "three squared."

### Think & Discuss

What do you think? How can Luke use Althea's idea to write $5 \cdot 5 \cdot 5 \cdot 5$ in a shorter form?

How can Althea's idea be used to write $2 \cdot 2 \cdot 2 \cdot 2 \cdot 2$ in a shorter form?

## ✅ Develop & Understand: B

**Write each multiplication in exponential form. Then write each product as a whole number.**

**7.** $6 \cdot 6$

**8.** $2 \cdot 2$

**9.** $7 \cdot 7$

**10.** $10 \cdot 10$

**11.** $5 \cdot 5 \cdot 5$

**12.** $3 \cdot 3 \cdot 3$

**13.** $4 \cdot 4 \cdot 4$

**14.** $2 \cdot 2 \cdot 2 \cdot 2$

**15.** $10 \cdot 10 \cdot 10 \cdot 10 \cdot 10$

**16.** $1 \cdot 1 \cdot 1 \cdot 1 \cdot 1 \cdot 1$

**Math Link**

The products $3 \cdot 3$ and $5 \cdot 5 \cdot 5 \cdot 5$ are written as repeated multiplications, and $3^2$ and $5^4$ are written in exponential form.

**Write each of the following as repeated multiplication. Then write each product as a whole number.**

**17.** $5^2$

**18.** $9^2$

**19.** $12^2$

**20.** $10^3$

**21.** $6^3$

**22.** $2^3$

**23.** $1^4$

**24.** $10^4$

**25.** $4^5$

**26.** $3^6$

In Exercises 1–6, you were given the rule and input to find the output. In the following exercises, you will be given the input and output. Your challenge is to find the missing rule.

## ✅ Develop & Understand: C

For Exercises 27–34, find the missing rule. Write each rule as repeated multiplication and in exponential form. Use ☐ to represent the factor when writing each rule.

**Real-World Link**

Earthquake strength can be measured using the Richter scale. Suppose an earthquake registers a 7.0 on the Richter scale. It would do considerably more damage than an earthquake registering a 5.0 because it is $10^2$, or 100, times stronger.

**27.**

9 ➡ ? ➡ 81

**28.**

2 ➡ ? ➡ 8

**29.**

4 ⟶ ? ⟶ 256

**30.**

7 ⟶ ? ⟶ 343

**31.**

4,000 ⟶ ? ⟶ 16,000,000

**32.**

2,000 ⟶ ? ⟶ 8,000,000,000

**33.**

15 ⟶ ? ⟶ 50,625

**34.**

1 ⟶ ? ⟶ 1

### Share & Summarize

Explain the difference between writing a number in exponential form and as repeated multiplication.

**Practice & Apply**

1. How many millimeters are in a kilometer?

2. In this exercise, you will figure out how far you would walk if you took 1 million steps, if you took 1 billion steps, and if you took 1 trillion steps.

   a. Measure or estimate the length of a single step you take.

   b. If you took 1 million steps, about how far would you walk? Give your answer in miles.

   c. If you took 1 billion steps, about how far would you walk? Give your answer in miles.

   d. If you took 1 trillion steps, about how far would you walk?

   e. Do you think you will walk 1 million steps in your lifetime? What about 1 billion steps? Explain your answers.

In Exercises 3–8, write each multiplication expression in exponential form. Then write each product as a whole number.

3. $9 \cdot 9$

4. $4 \cdot 4$

5. $12 \cdot 12$

6. $6 \cdot 6 \cdot 6$

7. $3 \cdot 3 \cdot 3 \cdot 3$

8. $5 \cdot 5 \cdot 5 \cdot 5 \cdot 5$

In Exercises 9–14, write each of the following as repeated multiplication. Then write each product as a whole number.

9. $7^2$

10. $11^2$

11. $2^3$

12. $8^3$

13. $6^5$

14. $1^7$

For Exercises 15 and 16, find the missing rule. Write each rule as repeated multiplication and in exponential form. Use □ when writing each rule.

15.

$4 \rightarrow \boxed{?} \rightarrow 1{,}024$

16.

$2{,}000 \rightarrow \boxed{?} \rightarrow 4{,}000{,}000$

*Connect* **&** *Extend*

**17. Economics** In this exercise, you will investigate what you could buy if you had $1 million. To answer these questions, it might help to look at advertisements in the newspaper.

   **a.** How many cars could you buy? Tell how much you are assuming each car costs.

   **b.** How many houses could you buy? Tell how much you are assuming each house costs.

   **c.** Make a shopping list of several items that total about $1 million. Give the price of each item.

**18. Astronomy** The average distance from Earth to the Sun is 93,000,000 miles. How many of you, stacked on top of yourself, would it take to reach to the Sun? Explain how you found your answer.

**19.** How many dollars is 1 million quarters?

**20.** Count the number of times your heart beats in 1 minute.

   **a.** At your current heart rate, approximately how many times has your heart beaten since your birth?

   **b.** How many times will your heart have beaten when you reach age 20?

   **c.** How many times will your heart have beaten when you reach age 80?

**21.** The number 400 can be written as the product $4 \cdot 10 \cdot 10$, or $4 \cdot 10^2$. Write each of the following products two ways, one with exponents and one without.

   **a.** 700

   **b.** 3,000

   **c.** 80,000

   **d.** 2,000,000

   **e.** 9,000,000,000

   **f.** 6,000,000,000,000

**22.** Refer to Exercise 21. What relationship do you see between each number and its products?

**23.** **In Your Own Words** Are the expressions $3 \cdot 4$ and $3^4$ equal? Why or why not?

**24.** **Preview** In Lesson 3.2, you will be working with patterns. Give the next three terms of each sequence.

   **a.** 1, 3, 5, 7, 9, ...

   **b.** 2, 6, 18, ...

   **c.** 800, 400, 200, ...

   **d.** 10, 15, 14, 19, 18, 23, 22, ...

*Mixed Review*

**25.** From the following list, identify each fraction that simplifies to $\frac{1}{2}$.

$$\frac{10}{20} \qquad \frac{15}{35} \qquad \frac{8}{14} \qquad \frac{2}{6} \qquad \frac{2}{14} \qquad \frac{3}{6} \qquad \frac{10}{12}$$

**26.** In the number 35,217, which digit is in the thousands place?

**27.** In the number 73.412, which digit is in the tenths place?

**28.** In the number 892,341.7, which digit is in the tens place?

**29.** Write 322 in words.

**30.** Write 10,010 in words.

**31.** List the following numbers in order from least to greatest.

$$\frac{3}{2} \qquad 0.\overline{3} \qquad \frac{2}{4} \qquad 1\frac{1}{5} \qquad 0.3$$

In Exercises 32 and 33, give your answer as a mixed number and as a fraction.

**32.** Evelyn ran 17 miles in four days. How many miles per day did she run?

**33.** Seven people order one large pizza. If a large pizza has 16 slices, what portion of the pizza will each person receive?

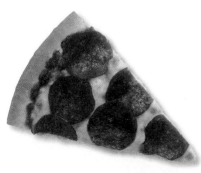

# Patterns

Patterns are everywhere. You can see patterns in wallpaper, fabric, buildings, flowers, and insects. You can hear patterns in music and song lyrics and even in the sound of a person's voice. You can follow patterns to catch a bus or a train or to locate a store with a particular address.

Patterns are an important part of mathematics. You use them every time you read a number, perform a mathematical operation, interpret a graph, or identify a shape. In this lesson, you will search for, describe, and extend many types of patterns.

## Materials

- toothpicks (optional)
- counters (optional)

### Explore

In this diagram, you can begin at "Start." Trace a path down the diagram, following the arrows, to any of the letters.

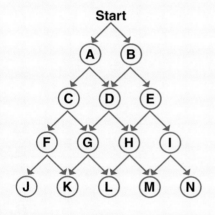

How many paths are there from Start to A? Describe each path.

How many paths are there from Start to D? Describe each path.

How many paths are there from Start to G? Describe each path.

There are four paths from Start to K. Describe all four.

Copy the diagram. Add another row of circles following the pattern of arrows and letters. How many paths are there from Start to S? Describe them.

On a new copy of the diagram, replace each letter with the number of paths from Start to that letter. For example, replace A with 1 and K with 4.

The triangle of numbers you just created is quite famous. You will learn more about the triangle and the patterns it contains in Investigation 1.

# Investigation 1 Patterns and Sequences

## Vocabulary

sequence

term

The number triangle that you created in Explore has fascinated mathematicians for centuries because of the many patterns it contains. Chinese and Islamic mathematicians worked with the triangle as early as 1100 A.D. Blaise Pascal, a French mathematician who studied it in 1653, called it the *arithmetic triangle*. It is now known as *Pascal's triangle* in his honor.

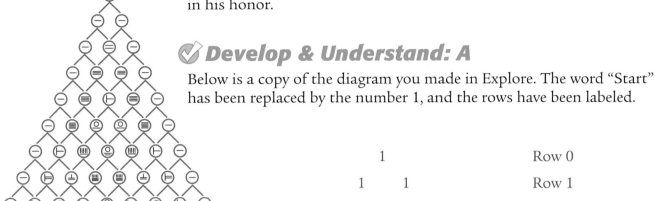

The number triangle is shown as it appears in *Precious Mirror of the Four Elements,* written by Chinese mathematician Chu Shih-Chieh in 1303.

## ✓ Develop & Understand: A

Below is a copy of the diagram you made in Explore. The word "Start" has been replaced by the number 1, and the rows have been labeled.

| | | | | | | | | | | | | |
|---|---|---|---|---|---|---|---|---|---|---|---|---|
| | | | | | 1 | | | | | | | Row 0 |
| | | | | 1 | | 1 | | | | | | Row 1 |
| | | | 1 | | 2 | | 1 | | | | | Row 2 |
| | | 1 | | 3 | | 3 | | 1 | | | | Row 3 |
| | 1 | | 4 | | 6 | | 4 | | 1 | | | Row 4 |
| 1 | | 5 | | 10 | | 10 | | 5 | | 1 | | Row 5 |

There are many patterns in this triangle. For example, each row reads the same forward as it does backward.

1. Describe as many patterns in the triangle as you can.

2. To add more rows to the triangle, you could count paths as you did in Explore. But that might take a lot of time. Instead, use some of the patterns you found in Exercise 1 to extend the triangle to Row 7. You may not be able to figure out all the numbers, but fill in as many as you can.

3. One way to add new rows to the triangle is to consider how each number is related to the two numbers just above it to the left and right. Look at the numbers in Rows 3 and 4. Describe a rule for finding the numbers in Row 4 from those in Row 3. Does your rule work for other rows of the triangle as well?

4. Use your rule from Exercise 3 to complete the triangle to Row 9.

Pascal's triangle has many interesting patterns in it. You have probably worked with other patterns in the form of puzzles like the following.

> Fill in the blanks.
>
> **Puzzle A:** 2, 5, 8, 11, ___, ___, ___
>
> **Puzzle B:** 16, 8, 4, 2, ___, ___, ___
>
> **Puzzle C:** 3, 2, 3, 2, ___, ___, ___
>
> **Puzzle D:** ★, ✳, ★, ✳, ___, ___, ___

To solve these puzzles, you need to find a pattern in the part of the list given and use it to figure out the next few items. Ordered lists like these are called **sequences**. Each item in a sequence is called a **term**. When patterns consist of shapes that build from one to the next, we sometimes refer to each term as a *stage*.

## Think & Discuss

Here is Puzzle A. Describe a rule you can follow to get from one term to the next.

$$2, 5, 8, 11, \underline{\phantom{x}}, \underline{\phantom{x}}, \underline{\phantom{x}}$$

According to your rule, what are the next three terms?

Now look at Puzzle B. Describe the pattern you see.

$$16, 8, 4, 2, \underline{\phantom{x}}, \underline{\phantom{x}}, \underline{\phantom{x}}$$

According to the pattern you described, what are the next three terms?

What pattern do you see in Puzzle C: 3, 2, 3, 2, ___, ___, ___?

According to the pattern, what are the next three terms?

Sequences do not always involve numbers. Look at Puzzle D, for example.

Describe the pattern, and give the next three terms.

In Puzzles A and B, each term is found by applying a rule to the term before it. In Puzzles C and D, the terms or stages follow a repeating pattern. In the following exercises, you will explore more sequences of both types.

5. The sequences in Parts a–e follow a repeating pattern. Give the next three terms or stages of each sequence.

   a.

   b. 3, 6, 9, 3, 6, 9, 3, 6, …

   c.

   d. 7, 1, 1, 7, 1, 1, 7, 1, 1, …

   e. $\frac{1}{2}, \frac{2}{3}, \frac{1}{2}, \frac{2}{3}, \frac{1}{2}, \frac{2}{3}, …$

6. In Parts a–e, each term or stage in the sequence is found by applying a rule to the one before it, the *preceding* term or stage. Give the next three terms or stages of each sequence.

   a. 3, 6, 9, 12, …

   b.

   c. 100, 98.5, 97, …

   d. 3, 5, 8, 12, …

   e. $\frac{1}{2}, \frac{1}{3}, \frac{1}{4}, \frac{1}{5}, …$

**7.** Below are two sequences, one made with toothpicks and the other with counters. You and your partner should each choose a different sequence. Do Parts a–c on your own using your sequence.

**Sequence A**

**Sequence B**

a. Make or draw the next three stages of your sequence.

b. How many toothpicks or counters will be in the tenth stage? Check by making or drawing the tenth stage.

c. Give a number sequence that describes the number of toothpicks or counters in each stage of your pattern.

d. Compare your answers to Parts a–c with those of your partner. What is the same about your answers? What is different?

**8.** Describe the pattern in each number sequence. Use the pattern to fill in the missing terms.

a. 5, 12, 19, 26, ___ , ___ , ___

b. 0, 9, 18, 27, ___ , ___ , ___

c. 125, 250, ___ , 1,000, ___ , ___ , 8,000

d. 1, 0.1, ___ , 0.001, ___ , 0.00001, ___

e. 4, 6, 9, 11, 14, 16, 19, ___ , ___ , ___

**9.** Consider this sequence of symbols.

$$\Delta, \Delta, \Delta, \Omega, \Omega, \Delta, \Delta, \Delta, \Omega, \Omega, \Delta, \Delta, \Delta, \Omega, \Omega, \dots$$

a. If this repeating pattern continues, what are the next six terms?

b. What is the 30th term?

c. How could you find the 100th term without drawing 100 symbols? What is the 100th term?

**Math Link**

The symbols in Exercise 9 are letters of the Greek alphabet. $\Delta$ is the letter *delta,* and $\Omega$ is the letter *omega.* Greek letters are used frequently in physics and advanced mathematics.

10. The sequence below is known as the *Fibonacci sequence* after the mathematician who studied it. The Fibonacci sequence is interesting because it appears often in both natural and manufactured things.

$$1, 1, 2, 3, 5, 8, 13, \ldots$$

a. Study the sequence carefully to see whether you can discover the pattern. Give the next three terms of the sequence.

b. Write instructions for continuing the Fibonacci sequence.

## Share & Summarize

1. The diagram from Explore on page 120 is repeated below. How is Pascal's triangle related to the number of paths from Start to each letter in this diagram?

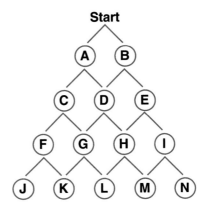

2. You discovered that each number in Pascal's triangle is the sum of the two numbers just above it. Explain what this means in terms of the number of paths to a particular letter in the diagram.

3. Describe some strategies you use when searching for a pattern in a sequence.

# Investigation 2 Order of Operations

## Vocabulary

order of operations

A *convention* is a rule people have agreed to follow because it is helpful or convenient for everyone to do the same thing. The rules, "When you drive, keep to the right" and "In the grocery store, wait in line to pay for your selections," are two conventions.

Reading across the page from left to right is a convention that English-speaking people have adopted. When you see the words "dog bites child," you know to read "dog" then "bites" then "child" and not "child bites dog." Not all languages follow this convention. For example, Hebrew is read across the page from right to left, and Japanese is read down the page from left to right.

To do mathematics, you need to know how to read mathematical expressions. For example, how would you read this expression?

$$5 + 3 \cdot 7$$

There are several possibilities.
- *Left to right:* Add 5 and 3 to get 8 and then multiply by 7. The result is 56.
- *Right to left:* Multiply 7 and 3 to get 21 and then add 5. The result is 26.
- *Multiply and then add:* Multiply 3 and 7 to get 21 and then add 5. The result is 26.

**Math Link**

*Evaluating* a mathematical expression means finding its value.

To communicate in the language of mathematics, people follow a convention for reading and evaluating expressions. The convention, called the **order of operations**, says that expressions should be evaluated in this order.
- Evaluate expressions inside parentheses.
- Simplify exponents.
- Do multiplications and divisions from left to right.
- Do additions and subtractions from left to right.

To evaluate $5 + 3 \cdot 7$, you multiply first and then add.

$$5 + 3 \cdot 7 = 5 + 21 = 26$$

If you want to indicate that the addition should be done first, you would use parentheses.

$$(5 + 3) \cdot 7 = 8 \cdot 7 = 56$$

## Example

These calculations follow the order of operations.

$$15 - 3 \times 4 = 15 - 12 = 3$$

$$1 + 4 \cdot (2 + 3) = 1 + 4 \cdot 5 = 1 + 20 = 21$$

$$3 + 6 \div 2 - 1 = 3 + 3 - 1 = 6 - 1 = 5$$

Another convention in mathematics involves the symbols used to represent multiplication. You are familiar with the $\times$ symbol. An asterisk or a small dot between two numbers also means to multiply. So, each of the following expressions means "three times four."

$$3 \times 4 \qquad 3 \cdot 4 \qquad 3 * 4$$

## ✅ Develop & Understand: A

In Exercises 1–4, use the order of operations to decide which of the expressions are equal.

1. $8 \cdot 4 + 6$     $(8 \cdot 4) + 6$     $8 \times (4 + 6)$

2. $2 + 8 \cdot 4 + 6$    $(2 + 8) \times (4 + 6)$    $2 + (8 \cdot 4) + 6$

3. $(10 - 4) \times 2$    $10 - (4 * 2)$     $10 - 4 * 2$

4. $24 \div 6 * 2$     $(24 \div 6) \times 2$     $24 \div (6 \cdot 2)$

5. Create a mathematical expression with at least three operations. Calculate the result. Then write your expression on a separate sheet of paper. Trade expressions with your partner. Evaluate your partner's expression, and have your partner check your result.

6. Most modern calculators follow the order of operations.

   a. Use your calculator to compute $2 + 3 \times 4$. What is the result? Did your calculator follow the order of operations?

   b. Use your calculator to compute $1 + 4 \times 2 + 3$. What is the result? Did your calculator follow the order of operations?

As a rule, the small dot will be used to indicate multiplication in this course.

### ✅ *Develop & Understand: B*

A middle school student council is holding a fundraiser for a local charity. Teachers and students are donating paperback and hard cover books to sell. The student council will sell books according to the following rule.

*Charge $2 for each paperback book and $3 for each hard cover book.*

7. The student council collected 410 paperbacks and 180 hard cover books. If the student council sells all of the donated books, how much will it collect from the book sale?

8. The student council treasurer uses a calculator to determine the fundraiser expenses. The calculator *does not* use the order of operations. Instead, the calculator evaluates the operations in the order they are entered.

   To figure out the book sale total, the treasurer enters the expression below. Will the result be correct, too little, or too much? Explain.

   $$410 \cdot 2 + 180 \cdot 3$$

9. Suppose the treasurer enters the calculation below instead. Will the result be correct, too little, or too much? Explain.

   $$2 \cdot 410 + 3 \cdot 180$$

A fraction bar is often used to indicate division. For example, the expressions below both mean "divide 10 by 2" or "10 divided by 2."

$$10 \div 2 \qquad \frac{10}{2}$$

Sometimes a fraction bar is used in more complicated expressions.

$$\frac{2 + 3}{4 + 4}$$

In expressions such as this, the bar not only means "divide," it also acts as a grouping symbol, grouping the numbers and operations above the bar and grouping the numbers and operations below the bar. It is as if the expressions above and below the bar are inside parentheses.

The expression $\frac{2 + 3}{4 + 4}$ means "Add 2 + 3. Next, add 4 + 4. Then divide the results." So, this expression equals $\frac{5}{8}$, or 0.625.

This more complete order of operations includes the fraction bar.
- Evaluate expressions inside parentheses and above and below fraction bars.
- Simplify exponents.
- Do multiplications and divisions from left to right.
- Do additions and subtractions from left to right.

## ✓ *Develop & Understand: C*

**Find the value of each expression.**

10. $\frac{2 + 2}{1 + 1}$

11. $2 + \frac{2}{1 + 1}$

12. Your calculator does not have a fraction bar to serve as a grouping symbol. So, you have to be careful when entering expressions like $\frac{2 + 2}{1 + 1}$.

   a. What result does your calculator give if you enter 2 + 2/1 + 1 or 2 + 2 ÷ 1 + 1? Can you explain why you get that result?

   b. What should you enter to evaluate $\frac{2 + 2}{1 + 1}$?

### *Share* & *Summarize*

Why is it important to learn mathematical conventions such as the order of operations?

# Investigation ③ Order of Operations with Exponents

Squaring is an operation, just like addition, subtraction, division, and multiplication. In Investigation 2, you learned about *order of operations,* a rule that specifies the order in which the operations in an expression should be performed. Below, the rule has been extended to include squares and other exponents.

---

**Order of Operations**

- Evaluate expressions inside parentheses and above and below fraction bars.
- Evaluate all exponents, including squares.
- Do multiplications and divisions from left to right.
- Do additions and subtractions from left to right.

---

## Think & Discuss

Evaluate each expression.

$$2 \cdot 11^2 \qquad (2 \cdot 11)^2$$

Explain how the order in which you performed the operations is different for the two expressions.

## ✓ Develop & Understand: A

1. Does $(3 + 5)^2$ have the same value as $3^2 + 5^2$? Explain.

2. Does $(5 \cdot 3)^2$ have the same value as $5^2 \cdot 3^2$? Explain.

3. Is $(5 \cdot x)^2$ equivalent to $5^2 \cdot x^2$? Explain.

4. This equation is *not* true.

$$2 \cdot 5 + 2^2 = 11^2 - 23$$

a. Show that the equation above is not true by finding the value of each side.

b. **Challenge** Place one pair of parentheses in the equation to make it true. Show that your equation is true by finding the value of each side.

5. Consider the four digits of the year in which you were born. Write at least three expressions using these four digits and any combination of parentheses, squaring, addition, subtraction, multiplication, and division. Use each digit only once in an expression. Evaluate each expression.

**Math Link**

The convention of using raised exponents was introduced by René Descartes, a French mathematician, philosopher, and scientist. Descartes invented the Cartesian coordinate system.

## ✓ *Develop & Understand: B*

In Exercises 6–13, you will compare squaring to doubling.

You and a partner will play the game *Square to a Million*. The object of the game is to get a number as close to 1 million as possible, without going over, using only the operation of squaring.

Here are the rules for the game.

- Player 1 enters a number greater than 1 into a calculator.
- Starting with Player 2, players take turns choosing to continue or to end the game. In either case, the player states his or her decision and then presses $x^2$ $\boxed{\text{ENTER}}$.

  *If the player chooses to continue the game* and the result is greater than or equal to 1 million, the player loses the round. If the result is less than 1 million, it is the other player's turn.

  *If the player chooses to end the game* and the result is greater than or equal to 1 million, the player wins. If it is less than 1 million, the player loses.

Play six games with your partner, switching roles for each round.

6. On your turn, how did you decide whether to continue or to end the game?

7. What is the greatest whole number whose square is less than 1 million?

8. What is the greatest whole number with which you could start, press $x^2$ twice, and get a number less than 1 million?

9. What is the greatest whole number with which you could start, press $x^2$ three times, and get a number less than 1 million?

10. What is the greatest whole number with which you could start, press $x^2$ four times, and get a number less than 1 million?

11. What would happen if you started the game with a positive number less than or equal to 1?

12. Imagine you are playing the game *Double to a Million*, in which you double the number in the calculator instead of squaring it. If you start with the given number, how many times will you have to double until you produce a number greater than or equal to 1 million?

    a. 50                          b. 5

    c. 1                           d. 0.5

13. For each part of Exercise 12, describe what would happen if you repeatedly squared the result instead of doubling it.

## Share & Summarize

1. Write an expression that involves parentheses, squaring, and at least two other operations. Explain how to use order of operations to evaluate your expression.

2. Copy the table. Fill in the missing information. The first row has been completed for you.

| Number | Double It Is the result greater than, less than, or equal to the original number? | Square It Is the result greater than, less than, or equal to the original number? | Which gives the greater result, squaring or doubling? |
|---|---|---|---|
| Between 0 and 1 | greater than | less than | doubling |
| 1 | | | |
| Between 1 and 2 | | | |
| 2 | | | |
| Greater than 2 | | | |

# Investigation ④ Find the Rule

A fun way to practice recognizing patterns and finding rules is to play a game called *What's My Rule*. In this game, one player thinks of a rule about numbers. The other players try to guess the rule.

## Example

Hannah, Jahmal, and Miguel were playing *What's My Rule*.

Now you will have a chance to play *What's My Rule*. As you play, try to come up with some strategies for finding the rule quickly.

## ✅ Develop & Understand: A

1. Play *What's My Rule* at least six times with your group. Take turns making up the rule. Do the following for each game you play.
   - Write down the name of the person who made up the rule.
   - Make a table showing the numbers the players guess and the results the rule gives for those numbers.
   - After a player correctly guesses the rule, write it down.

2. Work with your group to create a list of strategies for playing *What's My Rule*.

In the *What's My Rule* game, you try to guess a rule created by another student. Now you will play a rule-guessing game that does not require a partner.

To play, imagine a machine like the one that you used in Lesson 1. This machine has taken some *input* numbers, applied a rule to each one, and given the resulting *output* numbers. Your job is to guess the rule the machine used.

### Think & Discuss

Here are the outputs one machine gave for the inputs 6, 3, 10, and 11. What rule did the machine use?

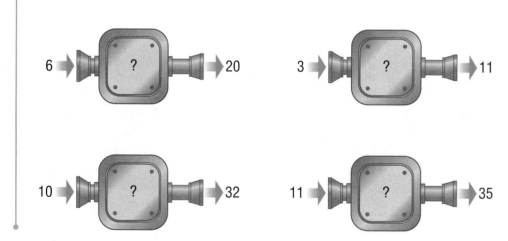

# ✅ Develop & Understand: B

Each table shows the outputs a particular machine produced for the given inputs. Find a rule the machine could have used. Check to make sure your rule works for all the inputs listed.

**3.**

| Input | 3 | 5 | 8 | 4 | 1 |
|-------|---|---|---|---|---|
| Output | 2 | 4 | 7 | 3 | 0 |

**4.**

| Input | 4 | 7 | 10 | 3 | 0 |
|-------|---|---|----|---|---|
| Output | 2 | 3.5 | 5 | 1.5 | 0 |

**5.**

| Input | 10 | 6 | 3 | 4 | 0 | 100 |
|-------|----|---|---|---|---|-----|
| Output | 23 | 15 | 9 | 11 | 3 | 203 |

## Share & Summarize

1. In one round of the *What's My Rule* game, the first clue was "2 gives 4." Write at least two rules that fit this clue.

2. The next clue in the same game was "3 gives 9." Write at least two rules that fit this clue. Do any of the rules you wrote for the first clue work for this clue as well?

3. The third clue was "10 gives 100." Give a rule that fits all three clues. How did you find the rule?

4. Describe some strategies that you use to find a rule for an input/output table.

**Practice & Apply**

**Real-World Link**

Pascal's triangle is named for Blaise Pascal (1623–1662).

**1.** Here are the first few rows of Pascal's triangle.

$$
\begin{array}{ccccccc}
& & & 1 & & & & \text{Row 0} \\
& & 1 & & 1 & & & \text{Row 1} \\
& 1 & & 2 & & 1 & & \text{Row 2} \\
1 & & 3 & & 3 & & 1 & \text{Row 3}
\end{array}
$$

  **a.** How many numbers are in each row shown?

  **b.** How many numbers are in Row 4? In Row 5? In Row 6?

  **c.** If you are given a row number, how can you determine how many numbers are in that row?

  **d.** In some rows, every number appears twice. Other rows have a middle number that appears only once. Will Row 10 have a middle number? Will Row 9? How do you know?

**2.** A certain row of Pascal's triangle has 252 as the middle number and 210 just to the right of the middle number.

  ...   ?   ?   ?   ?   252   210   ?   ?   ?   ...

  **a.** What is the number just to the left of the middle number? How do you know?

  **b.** What is the middle number two rows later? How do you know?

**Describe the pattern in each sequence. Use the pattern to find the next three terms.**

**3.** 3, 12, 48, 192, __, __, __

**4.** 0.1, 0.4, 0.7, 1.0, __, __, __

**5.** 2, 5, 4, 7, 6, 9, __, __, __

**6.** $\triangle, \infty, \triangle, \triangle, \infty, \triangle, \triangle, \triangle, \infty$, __, __, __

**7.** $-5, -4, -3, -2$, __, __, __

**8.** a, c, e, g, __, __, __

**Evaluate each expression.**

**9.** $3 + 3 \cdot 2 + 2$

**10.** $(3 + 3) \cdot (2 + 2)$

**11.** $(3 + 3) + 2 \div 2$

**12.** $\dfrac{7 + 6 - 2 \cdot 6}{11 - 5 \cdot 2}$

**Tell whether each expression was evaluated correctly using the order of operations. If not, give the correct result.**

**13.** $10 \cdot (6 - 5) + 7 = 80$

**14.** $54 - 27 \div 3 = 45$

**15.** $(16 - 4 \cdot 2) - (14 \div 2) = 5$

**16.** $100 - 33 \cdot 2 - (4 + 8) = 22$

**Find the value of each expression.**

**17.** $5 \cdot 3^2 - 2$

**18.** $2 \cdot (5^2 - 10)$

**19.** $3^2 - 2^2$

**20.** $7 + \dfrac{6^2}{3}$

**21.** Does $(1 + 3)^2$ have the same value as $1^2 + 3^2$? Explain.

**22.** Does $(4 - 2)^2$ have the same value as $4 - 2^2$? Explain.

**23.** Does $(11 \cdot 7)^2$ have the same value as $11^2 \cdot 7^2$? Explain.

**24. Challenge** Place one pair of parentheses in the equation below to make it true. Show that it is true by computing the value of each side.

$$22 - 7 - 5^2 \cdot 2 = 2 \cdot 3^2 - 4$$

**25.** Suppose you are playing *Square to a Million*. You chose the starting number 5, and your partner squared it. Now it is your turn. Should you continue or end the game? Explain.

**26.** Suppose you are playing *Square to a Million*. Your partner chose the starting number 1,001. Should you continue or end the game? Explain.

In Exercises 27–31, suppose you square the number. Without doing any calculations, tell whether the result will be *less than, greater than,* or *equal to* the original number.

**27.** 0.75      **28.** $\dfrac{2}{3}$      **29.** 1      **30.** 1.5      **31.** 5

**Find a rule that works for all the pairs in each input/output table. Use your rule to find the missing outputs.**

**32.**

| Input  | 0 | 1 | 2 | 5 | 8 | 12 |
|--------|---|---|---|---|---|----|
| Output | 4 | 5 | 6 | 9 |   |    |

**33.**

| Input  | 3 | 24 | 36 | 12 | 45 | 60 |
|--------|---|----|----|----|----|----|
| Output | 1 | 8  | 12 | 4  |    |    |

**34.**

| Input  | 2 | 10 | 16 | 22 | 32 | 44 |
|--------|---|----|----|----|----|----|
| Output | 0 | 4  | 7  | 10 |    |    |

**35.**

| Input  | 1 | 2  | 3  | 4  | 6 | 10 |
|--------|---|----|----|----|---|----|
| Output | 9 | 19 | 29 | 39 |   |    |

**Connect & Extend**

**36.** Some patterns in Pascal's triangle appear in unexpected ways. For example, look at the pattern in the sums of the rows.

| | | | | | | | | | | | | | |
|---|---|---|---|---|---|---|---|---|---|---|---|---|---|
| | | | | | 1 | | | | | | Row 0 | Sum = 1 |
| | | | | 1 | + | 1 | | | | | Row 1 | Sum = 2 |
| | | | 1 | + | 2 | + | 1 | | | | Row 2 | Sum = 4 |
| | | 1 | + | 3 | + | 3 | + | 1 | | | | |
| | 1 | + | 4 | + | 6 | + | 4 | + | 1 | | | |
| 1 | + | 5 | + | 10 | + | 10 | + | 5 | + | 1 | | |

  **a.** Find the sum of each row shown above.

  **b.** Describe the pattern in the row sums.

**37.** The pattern below involves two rows of numbers. If the pattern continued, what number would be directly to the right of 98? Explain how you know.

| 3 | | 6 | | 9 | | 12 | | 15 | | 18 |
|---|---|---|---|---|---|---|---|---|---|---|
| 1 | 2 | 4 | 5 | 7 | 8 | 10 | 11 | 13 | 14 | 16 | 17 |

**38.** Look at this pattern of numbers. If it continued, what number would be directly below 100?

$$1$$
$$2 \quad 3 \quad 4$$
$$5 \quad 6 \quad 7 \quad 8 \quad 9$$
$$10 \quad 11 \quad 12 \quad 13 \quad 14 \quad 15 \quad 16$$

**39.** Imagine that an ant is standing in the square labeled A on the grid below. The ant can move horizontally or vertically, with each step taking it one square from where it started.

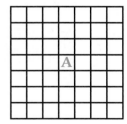

  **a.** On a copy of the grid, color each square, except the center square, according to the least number of steps it takes the ant to get there. Use one color for all squares that are one step away, another color for all squares that are two steps away, and so on.

  **b.** What shapes are formed by squares of the same color? How many squares of each color are there? What other patterns do you notice?

**40.** For this exercise, you may want to draw the shapes on graph paper.

   **a.** Find the next stage in this sequence.

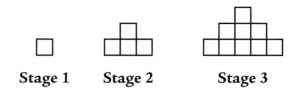

       **Stage 1**    **Stage 2**    **Stage 3**

   **b.** This table shows the number of squares in the bottom rows of stages 1 and 2. Copy and complete the table to show the number of squares in the bottom rows of the next two stages.

| Stage | Squares in Bottom Row |
|:-----:|:---------------------:|
| 1 | 1 |
| 2 | 3 |
| 3 |  |
| 4 |  |

   **c.** Look at your table carefully. Describe the pattern of numbers in the second column. Use your pattern to extend the table to show the number of squares in the bottom rows of stages 5 and 6.

   **d.** Predict the number of squares in the bottom row of stage 30.

   **e.** Now make a table to show the *total number of squares* in each of the first five stages.

| Stage | Total Number of Squares |
|:-----:|:-----------------------:|
| 1 | 1 |
| 2 | 4 |
| 3 |  |
| 4 |  |
| 5 |  |

   **f.** Look for a pattern in your table from Part e. Use the pattern to predict the total number of squares in stage 10.

**41. In Your Own Words** What is a pattern? Is every sequence of numbers a pattern? Is every sequence of shapes a pattern? Explain your answers.

In Exercises 42–45, tell whether each rule is a convention or a rule we cannot change.

**42.** Nine times a number is equal to the difference between ten times the number and the number.

**43.** In an expression involving only addition and multiplication and no parentheses, such as $2 \cdot 3 + 4 \cdot 5 + 6$, do the multiplication first.

**44.** $4 + 3 = 7$

**45.** Use a decimal point to separate the integer part of a number from the fractional part.

**46.** This computation gives the same result whether you compute correctly, using order of operations, or whether you do the computations from left to right.

$$16 - 6 \cdot 2 - 15 \div 5$$

**a.** Find the value of the expression both ways. Show that you get the same result.

**b.** Find another computation that you should *not* evaluate from left to right but that gives the correct result if you do.

**47.** Marcela receives $2 a week as an allowance. Her older brother Omar gets $3 each week. Her younger sister Stella receives $1 each week. All three children have asked their parents for larger allowances. Their parents have given them these choices.

• Option I: Add $1 to your current weekly allowance.
• Option II: Square your current weekly allowance.

Which option should each child choose? Why?

**48.** Randall squared a number and got 390,625. Without using your calculator, find the possible ones digits of his original number. Explain.

**49.** Marlene squared a number and got 15,376. Without using your calculator, find the possible ones digits of her original number. Explain.

**50.** Courtney squared a number and got 284,089. Without using your calculator, find the possible ones digits of her original number. Explain.

**51. In Your Own Words** Explain the order you would perform the operations in the expression. $3 + 5(1 + 4)^2$.

**52.** Not all input/output tables involve numbers. In this table, the inputs are words and the outputs are letters.

| Input | Alice | Justin | Darren | Jarvis | Jimmy | Mara |
|---|---|---|---|---|---|---|
| Output | i | s | r | r | | |

**a.** Complete the last two columns of the table.

**b.** What would be the output for your name?

**c.** Describe a rule for finding the output letter for any input word.

**d.** Are there input words that have no outputs? Explain your answer.

**53.** In this input/output table, the inputs are numbers and the outputs are letters.

| Input | 1 | 2 | 3 | 4 | 5 | 6 |
|---|---|---|---|---|---|---|
| Output | O | T | T | F | F | S |

**a.** What would be the outputs for the inputs 7 and 8?

**b.** Describe a rule for finding the output letter for any input number.

**54.** Camille was trying to find a relationship between the number of letters in a word and the number of different ways the letters can be arranged. She considered only words in which all the letters are different.

| Number of Letters | Example | Number of Arrangements |
|---|---|---|
| 1 | A | 1 (A) |
| 2 | OF | 2 (OF, FO) |
| 3 | CAT | 6 (CAT, CTA, ACT, ATC, TAC, TCA) |

**a.** Continue Camille's table, finding the number of arrangements of four different letters. You could use MATH as your example, since it has four different letters.

**b. Challenge** Predict the number of arrangements of five different letters. Explain how you found your answer.

**55. In Your Own Words** How is looking for a rule similar to looking for a pattern?

**56. Preview** In Lesson 3.3, you will be working with *variables*. You will substitute values for variables to find the value of an expression. For example, when $x = 0$, the expression $5x + 1$ equals $5(0) + 1$, or 1.

Determine the value of the expression $5x + 1$ for each value of $x$.

**a.** $x = 1$                 **b.** $x = 2$

**c.** $x = 10$             **d.** $x = 25$

**Mixed Review**

**Geometry** Tell what fraction of each figure is shaded.

**57.**

**58.**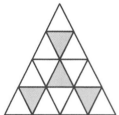

**Replace each ◯ with <, >, or = to make a true sentence.**

**59.** $\dfrac{7}{8} \bigcirc \dfrac{7}{10}$             **60.** $\dfrac{2}{7} \bigcirc \dfrac{2}{9}$

**61.** $\dfrac{9}{15} \bigcirc \dfrac{9}{12}$           **62.** $\dfrac{6}{5} \bigcirc 1.2$

**63.** $0.37 \bigcirc \dfrac{7}{20}$           **64.** $\dfrac{13}{18} \bigcirc \dfrac{7}{10}$

**65.** $0.0375 \bigcirc \dfrac{3}{80}$        **66.** $\dfrac{23}{24} \bigcirc \dfrac{24}{25}$

**67.** Write three fractions that are equivalent to $\dfrac{6}{10}$.

**Write each multiplication expression in exponential form. Then write each product as a whole number.**

**68.** $3 \cdot 3$           **69.** $4 \cdot 4 \cdot 4$         **70.** $2 \cdot 2 \cdot 2 \cdot 2 \cdot 2$

# Variables and Rules

## Vocabulary

**variable**

Every day, people are confronted with problems that they have to solve. Some of these situations involve such quantities as the amount of spice to add to a recipe, the cost of electricity, and interest rates. In some situations, it helps to have a way to record information without using many words. For example, both expressions in the box present the same idea.

> To find the circumference of a circle, multiply the radius length by $2\pi$.
>
> $$C = 2\pi r$$

While the statement on the top may be easier to read and understand at first, the statement on the bottom has several advantages. It is shorter and easier to write. It shows clearly how the quantities, radius and circumference, are related. It allows you to try different radius values and compute the corresponding circumferences.

In this lesson, you will see that by using a few simple rules, you can write powerful algebraic expressions and equations for a variety of situations.

### Think & Discuss

Shaunda, Kate, and Simon are holding bags of blocks. Isabel has just two blocks.

Suppose you know how many blocks are in each bag. How can you figure out how many blocks there are altogether?

If you know the number of blocks in each bag, it is possible to express the total number of blocks. For example, if there are 20 blocks in each bag, you can add as shown below.

$$20 + 20 + 20 + 2 = 62$$

Or, you can multiply and add.

$$3 \cdot 20 + 2 = 62$$

What if you do not know the number in each bag? First, notice that, in this situation, the number of bags and the number of loose blocks do not change, but the number of blocks in each bag can change. **Variables** represent unknown values and quantities that can change, or vary.

In algebra, letters are often used to represent variables. For example, you can let the letter $n$ stand for the number of blocks in each bag.

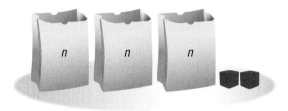

Now you can find the total number of blocks as you did before, by adding.

$$n + n + n + 2$$

Or, you can multiply and add.

$$3 \cdot n + 2$$

In algebra, the multiplication symbol between a number and a variable is usually not shown. So, $3 \cdot n + 2$ can be written $3n + 2$.

**Math Link**

Multiplication can be shown in several ways.

$3 \times n \qquad 3(n)$

$3 \cdot n \qquad 3 * n$

# Investigation ① Variables and Expressions

## Vocabulary

**algebraic expression**

In the bags-and-blocks situation above, you can think of $3n + 2$ as a rule for finding the total number of blocks when you know the number of blocks in each bag. Substitute the number in each bag for $n$. For example, for 100 blocks in each bag, the total number of blocks is as follows.

$$3n + 2 = 3 \cdot 100 + 2 = 302$$

Rules written with numbers and symbols, such as $n + n + n + 2$ and $3n + 2$, are called **algebraic expressions**.

As you progress in your study of mathematics, you will often work with algebraic expressions. Using bags and blocks is a good way to start thinking about expressions. Imagining the variable as a bag into which you can put any number of blocks can help you see how the value of an expression changes as the value of the variable changes.

## ✅ Develop & Understand: A

In these exercises, you will continue to explore the situation in which there are three bags, each containing the same number of blocks plus two extra blocks.

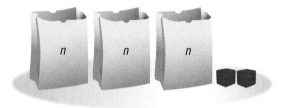

1. Copy and complete the table.

| Number of Blocks in Each Bag, $n$ | 0 | 1 | 2 | 3 | 4 | 5 |
|---|---|---|---|---|---|---|
| Total Number of Blocks, $3n + 2$ | | 5 | 8 | | | |

2. If $n = 7$, what is the value of $3n + 2$?

3. If $n = 25$, what is the value of $3n + 2$?

4. If there are 50 blocks in each bag, how many blocks are there altogether?

5. If there are 20 blocks altogether, how many blocks are in each bag?

6. Copy and complete the table.

| $n$ | 10 | 5 | 40 | 25 | 100 | | | 22 | | |
|---|---|---|---|---|---|---|---|---|---|---|
| $3n + 2$ | | 17 | | 77 | | 23 | 92 | | 128 | 3,143 |

7. Compare the tables in Exercises 1 and 6. Which table do you think was more difficult to complete? Why?

8. Could the total number of blocks in this situation be 18? Explain.

9. To represent the number of blocks in three bags plus two extra blocks with the expression $3n + 2$, you need to assume that all the bags contain the same number of blocks. Why?

10. The expression $3n + 2$ describes the total number of blocks in three bags, each with the same number of blocks plus two extra blocks.

   a. Describe a bags-and-blocks situation that can be represented by the expression $5n + 6$.

   b. Explain how the expression fits your situation.

You have explored the number of blocks in three bags plus two extra blocks. Now you will investigate some other bags-and-blocks situations.

## ✅ Develop & Understand: B

11. Here are five bags and four extra blocks.

   a. What is the total number of blocks if each bag contains three blocks? If each bag contains ten blocks?

   b. Using $n$ to represent the number of blocks in each bag, write an algebraic expression for the total number of blocks.

   c. Find the value of your expression for $n = 3$ and $n = 10$. Do you get the same answers you found in Part a?

12. Now suppose you have four bags, each with the same number of blocks plus two extra blocks.

   a. Draw a picture of this situation.

   b. Write an expression for the total number of blocks.

13. Write an expression to represent seven bags, each with the same number of blocks plus five extra blocks.

14. Write an expression to represent ten bags, each with the same number of blocks plus one extra block.

**15.** Any letter can be used to stand for the number of blocks in a bag. Match each expression below with a drawing.

$$2c + 4 \qquad 4m + 2 \qquad 4y + 5 \qquad 2f + 5$$

**a.**

**b.**

**c.**

**d.**

**16.** Rebecca wrote the expression $3b + 1$ to describe the total number of blocks represented in this picture.

**a.** What does the variable $b$ stand for in Rebecca's expression?

**b.** What does the 3 represent?

**c.** What does the 1 represent?

**d.** Complete the table for Rebecca's expression.

| $b$ | 1 | 2 | 3 | | 4 | | 100 |
|---|---|---|---|---|---|---|---|
| $3b + 1$ | | 7 | | 31 | | 76 | |

**17.** Brandi thought of the following situation.

"Imagine that the total number of blocks is two blocks less than three bags' worth. This is hard to draw, but I just described it easily in words. I can write it algebraically as $3n - 2$."

**a.** Describe a situation that $4n - 1$ could represent.

**b.** Describe a situation that $5x - 7$ could represent.

**c.** Describe a situation that $14 - 3p$ could represent.

**18.** Len has one more block than Blake. Ramon has one more block than Len. Patrick has two blocks.

**a.** If Len has six blocks, how many blocks does each boy have? How many do they have altogether?

**b.** If Len has 15 blocks, how many blocks does each boy have? How many do they have altogether?

**c.** If you know how many blocks Len has, how can you determine the total number of blocks without figuring out how many blocks each of the other boys has?

**d.** If the boys have 26 blocks altogether, how many does each boy have? Explain how you arrived at your answer.

**e.** Let $l$ stand for the number of blocks that Len has. Write an expression for the number each boy has. Then write an expression for the total number of blocks.

**f.** Let $b$ stand for the number of blocks that Blake has. Write an expression for the number each boy has. Then write an expression for the total number of blocks.

**g.** Your expressions for Parts e and f both tell how many blocks the group has. Yet, the expressions are different. Explain why.

## Share & Summarize

**1.** Make a bags-and-blocks drawing.

**2.** Write an expression that describes your drawing.

**3.** Explain how you know your expression matches your drawing.

# Investigation ② Variables, Sequences, and Rules

In Lesson 3.2, you explored sequences involving numbers, symbols, and shapes. Consider the following sequence.

The toothpick sequence can be represented by the following numerical sequence.

$$4, 7, 10, 13, 16, \ldots$$

Making a table is a good way to study the relationship between two quantities. This table shows the stage number and the number of toothpicks for the first five stages.

| Stage Number | 1 | 2 | 3 | 4 | 5 |
|---|---|---|---|---|---|
| Number of Toothpicks | 4 | 7 | 10 | 13 | 16 |

You can sometimes write a rule to show how two variables are related. Here is one possible rule relating the number of toothpicks to the stage number.

$$\text{number of toothpicks} = 3 \cdot \text{stage number} + 1$$

In the following exercises, you will make tables and find rules for variables in the toothpick sequence.

Math Link

An *arithmetic* sequence is created by adding each term by the same number to get the next term. Here are two arithmetic sequences.

1, 2, 3, 4, 5, …

4, 7, 10, 13, 16, …

## ✓ Develop & Understand: A

1. In Parts a–d, make a table showing the values of the given variable for the first four stages of the toothpick sequence shown above. The first table has been started for you.

   a. number of squares

   | Stage Number | 1 | 2 | 3 | 4 |
   |---|---|---|---|---|
   | Number of Squares | 1 | 2 | | |

   b. number of vertical toothpicks

   c. number of horizontal toothpicks

   d. number of rectangles (Hint: In each stage, count the squares, the rectangles made from two squares, the rectangles made from three squares, and so on.)

**2.** For each table from Exercise 1, try to find a relationship between the stage number and the other variable. Then write a rule to describe the relationship.

   **a.** number of squares =

   **b.** number of vertical toothpicks =

   **c.** number of horizontal toothpicks =

   **d. Challenge** number of rectangles =

**3.** To check your rules, you can test them for a particular stage number. Although this will not tell you for certain that a rule is correct, it is a good way to find mistakes. For each part of Exercise 2, use your rule to predict the value of the variable for stage 5. Then draw stage 5. Check your predictions.

**4.** Explain how you know that the rules that you wrote in Parts a–c of Exercise 2 will work for every stage.

As you have seen, letters are often used to represent variables. For example, consider this rule.

$$\text{number of toothpicks} = 3 \cdot \text{stage number} + 1$$

If you use the letter $n$ to represent the stage number and the letter $t$ to represent the number of toothpicks, you can write the rule as shown.

$$t = 3 \cdot n + 1$$

This rule is shorter than the original rule.

**Math Link**

Remember to follow the order of operations when simplifying expressions.

- Evaluate expressions inside parentheses and above and below fraction bars.

- Do multiplications and divisions from left to right.

- Do additions and subtractions from left to right.

When a number is multiplied by a variable, the multiplication symbol is often not shown. So, you can write the rule above in an even shorter form.

$$t = 3n + 1$$

You can use any letter to represent a variable, as long as you say what the letter represents. For example, you could let $w$ represent the stage number and $z$ represent the number of toothpicks and write the rule as $z = 3w + 1$.

A single rule can usually be written in many ways. Here are six ways to write the rule for the number of toothpicks in a stage.

$$t = n \cdot 3 + 1 \qquad t = (n * 3) + 1 \qquad t = 1 + 3n$$

$$t = 1 + (3 \cdot n) \qquad t = 1 + n \cdot 3 \qquad t = 3 \times n + 1$$

None of the rules above need parentheses because the order of operations states to multiply before you add. However, it is not incorrect to include them. Since some rules do need parentheses, be careful when you write your rules.

# ⊘ Develop & Understand: B

**5.** Rewrite your rules from Exercise 2 in a shorter form by using *n* for the stage number and a different letter for the other variable. Make sure to state what variable each letter represents.

**6.** Consider this toothpick sequence.

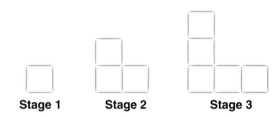

Stage 1      Stage 2      Stage 3

**a.** Choose a variable other than the stage number.

**b.** Create a table showing the value of your variable for each stage.

| Stage Number, *n* | 1 | 2 | 3 | 4 |
|---|---|---|---|---|
| Your Variable | | | | |

**c.** Try to find a rule that connects the stage number and your variable. Write the rule as simply as you can, using *n* to represent the stage number and a different letter to represent your variable.

**7.** Complete Parts a–c of Exercise 6 for this sequence of toothpicks.

Stage 1      Stage 2      Stage 3      Stage 4

**8.** Complete Parts a–c of Exercise 6 for this sequence of dots.

Stage 1      Stage 2      Stage 3      Stage 4

**9.** Consider the rule $t = 4 \cdot n + 2$, where *n* represents the stage number and *t* represents the number of toothpicks in a sequence.

**a.** Write the first four numbers in the sequence.

**b.** Draw a toothpick sequence that fits the rule.

**10.** Consider the rule $d = 3 \cdot (n + 1)$, where *n* represents the stage number and *d* represents the number of dots in a sequence.

**a.** Write the first four numbers in the sequence.

**b.** Draw a dot sequence that fits the rule.

---

### Math Link

Remember when a number is multiplied by a quantity in parentheses, the multiplication symbol is often left out. So, $2 \cdot n$ can be written as $2n$, and $3 \cdot (n + 1)$ can be written $3(n + 1)$.

1. Draw a toothpick or dot sequence. Make sure your sequence changes in a predictable way.

2. Name two variables in your sequence.

3. For each variable you named, try to write a rule relating the stage number to the variable. Use letters to represent the variables. Tell what each letter represents.

# Investigation 3 Equivalent Rules

Sometimes people can write different rules for the same pattern. You and your classmates may have written different rules in the last investigation.

Consider this toothpick sequence.

| Stage 1 | Stage 2 | Stage 3 | Stage 4 |

**Math Link**

It is not enough to show that a rule works in a few specific cases. Try to explain why it works based on how the terms are built.

Rosita and Conor wrote rules for the number of toothpicks in each stage. Both students used $n$ to represent the stage number and $t$ to represent the number of toothpicks.

Rosita's rule: $t = 3 + 2 \cdot (n - 1)$        Conor's rule: $t = 1 + 2n$

### Think & Discuss

Use the two rules to find the number of toothpicks in stage 10. Check your results by drawing stage 10 and counting toothpicks.

Show that both rules give the same result for stage 20 and for stage 100.

Do you think the rules will give the same result for every stage?

One way to show that the two rules will give the same result for every stage is to explain why both rules must work for any stage in the sequence.

## *Example*

Rosita explains why her rule works.

Conor explains why his rule works.

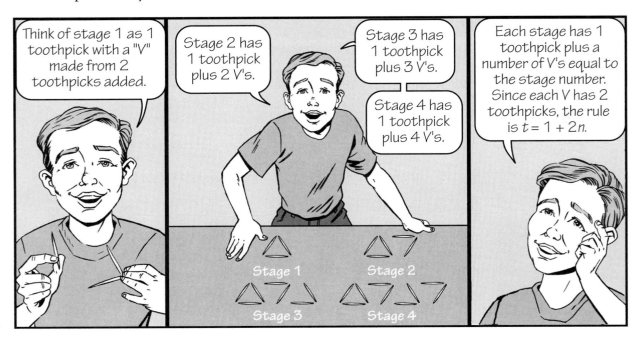

Rosita's and Conor's rules both correctly describe the toothpick sequence, so they *will* give the same result for every stage. Two rules that look different but describe the same relationship are said to be *equivalent*.

# Develop & Understand: A

1. Consider this sequence.

Stage 1　　　　Stage 2　　　　Stage 3

Mariana and Kendra wrote equivalent rules for this sequence. Both students used $n$ to represent the stage number and $t$ to represent the number of toothpicks.

Mariana's rule: $t = 2 \cdot n + 4$　　　Kendra's rule: $t = 2 \cdot (n + 2)$

Mariana used diagrams to explain why her rule is correct.

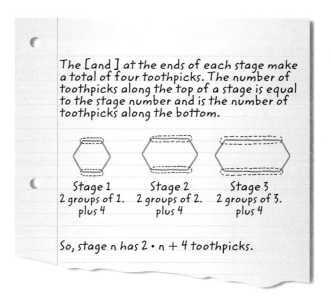

The [ and ] at the ends of each stage make a total of four toothpicks. The number of toothpicks along the top of a stage is equal to the stage number and is the number of toothpicks along the bottom.

Stage 1　　　Stage 2　　　Stage 3
2 groups of 1.　2 groups of 2.　2 groups of 3.
plus 4　　　　plus 4　　　　plus 4

So, stage n has 2 · n + 4 toothpicks.

Use diagrams to help explain why Kendra's rule is correct.

2. Consider this sequence.

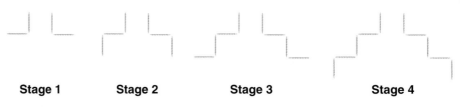

Stage 1　　　　Stage 2　　　　Stage 3　　　　Stage 4

Nadia and Uma wrote equivalent rules for this sequence. Both girls used $n$ to represent the stage number and $t$ to represent the number of toothpicks.

Nadia's rule: $t = 2n + 2$　　　Uma's rule: $t = 2 \cdot (n + 1)$

The Mayan temple of Kukulcan is on the Yucatan Peninsula, Mexico.

**a.** Copy and complete the table to show that both rules work for the first five stages of the sequence.

| Stage Number, *n* | 1 | 2 | 3 | 4 | 5 |
|---|---|---|---|---|---|
| Number of Toothpicks, *t* | | | | | |
| $2n + 2$ | | | | | |
| $2 \cdot (n + 1)$ | | | | | |

**b.** Use words and diagrams to explain why Nadia's rule is correct.

**c.** Use words and diagrams to explain why Uma's rule is correct.

**3.** Consider this sequence.

Stage 1      Stage 2      Stage 3

**a.** Write two equivalent rules for the number of toothpicks in each stage.

**b.** Use words and diagrams to explain why each rule is correct.

**4.** Consider this sequence.

Stage 1      Stage 2      Stage 3

**a.** Write two equivalent rules for the number of dots in each stage.

**b.** Use words and diagrams to explain why each rule is correct.

5. Conor, Jahmal, and Rosita each wrote a rule for this sequence.

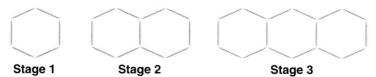

**Stage 1**         **Stage 2**         **Stage 3**

Conor's rule:
  $t = 5n + 1$, where $t$ is the number of toothpicks and $n$ is the stage number

Jahmal's rule:
  $t = 3 \cdot n + 3$, where $t$ is the number of toothpicks and $n$ is the stage number

Rosita's rule:
  $t = 6h - (h - 1)$, where $t$ is the number of toothpicks and $h$ is the number of hexagons

Determine whether each rule correctly describes the sequence. If it does, explain how you know it is correct. If it does not, explain why it is incorrect.

6. Consider strips of T-shapes like this one in which a strip can have any number of T-shapes.

a. Find a rule that connects the number of toothpicks in a strip to the number of T-shapes.

b. Explain why your rule is correct. Use diagrams if they help you to explain.

## Share & Summarize

1. How can you show that two rules for a sequence are equivalent?

2. Suppose two different rules give the same number of dots for stage 4 of a sequence. Can you conclude that the two rules are equivalent? Give an example to support your answer.

# *Inquiry*
## Investigation 4 Crossing a Bridge

While walking at night, a group of eight hikers, made up of six adults and two children, arrives at a rickety wooden bridge. A sign says the bridge can hold a maximum of 200 pounds. The group estimates that this means the following.

- one child can cross alone
- one adult can cross alone
- two children can cross together

Anyone crossing the bridge will need to use a flashlight. Unfortunately, the group has only one flashlight.

### Try It Out

1. Find a way to get all the hikers across the bridge in the fewest number of trips. Count one trip each time one or two people walk across the bridge.

   **a.** Describe your plan using words, drawings, or both.

   **b.** How many trips will it take for everyone in the group to get across the bridge?

2. A second group of hikers approaches the bridge. This group has ten adults, two children, and one flashlight.

   **a.** Can everyone in this group get across the bridge? If so, describe how.

   **b.** What is the least number of trips it will take for this group to cross the bridge?

*Go on*

**3.** In Parts a–c, find the least number of trips it would take for the group to get across the bridge. Assume each group has only a single flashlight.

  **a.** 8 adults and 2 children

  **b.** 1 adult and 2 children

  **c.** 100 adults and 2 children

**4.** Suppose a group has two children, *a* adults, and one flashlight. Write a rule that relates these two variables, the least number of trips *t* needed to get everyone across and the number of adults.

**5.** Could a group with 15 adults, one child, and one flashlight cross the bridge? Explain.

## Try It Again

Now you will explore how the number of children in a group affects the number of trips needed for the group to cross the bridge. You will start by thinking about groups with no adults at all.

**6.** Tell how many trips it would take to get each of these groups across the bridge. (Hint: First figure out a method for getting everyone across and then think about how many trips it would take.)

  **a.** 2 children          **b.** 3 children

  **c.** 6 children          **d.** 10 children

**7.** Look for a pattern relating the number of children to the number of trips.

  **a.** Write rule that relates the number of trips *t* to the number of children in the group *c*. Assume the group has no adults, more than one child, and just one flashlight.

  **b.** Part a states that the group must have more than one child. How many trips would it take a single child to get across the bridge? Does your rule give the correct result for one child?

8. Tell how many trips it would take to get each of the following groups across the bridge.

   **a.** 6 adults and 2 children   **b.** 6 adults and 3 children

   **c.** 6 adults and 6 children   **d.** 8 adults and 10 children

9. **Challenge** Write a rule that relates the number of trips to the number of children and adults in the group. Assume the group has at least two children and exactly one flashlight.

## What Did You Learn?

A second bridge has the following weight restriction.
   • one adult can cross alone
   • one child can cross alone
   • one adult and one child can cross together
   • two children can cross together

10. With these new rules, a group might be able to use a different method to get across the bridge.

    **a.** Find a way to get a group of eight adults and two children across the bridge in the fewest trips. Describe your method using words, pictures, or both.

    **b.** How many trips does it take for everyone to get across the bridge?

    **c.** How does the number of trips needed to get the group across this bridge compare to the number of trips needed for the first bridge? See your answer for Part a of Exercise 3.

11. A group of 15 adults and one child could not get across the first bridge. Could this group cross the second bridge? If so, explain how. Tell how many trips it would take.

12. Tell how many trips it would take each of these groups to get across the second bridge.

    **a.** 3 adults and 1 child

    **b.** 5 adults and 1 child

    **c.** 100 adults and 1 child

13. Write a rule that relates the number of trips needed to cross the second bridge to the number of adults in the group. Assume the group has one child and one flashlight.

# Investigation 5 Use Inputs and Outputs

In this investigation, you will play the game *What's My Rule*. Here is how you play.

- One player, the rule-maker, thinks of a secret rule for calculating an output number from a given input number. An example follows.

    *To find the output, add 3 to the input and multiply by 4.*

- The other players take turns giving the rule-maker input numbers. For each input, the rule-maker calculates the output and says the result out loud.

- By comparing each input to its output, the players try to guess the secret rule. The first player to guess the rule correctly wins.

In the *What's My Rule* game, the input and output are variables. In this investigation, you will play *What's My Rule* using letters to represent these variables. For example, if you let $i$ represent the input and $o$ represent the output, you can write the rule above as follows.

$$o = (i + 3) \cdot 4$$

## Think & Discuss

Rosita, Jahmal, and Althea are playing *What's My Rule*. Rosita's secret rule is $a = 3b + 4$, where $a$ is the output and $b$ is the input.

- Jahmal guesses that the rule is $a = 4 + 3b$, where $a$ is the output and $b$ is the input.

- Althea guesses that the rule is $x = 3y + 4$, where $x$ is the output and $y$ is the input.

Rosita is not sure whether the rules Jahmal and Althea wrote are the same as her secret rule. Tell whether each rule is correct. Explain how you know.

# ✅ *Develop & Understand: A*

1. Play *What's My Rule* with your group, using these added rules.
   - The rule-maker should write the secret rule with symbols, using letters for the variables.
   - The other players should make a table to keep track of the inputs and outputs.
   - When a player guesses the rule, he or she should write it with symbols, using letters for the variables.

   Take turns being the rule-maker so everyone has a chance. As you play, you may have to decide whether a guessed rule is equivalent to the secret rule even though it looks different.

   After your group has played several games, write a paragraph describing what you learned while playing. In your paragraph, you might discuss the following.
   - strategies you used to help you guess the rule
   - a description of what makes a rule easy to guess and what makes a rule difficult to guess
   - strategies you used to decide whether two rules were equivalent even when they looked different

# ✅ *Develop & Understand: B*

These tables were made during games of *What's My Rule*. Two rules are given for each table. Determine whether each rule could be correct. Explain how you know.

2.

| $q$ | 1 | 2 | 3 | 10 |
|---|---|---|---|---|
| $p$ | 7 | 11 | 15 | 43 |

$p = q + q + q + 4$

$p = 4 \cdot q + 3$

3.

| $s$ | 1 | 2 | 5 | 10 |
|---|---|---|---|---|
| $t$ | 10 | 20 | 50 | 100 |

$t = 4 \cdot s + 6 \cdot s$

$t = 10s$

4.

| $k$ | 0 | 2 | 5 | 10 |
|---|---|---|---|---|
| $j$ | 1 | 17 | 101 | 402 |

$j = 5 \cdot k^2 + 1$

$j = 5 \cdot k \cdot k + 1$

In the following exercises, you will try to figure out the rules for some *What's My Rule* games.

**Math Link**

Remember an exponent tells how many times a number is multiplied together. For example, $4^2 = 4 \cdot 4$ and $t^4 = t \cdot t \cdot t \cdot t$.

## ✅ Develop & Understand: C

These tables were made during games of *What's My Rule*. In each table, the values in the top row are the inputs and the values in the bottom row are the outputs.

Write a rule for each table, using the given letters to represent the variables.

**5.**

| *a* | 2 | 5 | 3 | 6 | 1 |
|-----|---|---|---|---|---|
| *b* | 9 | 21 | 13 | 25 | 5 |

**6.**

| *y* | 4 | 5 | 6 | 1 | $\frac{3}{5}$ |
|-----|----|----|----|---|---------------|
| *z* | 18 | 23 | 28 | 3 | 1 |

**7.**

| *w* | $\frac{12}{7}$ | 11 | 19 | 4 | 7 |
|-----|----------------|-----------------|------------------|---------------|------------------|
| *g* | $\frac{2}{7}$ | $1\frac{5}{6}$ | $3\frac{1}{6}$ | $\frac{2}{3}$ | $1\frac{1}{6}$ |

**8.**

| *q* | 10 | 5.5 | 1 | 2 | 3 |
|-----|----|------|-----|---|-----|
| *p* | 6 | 3.75 | 1.5 | 2 | 2.5 |

**9.**

| *c* | 100 | 42 | 17 | 1 | 0.3 |
|-----|--------|-------|-----|---|------|
| *d* | 10,000 | 1,764 | 289 | 1 | 0.09 |

**10.**

| *s* | 1 | 3.1 | 10 | 5 | 6.5 |
|-----|---|-------|-----|----|-------|
| *t* | 3 | 11.61 | 102 | 27 | 44.25 |

## Share & Summarize

Jade and Diego were playing *What's My Rule*. Diego's secret rule was "To get the output, multiply the input by itself and subtract 1." Jade guessed that the rule was "Subtract 1 from the input, and multiply the result by itself."

**1.** Write both rules with symbols. Use *m* to represent the output and *n* to represent the input.

**2.** Is Jade's rule equivalent to Diego's rule? Explain.

# Investigation  Translate Words into Symbols

Writing a rule for a real-life situation can be difficult, even when the situation is fairly simple. It is easy to make a mistake if you do not pay close attention to the details.

## Real-World Link

The body and legs of a tarantula are covered with hairs. Each hair has a tiny barb on the end. When a tarantula is being attacked, it can rub its hairs on its attacker, causing itching and even temporary blindness.

### Think & Discuss

A spider has eight legs. If $S$ represents the number of spiders and $L$ represents the number of legs, which of the following rules is correct? How do you know?

$$S = 8 \cdot L \qquad L = 8 \cdot S$$

In the spider situation, it is easy to confuse the two rules. The example shows how Luke thought about the situation.

## Example

Creating a table and looking for a pattern can make finding a rule a little easier. Notice that, after Luke wrote his rule, he checked it by testing a value for which he knew the answer. It is a good idea to test a value whenever you write a rule. Although this will not guarantee that your rule is correct, it is a helpful way to uncover mistakes.

**Lesson 3.3** Variables and Rules **163**

### ✓ *Develop & Understand: A*

1. To tile his bathroom, Mr. Drury needs twice as many blue tiles as white tiles.

   a. What are the two variables in this situation?

   b. Make a table of values for the two variables.

   c. Look for a pattern in your table. Describe how to calculate the values of one variable from the values of the other.

   d. Write a rule for the relationship between the two variables. Use letters to represent the variables. Tell what each letter represents. Be sure to check your rule by testing a value.

In Exercises 2–6, complete Parts a–d of Exercise 1.

2. In packages of Cool Breeze mints, there are four green mints for every pink mint.

3. In a toothpick sequence, the total number of toothpicks in a term is four more than twice the number of vertical toothpicks in the term.

4. In a factory, each assembly worker earns one seventh as much money as his or her manager.

5. A community theater charges $3 less for a child's ticket than for an adult's ticket.

6. A pet store always carries six times as many fish as hamsters.

## ✅ Develop & Understand: B

Below are more situations involving rules. Make a table whenever you feel it will help you better understand. Also, be sure to test all your rules.

**Real-World Link**

The first breakfast cereal, known as Granula, was created in the 1860s. The cereal consisted of heavy bran nuggets and had to be soaked overnight before it could be chewed.

**7.** Nick and his friends collect the prizes hidden in boxes of Flako cereal. Joel has twice as many prizes as Nick. Ruben has three more prizes than Nick. Andrea has half as many prizes as Nick.

   **a.** If Nick has six prizes, how many prizes do the other friends have?

   **b.** Write a rule for the relationship between the number of prizes Joel has $j$ and the number of prizes Nick has $n$.

   **c.** Write a rule for the relationship between the number of prizes Ruben has $r$ and the number Nick has $n$.

   **d.** Write a rule for the relationship between the number of prizes Andrea has $a$ and the number Nick has $n$.

**8.** Suppose Germaine has two more prizes than Joel.

   **a.** Write a rule for the relationship between the number of prizes Germaine has $g$ and the number Joel has $j$.

   **b.** If Nick has 19 prizes, how many prizes does Joel have? How many does Germaine have?

   **c.** Describe in words the relationship between the number of prizes Germaine has and the number Nick has.

   **d.** Write a rule for the relationship between the number of prizes Germaine has $g$ and the number Nick has $n$.

### *Share* & *Summarize*

**1.** How can making a table help you find a rule for a situation?

**2.** Once you have written a rule, how can you test it to check for mistakes?

# On Your Own Exercises

**Lesson 3.3**

**Practice**  **Apply**

For each picture, write an expression for the total number of blocks. Assume each bag contains the same number of blocks.

**1.**

**2.**

**3.**

**4.**

**5.** Consider the expression $4n + 5$.

   **a.** Draw a bags-and-blocks picture for this expression.

   **b.** Copy and complete the table.

| n | 0 | 1 | 2 | 3 | 26 | 66 | |
|---|---|---|---|---|----|----|---|
| 4n + 5 | | 9 | | | | | 321 |

   **c.** If $n = 7$, what is the value of $4n + 5$?

   **d.** If $n = 25$, what is the value of $4n + 5$?

**6.** A particular bags-and-blocks situation can be represented by the expression $5n + 3$. What is the value of $5n + 3$ if there are 38 blocks in each bag?

**7.** Consider this sequence of toothpick "houses."

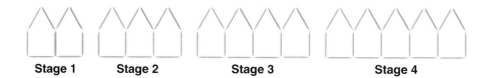

| Stage 1 | Stage 2 | Stage 3 | Stage 4 |

   **a.** Write a number sequence for the number of houses in each stage.

   **b.** Write a rule that connects the number of houses to the stage number. Use letters to represent the variables. Tell what each letter represents.

**8.** Consider this dot sequence.

**Stage 1**     **Stage 2**     **Stage 3**     **Stage 4**

  **a.** Write a number sequence for the number of dots in each stage.

  **b.** Write a rule that connects the number of dots to the stage number. Use letters to represent the variables. Tell what each letter represents.

In Exercises 9 and 10, give the first four numbers in the sequence. Then draw a sequence of toothpicks or dots that fits the rule.

**9.** $t = 4 \cdot (k + 1)$, where $t$ represents the number of toothpicks and $k$ represents the stage number

**10.** $d = 3 \cdot p - 2$, where $d$ represents the number of dots and $p$ represents the stage number

**11.** Consider this sequence.

**Stage 1**   **Stage 2**     **Stage 3**       **Stage 4**

  **a.** Write two equivalent rules for the number of toothpicks in each stage.

  **b.** Use words and diagrams to explain why each of your rules is correct.

**12.** Three students wrote rules for this dot sequence. Determine whether each rule correctly describes the pattern. If it does, explain how you know it is correct. If it does not, explain why it is not correct.

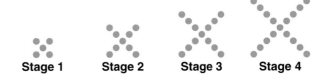

**Stage 1**     **Stage 2**     **Stage 3**     **Stage 4**

  **a.** Ilsa's rule: $d = 5 \cdot m$, where $d$ is the number of dots and $m$ is the stage number

  **b.** Mattie's rule: $d = 1 + 4k$, where $d$ is the number of dots and $k$ is the stage number

  **c.** Mauricio's rule: $d = 4 \cdot (j - 1) + 5$, where $d$ is the number of dots and $j$ is the stage number

**13.** When tiling a walkway, a particular contractor surrounds each pair of blue tiles with white tiles as shown at right. Four copies of this design are put together below.

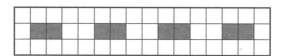

**a.** Copy and complete the table to show the number of white tiles needed for each given number of blue tiles.

| Blue Tiles | 2 | 4 | 6 | 8 | 20 | 100 |
|---|---|---|---|---|---|---|
| White Tiles | 10 | | | | | |

**b.** Find a rule that describes the connection between the number of white tiles and the number of blue tiles. Use letters for the variables in your rule. Tell what each letter represents.

**c.** Explain how you know your rule is correct. Use diagrams if necessary.

**14.** Consider this sequence of cubes.

  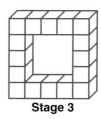

Stage 1          Stage 2          Stage 3

**a.** Find a rule for the number of cubes in each stage. You may want to make a table first. Use letters for the variables in your rule. Tell what each letter represents.

**b.** Explain how you know your rule is correct. Use diagrams if necessary.

**15.** In a round of the *What's My Rule* game, Hiam's secret rule was $a = b^2 \cdot 3$, where $b$ is the input and $a$ is the output. Complete the table to show the outputs for the given inputs.

| b | 8 | 0 | 15 | 7 | 4 |
|---|---|---|---|---|---|
| a | | | | | |

The tables below were made during games of *What's My Rule*. The values in the top row are inputs, and the values in the bottom row are outputs. Write a rule for each table, using the given letters to represent the variables.

**16.**

| *f* | 11 | 4 | 1 | 7 | 2 |
|-----|----|----|----|----|----|
| *g* | 43 | 22 | 13 | 31 | 16 |

**17.**

| *j* | 12 | 9 | 2 | 16 | 23 |
|-----|----|----|----|----|----|
| *k* | 6 | 5.25 | 3.5 | 7 | 8.75 |

The tables in Exercises 18 and 19 were made during games of *What's My Rule*. Two rules are given for each table. Determine whether each rule could be correct, and explain how you know.

**18.**

| *s* | 2 | 5 | 11 | 7 | 10 |
|-----|----|----|----|----|----|
| *t* | 5 | 14 | 32 | 20 | 29 |

$s = (t - 1) \cdot 3$

$t = 2s + (s - 1)$

**19.**

| *p* | 2 | 5 | 1 | 4 | 3 |
|-----|----|----|----|----|----|
| *m* | 10 | 127 | 3 | 66 | 29 |

$m = 2 + p \cdot p \cdot p$

$m = p^3 + 2$

**20. Economics** Tickets for a school play cost $3.75 each. Write a rule connecting the total cost in dollars *c* and the number of tickets bought *t*. Make a table if necessary. Check your rule by testing a value.

**21.** The cooking time for a turkey is 18 minutes for every pound plus an extra 20 minutes.

**a.** How long will it take to cook a 12-pound turkey?

**b.** Write a rule for this situation, using *m* for the number of minutes and *p* for the number of pounds. Make a table if necessary. Check your rule by testing a value.

**22.** Three students wrote rules for the relationship between the number of eyes and the number of noses in a group of people. Each student used *e* to represent the number of eyes and *n* to represent the number of noses. Which of the rules are correct? Explain how you decided.

Miguel's rule: $2 \cdot e = n$

Althea's rule: $n \times 2 = e$

Hannah's rule: $n = e \div 2$

**23.** Alma and her friends are training for a marathon. Today, Alma ran *a* miles. In Parts a–d, write a rule expressing the relationship between the number of miles a friend ran and the number of miles Alma ran. Be sure to state what the letters in your rules represent. Make a table if necessary. Check your rule by testing a value.

    **a.** Juan ran twice as far as Alma.

    **b.** Kai ran 3 miles more than Alma.

    **c.** Toshio ran $\frac{2}{3}$ as far as Alma.

    **d. Challenge** Melissa ran three times as far as Toshio. Remember, your rule should relate Melissa's distance to Alma's distance.

**24.** In football, a team receives six points for a touchdown, one point for making the kick after a touchdown, and three points for a field goal.

    **a.** If a team scores three touchdowns, makes two of the kicks after the touchdowns, and scores two field goals, what is the team's total score?

    **b.** Write a rule for a team's total score *S* if the team gets *t* touchdowns, makes *p* kicks after touchdowns, and scores *g* field goals. Be sure to check your rule by testing it for a specific case.

**Connect & Extend**

**25.** Consider this expression.

$$3n - 2$$

    **a.** Why is it difficult to draw a picture of bags and blocks for this expression?

    **b.** Copy and complete the table.

| *n* | 15 | 24 | | 38 | 45 | 60 |
|---|---|---|---|---|---|---|
| **3n − 2** | | | 88 | | | |

**26.** Orlando and Tate are packing the 27 prizes left in their booth after the school fair. They have four boxes, and each box holds eight prizes.

  **a.** How many boxes can they fill completely?

  **b.** After they fill all the boxes they can, will they have any prizes left to fill another box? If so, how many prizes will be in that box?

  **c.** How many empty boxes will there be, if any?

**27.** Claudio has two bags and a box. Each bag contains the same number of blocks. The box contains ten more blocks than a bag contains.

  **a.** Draw a sketch of this situation. Label each part of your sketch with an expression showing how many blocks that part contains.

  **b.** Write an expression for the total number of blocks Claudio has.

  **c.** If Claudio has a total of 49 blocks, how many blocks are in each bag?

**28.** Here are the first and fifth stages of a toothpick sequence.

**Stage 1**       **Stage 5**

  **a.** What might stages 2, 3, and 4 look like?

  **b.** Write a rule that connects the stage number and the number of toothpicks in your sequence. Use letters to represent the variables.

**29.** Each stage of this sequence is made from a one-inch straw cut into equal-sized pieces.

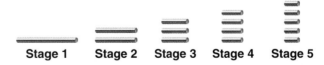

**Stage 1**   **Stage 2**   **Stage 3**   **Stage 4**   **Stage 5**

  **a.** How many pieces of straw will be in stage 10? What fraction of an inch will the length of each piece be?

  **b.** Write a rule that connects the number of straw pieces to the stage number.

  **c.** Write a rule that connects the length of each straw piece in a stage to the stage number.

## Math Link

The *perimeter* of a figure is the distance around it.

**30. Geometry** Gage and Tara wrote formulas for the perimeter of a rectangle. Both students used $P$ for the perimeter, $L$ for the length, and $W$ for the width.

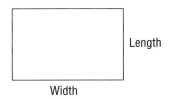

Length

Width

Gage's formula: $P = 2 \cdot L + 2 \cdot W$

Tara's formula: $P = 2 \cdot (L + W)$

**a.** Tell whether each formula is correct. If it is correct, draw diagrams showing why it is correct. If it is not correct, explain what is wrong.

**b.** Gage said, "I wonder whether I could just write my formula as $P = 2 \cdot L \cdot W$." Does this formula give the correct perimeter for a rectangle? Explain how you know.

**31.** This table shows values of the variables $m$ and $n$.

| m | 6 | 0 | 11 | 7 | 3 |
|---|---|---|----|---|---|
| n | 78 | 0 | 143 | 91 | 39 |

**a.** Complete this rule for the relationship between $m$ and $n$.

$$n = \underline{\hspace{3cm}}$$

**b.** Complete this rule for the relationship between $m$ and $n$.

$$m = \underline{\hspace{3cm}}$$

**c.** Explain how your two rules describe the same relationship.

**32.** Danilo and his brother Adan are having a walking race. Since Adan is younger, Danilo gives him a head start. The table shows the number of minutes after the start of the race that each boy has reached the given distance. The boys are walking at a steady pace, and all the blocks are about the same length.

| Number of Blocks | 1 | 4 | 6 | 10 |
|---|---|---|---|---|
| Adan's Time (min) | 3 | 12 | 18 | 30 |
| Danilo's Time (min) | 22 | 28 | 32 | 40 |

**a.** Write a rule for finding the number of minutes $M$ that it takes Adan to reach $N$ number of blocks.

**b.** Write a rule for finding the number of minutes $M$, after the start of the race, that it takes Danilo to reach $N$ number of blocks.

**c.** If both boys stop walking an hour after the race began, will Danilo catch up to Adan? Explain.

## Real-World Link

In race walking, a competitor's leading leg must be straight from the time his heel hits the ground until his leg is under his hip, and one foot must be in contact with the ground at all times.

**33.** In a children's story, peacocks and rabbits lived in a king's garden. A peacock has two legs, and a rabbit has four legs.

   **a.** Complete the table to show the total number of legs in the garden for the given numbers of peacocks and rabbits.

| Peacocks | 2 | 4 | 6 | 8 | 10 |
|----------|---|---|---|---|-----|
| Rabbits | 3 | 6 | 9 | 12 | 15 |
| Legs | | | | | |

   **b.** Describe how you calculated the total number of legs for each group of animals.

   **c.** Use letters and symbols to write a rule to calculate the total number of legs in the garden if you know the number of rabbits and the number of peacocks. Tell what variable each letter represents.

**34. In Your Own Words** Write a paragraph explaining what you have learned about writing rules for real-life situations. Be sure to discuss the following.
   • strategies for making rule-writing easier
   • how to tell whether a rule is correct

*Mixed Review*

**35.** Order the following 40-meter-dash times from slowest to fastest.

   4.52    5.01    4.82    5.1    5.26    6.00    5.24

**Write each fraction or mixed number as a decimal.**

**36.** $\frac{3}{5}$          **37.** $\frac{132}{10,000}$          **38.** $\frac{7}{9}$

**39.** $3\frac{17}{20}$          **40.** $\frac{72}{2,500}$          **41.** $\frac{173}{12}$

**Geometry** Find each missing angle measure in Exercises 42–43.

**42.**           **43.**

**44.** Write the next three terms in the sequence 1, 3, 4, 7, 11, 18, 29, ...

# Apply Properties

In previous lessons, you learned to use patterns to write mathematical rules. In this lesson, you will see that certain number patterns can be expanded to more general statements. These types of statements are called **properties**.

The page numbers of a newspaper form a pattern that will help you find the sum of any number of counting numbers.

## Vocabulary

**properties**

### Real-World Link

Carl Friedrich Gauss (1777–1855) made many contributions to the fields of mathematics, physics, and astronomy. His accomplishments included predicting the orbit of the asteroid Ceres.

### Explore

Each person in your group should take one sheet from the same section of a newspaper.

Notice that your sheet contains four printed pages, two on each side. Write down the pair of page numbers on one side of the sheet and the pair of page numbers on the other side.

Compare the two page numbers on one side with the two on the other side. Describe any patterns you notice.

Next, compare your two pairs of numbers with those of the other students in your group. Describe any patterns that fit every pair of numbers.

Now work with your group to find the following sum.

*A section of the newspaper has 48 pages that are numbered from 1 to 48. Without adding all the page numbers in the section, find the sum of all the page numbers. Explain how you found your answer.*

At a very young age, German mathematician Carl Friedrich Gauss used the strategy you discovered using the newspaper pattern to find the sum of the first 100 counting numbers.

Gauss' strategy involved rearranging and regrouping numbers. In this lesson, you will work with several properties, including the commutative and associative properties, which can be used to simplify computations.

# Investigation 1 Commutative and Associative Properties

## Vocabulary

**associative property**

**commutative property**

**identity element**

**inverse element**

In this chapter, you have studied a number of patterns. In this investigation, you will continue to use patterns to explore several important mathematical relationships, or properties.

### Think & Discuss

Are there everyday situations in which the order that you perform certain actions makes a difference?

When you are getting dressed for school, can you put on your shoes before your socks? When you are eating cereal, does it matter if you put the milk in your bowl before the cereal?

Think about situations that occur in your daily life. Give an example of a situation in which the order of your actions makes a difference. Give an example of a situation in which the order of your actions does not make a difference.

### Develop & Understand: A

Consider the following table, where ☆ is an operation on the given shapes. For example, ⬯ ☆ ▢ equals ⬯.

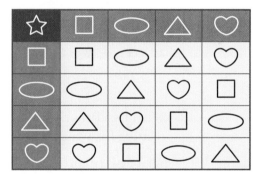

1. Perform the following operations. The first shape comes from the table's columns and the second shape comes from the table's rows.

   a. △ ☆ ⬯

   b. ⬯ ☆ △

   c. ▢ ☆ ▢

   d. ▢ ☆ ⬯

   e. ▢ ☆ △

   f. ▢ ☆ ♡

   g. ⬯ ☆ ♡

   h. ♡ ☆ ⬯

2. Describe at least one relationship you see in the table.

The **commutative property** states that the order that any two numbers are combined does not make a difference.

## ☑ Develop & Understand: B

Refer to the table on page 175 for Exercises 3–5.

3. Do you think the ☆ operation is commutative? How can you tell?

Number systems have **identity elements**. When an operation is performed on a number and an identity element, the result is the original number.

4. Examine your answers to Parts c–f in Exercise 1. Which shape acts like an identity element in the ☆ operation? How do you know?

Number systems also have **inverse elements**. When the identity element is the result of an operation being performed on a pair of numbers, the numbers are said to be inverses.

5. Examine your answers to Parts g and h in Exercise 1. Which shapes appears to be inverses in the ☆ operation? How do you know?

### Explore

Think of two numbers $x$ and $y$. Let $x$ represent the first number and $y$ represent the second number.

Determine which of the following statements are true.

a. $x + y = y + x$          b. $x - y = y - x$

c. $x \cdot y = y \cdot x$          d. $x \div y = y \div x$

With two other pairs of numbers, repeat Parts a–d. Based on your results, for which operations does the commutative property hold?

Write two expressions that are commutative. Then write two expressions that are not commutative.

**Real-World Link**

Many students commute to and from school daily. Suppose you reverse the path you take to school for your return trip. Your distances to and from school will be the same.

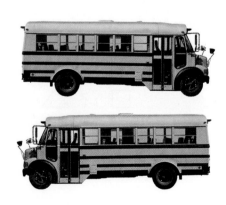

Without using a calculator, work with a partner to find the following sums and products. See if you can find a shortcut for the calculator.

$25 + 75 + 29$          $16 + 81 + 84$

$22 + 41 + 8$          $4 \cdot 23 \cdot 25$

$20 \cdot 5 \cdot 8$          $5 \cdot 18 \cdot 2$

The ☆ table from page 175 is shown below. Use the ☆ table to perform the following operations. An example is provided below the table.

## Math Link

The order of operations states you to simplify expressions inside grouping symbols, such as parentheses, first.

6. $(\square \, ☆ \, \bigcirc) \, ☆ \, \triangle$

7. $\square \, ☆ \, (\bigcirc \, ☆ \, \triangle)$

8. $(\heartsuit \, ☆ \, \triangle) \, ☆ \, \heartsuit$

9. $\heartsuit \, ☆ \, (\triangle \, ☆ \, \heartsuit)$

10. $(\triangle \, ☆ \, \square) \, ☆ \, \triangle$

11. $\triangle \, ☆ \, (\square \, ☆ \, \triangle)$

The **associative property** states that numbers can be grouped differently when they are combined and the result will be the same.

In other words, $(1 + 2) + 3 = 1 + (2 + 3)$.

## ✅ Develop & Understand: C

12. Do you think addition is associative? Why or why not?

13. Do you think multiplication is associative? Why or why not?

14. Do you think the ☆ operation is associative? How can you tell?

## ✅ Develop & Understand: D

By applying the associative property, calculations can be made easier by grouping compatible numbers. Consider the following examples.

$$15 + (85 + 24) = (15 + 85) + 24 = 100 + 24 = 124$$

$$(7 \cdot 2) \cdot 5 = 7 \cdot (2 \cdot 5) = 7 \cdot 10 = 70$$

Without using your calculator, use the associative property and compatible numbers to write an equivalent expression to find the following sums and products.

15. $29 + (21 + 88)$

16. $(12 + 9) + 91$

17. $(45 + 63) + 7$

18. $24 + (6 + 133)$

19. $(18 \cdot 5) \cdot 20$

20. $5 \cdot (2 \cdot 129)$

21. $(7 \cdot 25) \cdot 4$

22. $50 \cdot (2 \cdot 37)$

### Share & Summarize

How are the commutative and associative properties alike? How are they different?

Explain how you can use the commutative and associative properties and compatible numbers to find the sum of $(81 + 24) + 19$.

# Investigation ② Distributive Property

## Vocabulary

**distributive property**

In previous lessons, you have used different rules to represent the same situation. In this investigation, you will use different groupings to represent the same value.

One way to find different rules is to look at different groupings of quantities. For example, you can think of the diagram below as a single rectangular array of dots or as two rectangular arrays put together.

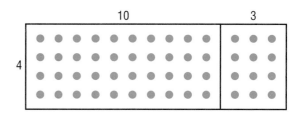

These two ways of thinking about the diagram lead to two ways of calculating the total number of dots in the diagram.

- The total number of dots can be found by noting that there are four rows with $10 + 3$ dots each.

$$4(10 + 3)$$

- The total number of dots can be found by adding the number of dots in the left rectangle to the number of dots in the right rectangle.

$$4 \cdot 10 + 4 \cdot 3$$

Both $4(10 + 3)$ and $4 \cdot 10 + 4 \cdot 3$ describe the total number of dots in the diagram.

$$4(10 + 3) = 4 \cdot 10 + 4 \cdot 3$$

## ✓ Develop & Understand: A

1. Describe two ways to find the number of dots in this diagram. Write an expression for each method.

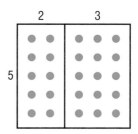

**2.** Create a dot diagram to show that $3(4 + 5) = 3 \cdot 4 + 3 \cdot 5$.

In the diagrams below, the dots are not shown, but the total number of dots is given. Labels indicate the number of rows and columns. Use the clues to determine the value of each variable.

**3.**

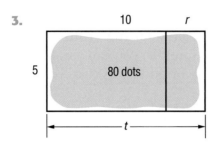

**4.**

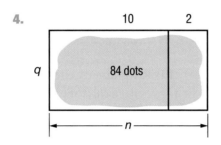

**5.** Jane wrote $5(10 + 6) = 5 \cdot 10 + 6$ in her notebook.

   **a.** Find the value of the expression on each side to show that Jane's statement is incorrect.

   **b.** Make a dot diagram you could use to explain to Jane why her statement does not make sense.

**6.** Describe two ways to find the number of dots in this diagram. Write an expression for each method.

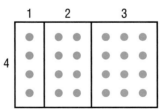

Dot diagrams help you see how different groupings can give equivalent expressions. In this lesson, you have seen many pairs of equivalent expressions like $3(10 + 9)$ and $3 \cdot 10 + 3 \cdot 9$, or $2(n - 3)$ and $2n - 2 \cdot 3$.

When you rewrite an expression like $3(10 + 9)$ or $3(20 - 1)$ as a sum or difference of products, you are using the **distributive property**.

### Think & Discuss

What does it mean to distribute something? A dictionary might help.

What is being distributed in the following equations?

$$3(10 + 9) = 3 \cdot 10 + 3 \cdot 9$$

$$5(8 + 4) = 5 \cdot 8 + 5 \cdot 4$$

$$9(3 + 1) = 9 \cdot 3 + 9 \cdot 1$$

$$7(4 - 2) = 7 \cdot 4 - 7 \cdot 2$$

$$10(25 - 21) = 10 \cdot 25 - 10 \cdot 21$$

At the beginning of this lesson, you read about shortcuts that can be used to compute numbers mentally. Such shortcuts are examples of the distributive property.

| Shortcut in Words | Shortcut in Symbols |
|---|---|
| It is $4 \cdot 20$ plus $4 \cdot 4$. | $4 \cdot 24 = 4(20 + 4) = 4 \cdot 20 + 4 \cdot 4$ |
| It is 4 less than $4 \cdot 25$. | $4 \cdot 24 = 4(25 - 1) = 4 \cdot 25 - 4 \cdot 1$ |

Sometimes a calculation can be simplified by using the distributive property in reverse.

### Example

The example shows a shortcut for calculating $12 \cdot 77 + 12 \cdot 23$.

$$12 \cdot 77 + 12 \cdot 23 = 12(77 + 23)$$
$$= 12(100)$$
$$= 1{,}200$$

## ✅ *Develop & Understand: B*

**Use the distributive property to help you do each calculation mentally. Write the grouping that shows the method you used.**

**7.** $5 \cdot 17$

**8.** $6 \cdot 41$

**9.** $4 \cdot 19$

**10.** $7 \cdot 27$

**11.** $6 \cdot 45$

**12.** $9 \cdot 38$

**Copy each equation, inserting parentheses if needed to make the equation true.**

**13.** $4 \cdot 8 + 3 = 44$

**14.** $4 \cdot 8 + 3 = 35$

**15.** $3 \cdot 7 + 4 = 25$

**16.** $3 \cdot 7 + 4 = 33$

**Find a shortcut for doing each calculation. Use parentheses to show your shortcut.**

**17.** $9 \cdot 2 + 9 \cdot 8$

**18.** $19 \cdot 2 + 19 \cdot 8$

**19.** $12 \cdot 4 + 12 \cdot 6$

**20.** $7 \cdot \dfrac{3}{5} + 3 \cdot \dfrac{3}{5}$

You have been rewriting expressions as sums or differences of products using two versions of the *distributive property.* Each version has its own name.

When addition is involved, you use the *distributive property of multiplication over addition.* The general form of this property states that for any numbers *n, a,* and *b,*

$$n(a + b) = na + nb.$$

The distributive property you have used to write an expression as a difference of products is the *distributive property of multiplication over subtraction.* The general form of this property states that for any numbers *n, a,* and *b,*

$$n(a - b) = na - nb.$$

Each of these more specific names mentions two operations, multiplication and either addition or subtraction. You distribute the number that multiplies the sum or difference to each part of the sum or difference.

In the next exercise set, you will explore whether distribution works for several combinations of operations.

# ✅ Develop & Understand: C

21. The expressions in the following statement involve division rather than multiplication.

$$\frac{a + b}{c} = \frac{a}{c} + \frac{b}{c}$$

Choose some values for $a$, $b$, and $c$. Test the statement to see whether it is true. For example, you might try $a = 2$, $b = 5$, and $c = 7$. Try several values for each variable. Do you think the statement is true for all values of $a$, $b$, and $c$?

22. The expressions in the statement below are like those in Exercise 21, but they involve multiplication rather than addition.

$$\frac{ab}{c} = \frac{a}{c} \cdot \frac{b}{c}$$

Choose some values for $a$, $b$, and $c$. Test the statement to see whether it is true. Try several values for each variable. Do you think the statement is true for all values of $a$, $b$, and $c$?

23. Choose some values for $a$ and $b$, and test this statement.

$$(a + b)^2 = a^2 + b^2$$

Do you think the statement is true for all values of a and b?

## Share & Summarize

1. Make a dot diagram to show that $6 \cdot 3 + 6 \cdot 2 = 6(3 + 2)$.

2. Give examples of calculations that look difficult but are easy to do mentally by using the distributive property. For each example, explain how the distributive property can be used to simplify the calculation.

# Investigation ③ Clock Systems

## Vocabulary

additive identity

additive inverse

identity property of multiplication

In this investigation, you will study a new number system so you can think more about the number system with which you are familiar. The number system you are going to explore is a clock system.

### ✅ Develop & Understand: A

Our method of telling time uses hours from 1 through 12. A number system based on the way we tell time can be called the clock-12 system.

**Real-World Link**

Atomic clocks tell time based on the frequencies of electromagnetic waves emitted by certain atoms. They can be accurate to within one second in more than a million years.

1. Suppose it is now 5 o'clock.

   a. What time will it be in 2 hours? In 5 hours? In 7 hours? How did you find your answers?

   b. What time will it be in 8 hours? In 10 hours? In 12 hours? How did you find your answers?

   c. What time will it be in 20 hours? In 30 hours? How did you find your answers?

2. Brianna says she can find the time $h$ in hours after 5 o'clock by computing $5 + h$ and then subtracting 12 until her answer is between 1 and 12. Does her method work? Explain.

3. Your answers to Parts a, b, and c in Exercise 1 should have been a number from 1 to 12. Explain why.

Suppose the time is 9 o'clock. To find the time in five hours, you would first add 9 and 5 to get 14. Then, you would subtract 12 to find that the time will be 2 o'clock. So, in the clock-12 system, $9 + 5 = 2$.

4. Write an addition equation to represent the following times.

   a. It is now 3 o'clock. What time will it be in 7 hours?

   b. It is now 2 o'clock. What time will it be in 12 hours?

   c. It is now 10 o'clock. What time will it be in 6 hours?

In Exercise 4, you wrote addition equations to represent times in the clock-12 system. For example, in Part a, you wrote $10 + 6 = 4$.

## ✓ Develop & Understand: B

5. Copy and complete the addition table to show *all* the possible sums in the clock-12 system.

| + | 12 | 1 | 2 | 3 | 4 | 5 | 6 | 7 | 8 | 9 | 10 | 11 |
|----|----|----|----|----|----|----|----|----|----|----|----|----|
| 12 | | | | | | | | | | | | |
| 1 | | | | | | | | | | | | |
| 2 | | | | | | | | | | | | |
| 3 | | | | | | | | | | | | |
| 4 | | | | | | | | | | | | |
| 5 | | | | | | | | | | | | |
| 6 | | | | | | | | | | | | |
| 7 | | | | | | | | | | | | |
| 8 | | | | | | | | | | | | |
| 9 | | | | | | | | | | | | |
| 10 | | | | | | | | | | | | |
| 11 | | | | | | | | | | | | |

6. Examine the table. Is addition commutative in the clock-12 system? How do you know?

7. Examine the table again. What happens when you add 12 to any number?

In the clock-12 system, the sum of twelve and any number equals the original number. For example, $2 + 12 = 2$ and $12 + 3 = 3$. So, 12 is the **additive identity** in the clock-12 system.

**Math Link**

The product of one and any number is the original number. For example, $5 \cdot 1 = 5$. This relationship is an example of the **identity property of multiplication**.

**Math Link**
The whole numbers are the counting numbers plus zero, (0, 1, 2, 3, ... ).

In the whole number system, zero is the identity element for addition, or *additive identity*. The sum of zero and any number equals the original number. For example, $2 + 0 = 2$ and $0 + 3 = 3$.

## ✓ Develop & Understand: C

Consider your table from page 185. Cross out the 12s in your table and replace them with 0s.

8. Find the diagonal of 0s in your table. Which pairs of numbers add to zero?

In any number system, pairs of numbers whose sum is the *additive identity* are known as **additive inverses**. In the clock-12 system, 2 and 10 are additive inverses.

9. What is the additive inverse of 5 in the clock-12 system?

10. What is the additive inverse of 6 in the clock-12 system?

### Share & Summarize

How is the clock-12 system like the whole number system? How are the two systems different?

Explain why $6 + 12 = 6$ is a true statement in the clock-12 system.

**Real-World Link**
The first clocks were used in ancient Egypt. They were shadow clocks, an early version of a sundial.

**Practice & Apply**

For Exercises 1–6, determine which of the following expressions are commutative. If the expression is commutative, use the commutative property to write an equivalent expression. If the expression is not commutative, write "not commutative."

**1.** $7 \cdot 9$

**2.** $3 + 10$

**3.** $8 \div 4$

**4.** $18 + 6$

**5.** $12 \cdot 6$

**6.** $30 - 14$

For Exercises 7–12, without using your calculator, use compatible numbers and the associative property to write an equivalent expression and find the following sums and products.

**7.** $32 + (18 + 12)$

**8.** $(31 + 12) + 88$

**9.** $(21 + 34) + 6$

**10.** $91 + (9 + 57)$

**11.** $(31 \cdot 2) \cdot 50$

**12.** $2 \cdot (5 \cdot 8)$

**Use the clues on each dot diagram to find the unknown values.**

**13.**

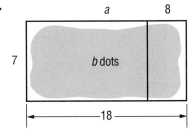

**14.**

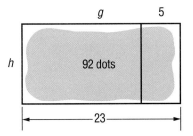

**Copy each equation, inserting parentheses when needed to make the equation true.**

**15.** $5 \cdot 2 + 3 = 25$

**16.** $12 + 3 \cdot 7 = 105$

**17.** $11 + 8 \cdot 4 = 43$

**18.** $0.2 + 0.2 \cdot 0.2 = 0.08$

**Use the distributive property to help you do each calculation mentally. Write the grouping that shows the method you used.**

**19.** $17 \cdot 2 + 17 \cdot 8$

**20.** $16 \cdot 4 - 4 \cdot 4$

**21.** $11 \cdot 5 + 5 \cdot 9$

**22.** $\frac{20}{87} + \frac{80}{87}$

**Decide whether each equation is true for all values of the variable. Justify your answers.**

**23.** $6(W + 2) = 6W + 2$

**24.** $(y + 176) \div 8 = \frac{y}{8} + \frac{176}{8}$

**25.** $2.5(B + 12) = 2.5B + 30$

**26.** $(a + 3) \cdot 7 = a \cdot 7 + 3$

**27.** This clock face is for the clock-6 system.

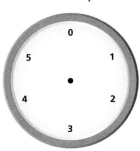

The clock-6 system works in a similar way to the clock-12 system. Consider the following examples.

$$1 + 3 = 4 \qquad 2 + 4 = 0 \qquad 3 + 5 = 2$$

**a.** Create an addition table for the clock-6 system.

| + | 0 | 1 | 2 | 3 | 4 | 5 |
|---|---|---|---|---|---|---|
| **0** | | | | | | |
| **1** | | | | | | |
| **2** | | | | | | |
| **3** | | | | | | |
| **4** | | | | | | |
| **5** | | | | | | |

**b.** Which pairs of numbers add to 0 in the clock-6 system?

**28.** Find each sum in the clock-6 system.

    **a.** $2 + 3$

    **b.** $5 + 1$

    **c.** $4 + 5$

    **d.** $(3 + 4) + (0 + 1)$

    **e.** What is the additive inverse of 4 in the clock-6 system?

**Connect & Extend**

**29.** An auditorium has 45 rows with $n$ seats in each row. Suppose a second auditorium has $n$ rows with 45 seats in each row. Would the two auditoriums have the same number of seats? Explain your answer.

**30.** Does the associative property hold for division? Provide at least one example to support your conjecture.

**31.** In this exercise, you will use the commutative and associative properties to explore the volume of two rectangular prisms made from cubes. To find the volume of a rectangular prism, you would multiply the length times the width times the height.

**a.** Find the volume of the following rectangular prism.

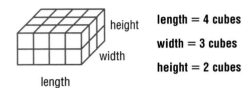

length = 4 cubes

width = 3 cubes

height = 2 cubes

**b.** Find three different measurements for the length, width, and height of a rectangular prism that has the same volume as the rectangular prism in Part a.

**c.** Find the volume of a rectangular prism that has a length of 9 cubes, a width of 2 cubes, and a height of 2 cubes.

**d.** Find three different measurements for the length, width, and height of a rectangular prism that has the same volume as the rectangular prism in Part c.

**32.** This dot diagram is missing information, and many sets of numbers will work. Find at least three sets of values for *a*, *b*, and *c*.

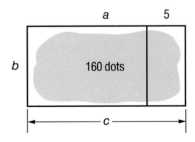

**33.** You have learned that multiplication distributes over addition. Do you think addition distributes over multiplication? That is, does $a + (b \cdot c) = (a + b) \cdot (a + c)$? Support your idea with numerical examples.

**34.** Marcus says he knows a shortcut for multiplying by 99 in his head. He claims he can mentally multiply any number by 99 within 5 seconds.

**a.** Find Marcus' shortcut for multiplying by 99.

**b.** Using symbols, explain why his shortcut works. (Hint: His shortcut uses the distributive property.)

**35.** For each *What's My Rule* table, fill in the missing numbers and find the rule. (Hint: Consider exponents when you are looking for the rules.)

**a.**

| Input | 0 | 1 | 2 | 3 | 5 | | 10 | |
|---|---|---|---|---|---|---|---|---|
| Output | 0 | 1 | 4 | 9 | | 64 | | 625 |

**b.**

| Input | 0 | 1 | 2 | 3 | 4 | | 7 | |
|---|---|---|---|---|---|---|---|---|
| Output | 3 | 4 | 7 | | | 28 | | 103 |

**36. Challenge** In Investigation 3, you learned that pairs of numbers whose sum is the identity element are additive inverses. In the clock-12 system, 12 is the additive identity element. Two additive inverses are 10 and 2.

In the whole number number system, 0 is the additive identity element. To find additive inverses for whole numbers, you use negative numbers. Think about a number line. Just as numbers start from zero and increase to the right of zero, there are numbers that start from zero and decrease to the left of zero.

**a.** Copy the following number line. Find the missing values.

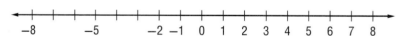

**b.** Use your number line to find each additive inverse.

**i.** 1              **ii.** 5              **iii.** 7

**iv.** −2              **v.** −6              **vi.** −4

**37. In Your Own Words** Assume you are talking to a student two years younger than you are. Explain why $3(b + 4) = 3b + 12$, not $3b + 4$. You might want to draw a picture to help explain the idea.

## Mixed Review

**Find each sum or difference without using a calculator.**

**38.** $165.7 + 47.5$      **39.** $3.179 - 0.238$      **40.** $976,556 + 0.002$

**41.** $87.78 + 94.76$      **42.** $10.0101 + 1.101$      **43.** $9.02 - 7.34$

**44.** Order these numbers from least to greatest

$$6\frac{1}{4} \qquad 6.2 \qquad \frac{26}{4} \qquad 6.05$$

**Convert each fraction to a decimal and each decimal to a fraction or mixed number.**

**45.** 0.36              **46.** $\frac{17}{20}$              **47.** 1.4

# Review & Self-Assessment

## Chapter Summary

In this chapter, you worked with expressions that helped you develop a sense of large numbers, like a million, a billion, and a trillion. You also learned how to write repeated multiplication expressions using exponents.

Next, you explored patterns and rules. You began by searching for patterns in Pascal's triangle and in sequences. You looked for ways to describe and extend the patterns you found. You then followed common rules and rules for creating sequences. You wrote rules for others to follow. You also learned about the *order of operations,* a convention for evaluating and writing mathematical expressions.

In an *algebraic expression,* symbols, usually letters, are used as variables. *Variables* can be quantities that change or unknown quantities. By investigating different values for a variable, you can explore what happens in a situation as the variable changes.

You found that the same situation can often be described with several *equivalent expressions.* Then, you focused on writing rules connecting two quantities, such as the term number and the number of toothpicks in the term, and the inputs and outputs in a game of *What's My Rule.*

You used the commutative and associative properties to simplify calculations. Next, you saw that you could change expressions into equivalent expressions by using the *distributive property* to *expand* and *factor* expressions. Lastly, you used a 12-clock system to explore *identities* and *inverses.*

## Vocabulary

additive identity

additive inverse

algebraic expression

associative property

commutative property

distributive property

exponent

factor

identity element

identity property of multiplication

input

inverse element

order of operations

output

property

sequence

term

variable

## Strategies and Applications

The questions in this section will help you review and apply the important ideas and strategies developed in this chapter.

### Developing a sense of large numbers

**1.** What repeater machine will stretch a stick one million miles long into a stick one billion miles long?

**2.** What repeater machine will stretch a stick one million miles long into a stick one trillion miles long?

**3.** Which is worth more, a billion $1 bills or a 100 million $1 bills?

### Understanding Exponents

**Write each multiplication in exponential form. Then write each product as a whole number.**

**4.** $7 \cdot 7 \cdot 7$

**5.** $10 \cdot 10 \cdot 10 \cdot 10$

### Recognizing, describing, and extending patterns

**6.** Use your calculator to help you complete this table.

| Number of 3s | Expression | Whole Number |
|---|---|---|
| 1 | 3 | 3 |
| 2 | 3 · 3 | 9 |
| 3 | 3 · 3 · 3 | |
| 4 | 3 · 3 · 3 · 3 | |
| 5 | 3 · 3 · 3 · 3 · 3 | |
| 6 | 3 · 3 · 3 · 3 · 3 · 3 | |
| 7 | 3 · 3 · 3 · 3 · 3 · 3 · 3 | |
| 8 | 3 · 3 · 3 · 3 · 3 · 3 · 3 · 3 | |

**a.** Look at the ones digits of the products. What pattern do you see?

**b.** Predict the ones digits of the product of nine 3s and the product of ten 3s. Use your calculator to check your predictions.

**c.** What is the ones digit of the product of twenty-five 3s? Explain.

### Following common rules and rules for sequences

**7.** Lakita works as a word processor. She charges customers according to this rule.

*Charge $7.50 for the project plus $2 per page.*

**a.** Kaylee hired Lakita to type an eight-page term paper. How much did Lakita charge him?

**b.** Ms. Thompson hired Lakita to type a business report. Lakita charged her $67.50 for the job. How many pages were in the report?

**c.** Lakita thinks she might get more customers if she does not charge the fixed rate of $7.50. She decides to use this new rule.

*Charge $2.50 per page.*

How much more or less would Kaylee and Ms. Thompson have been charged if Lakita had used this new rule?

**8.** Consider this starting stage and rule.

*Starting stage:* ▲

*Rule:* Add three triangles to the preceding stage.

**a.** Give the first four stages of two sequences that fit this rule.

**b.** Rewrite the rule so that only one of your sequences is correct.

## Applying the order of operations

**9.** Start with this string of numbers.

3 4 6 2 4 3

**a.** Copy the string of numbers. Create a mathematical expression by inserting operation symbols ($+$, $-$, $\times$, $\div$) and parentheses between the numbers. Evaluate your expression.

**b.** Copy the string two more times. Create and evaluate two more mathematical expressions so that each of your three expressions gives a different result.

## Writing rules that connect two quantities

**10.** Here are the first three stages of a sequence made from squares.

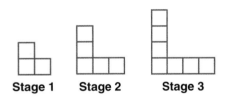

Stage 1      Stage 2      Stage 3

**a.** Figure out how many squares are in each of the first five stages. Record your results in a table.

**b.** How many squares are needed for stage 100?

**c.** Write a rule that connects the number of squares to the stage number. Use your rule to predict the number of squares in stages 6 and 7. Check your predictions by drawing those stages. If your rule does not work, revise it until it does.

**d.** Explain why your rule will work for any stage number.

## Applying Properties

**Without using a calculator, use the commutative property to find the following sums and products.**

**11.** $32 + 15 + 18$

**12.** $5 \cdot 16 \cdot 20$

**13.** $2 \cdot 11 \cdot 5$

**14.** $57 + 62 + 38$

**Without using your calculator, use the associative property to write an equivalent expression to find the following sums and products.**

**15.** $(8 \cdot 4) \cdot 25$

**16.** $15 + (15 + 28)$

**17.** $25 + (75 + 92)$

**18.** $50 \cdot (2 \cdot 135)$

## Demonstrating Skills

**Describe a rule for creating each sequence. Give the next three terms.**

**19.** 2, 5, 8, 11, 14,...

**20.** 1, 4, 2, 5, 3, 6, 4,...

**Evaluate each expression.**

**21.** $6 + 4 - 5 \div 5$

**22.** $5 \cdot (4 + 5) + 3$

**23.** $2 + \dfrac{7 \cdot 4}{5 + 2}$

**24.** $2 \cdot 3 + 2 \cdot 3 + 2$

For Exercises 25 and 26, use the fact that a dollar bill is approximately six inches long.

**25.** If you lined up one million dollar bills, how far would they reach? Give your answer in miles.

**26.** Approximately how many dollars would you need to lay end to end in order to reach from Earth to the moon 238,855 miles away?

**Write each of the following as repeated multiplication. Then write each product as a whole number.**

**27.** $11^2$

**28.** $5^3$

**29.** $3^2$

**Use the distributive property to help you do each calculation mentally. Write the grouping that shows the method you used.**

**30.** $4 \cdot 29$

**31.** $5 \cdot 13$

**Copy each equation, inserting parentheses if needed to make the equation true.**

**32.** $7 \cdot 10 + 5 = 105$

**33.** $15 \cdot 2 + 8 = 38$

**34.** Claire, Danny, and Jena each bought a $5 ticket and a $3 program at the school play.

   **a.** Write an equation using the distributive property to determine how much total money was spent by the three students.

   **b.** Determine how much total money was spent.

**Find each sum in the clock-12 system.**

**35.** $11 + 4$

**36.** $2 + 8$

**37.** $7 + 12$

**38.** $9 + 3$

**Writing and interpreting rules for sequences and input/output tables**

**39.** Consider this toothpick pattern.

| Stage 1 | Stage 2 | Stage 3 |

**a.** Choose a variable other than the stage number.

**b.** Create a table showing the value of your variable for each stage.

**c.** Try to find a rule that connects the stage number and your variable. Write the rule as simply as you can, using *n* to represent the stage number and another letter to represent your variable.

**40.** Write two different rules to express the relationship between a and b.

| *a* | 0 | 1 | 2 | 3 | 4 |
|---|---|---|---|---|---|
| *b* | 3 | 6 | 9 | 12 | 15 |

## Test-Taking Practice

**SHORT RESPONSE**

**1** Mr. Brunney works as a plumber. He charges customers according to this rule.

*Charge $50.00 for the work plus $25.00 per hour.*

Maxim hired Mr. Brunney to fix a leak in his kitchen sink. Mr. Brunney charged him $125.00 for the job. How many hours did Mr. Brunney work?

***Show your work.***

***Answer*** _____

**MULTIPLE CHOICE**

**2** Which algebraic expression represents the sequence 4, 7, 12, 19, ...?

**A** $3n$

**B** $n(n + 3)$

**C** $n^2 + 3$

**D** $3n^2$

**3** Which operation should you do first when evaluating the expression below?

$$32 - 6 + 9 \div 3 \cdot 2$$

**E** add

**F** subtract

**G** multiply

**H** divide

# Fraction and Decimal Operations

## Real-Life Math

**The House that Fractions Built** Did you know that planning and building a house requires calculations with fractions and decimals? Architects and contractors need to know how to add, subtract, multiply, and divide fractions and mixed numbers to make and read blueprints.

**Think About It** Blueprints are detailed drawings of the floorplan, front, back, and side views of a building. They are drawn to a particular scale. Suppose $\frac{1}{4}$ inch on a blueprint represents 1 foot on the actual house. What dimensions could you use on the drawing to represent a floor with dimensions 12 feet by 16 feet?

**Contents in Brief**

**Math Online** >
Take the **Chapter Readiness Quiz** at glencoe.com.

# Dear Family,

Many real-life trades, such as construction, cooking, and sewing, use fractions. In this chapter, your student will add, subtract, multiply, and divide fractions while exploring real-world situations. They will use their fraction knowledge to explore data, using measures such as the mean, median, and mode.

## Key Concept—Adding Fractions

Suppose you have two boards of lengths $\frac{3}{4}$ yard and $\frac{1}{8}$ yard. How much total lumber do you have? The sum $\frac{7}{8}$ yard is found by adding $\frac{3}{4}$ yard and $\frac{1}{8}$ yard.

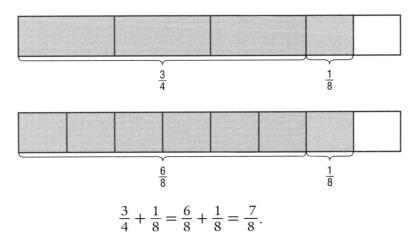

$$\frac{3}{4} + \frac{1}{8} = \frac{6}{8} + \frac{1}{8} = \frac{7}{8}.$$

Since we live in a computer age, it is becoming more and more likely that a number is represented as a decimal. For example, $\frac{7}{8}$ yard of lumber is the same as 0.875 yard. Your student will also be multiplying and dividing decimals.

## Chapter Vocabulary

| | |
|---|---|
| mean | outlier |
| median | range |
| mode | reciprocal |

## Home Activities

- Work with measurements, customary or metric, when you are measuring, sewing, or doing carpentry work.
- In the grocery store, find the unit cost, like the cost of 1 ounce, 1 liter, or 1 piece, of different brands or different packaging for the same item. Decide which is the better value.
- Figure out the correct quantities when doubling or halving a recipe.

# Add and Subtract Fractions

You know how to compare fractions and how to find equivalent fractions. Now you will begin to think about adding and subtracting fractions. You already know how to add and subtract fractions with the same denominator.

## Think & Discuss

Solve each problem in your head.

$$\frac{1}{5} + \frac{2}{5} = \underline{\qquad} \qquad \frac{5}{7} - \frac{2}{7} = \underline{\qquad} \qquad \frac{1}{2} + \frac{1}{2} = \underline{\qquad}$$

$$\frac{1}{3} + \underline{\qquad} = 1 \qquad \underline{\qquad} + \frac{2}{6} = 1 \qquad \frac{5}{8} + \underline{\qquad} = 1$$

$$1 - \frac{1}{3} = \underline{\qquad} \qquad 1 - \frac{3}{8} = \underline{\qquad} \qquad 1 - \frac{1}{6} = \underline{\qquad}$$

Explain how to add or subtract fractions with the same denominator.

Explain how to find a fraction that adds to another fraction with the same denominator to produce a 1.

Explain how to subtract a fraction from 1.

# Investigation 1 Use Fraction Models

## Materials

- set of fraction pieces and whole square

You can use models to add and subtract fractions.

The square represents the whole, or 1.

The rectangles represent fractions of the whole, or fraction pieces.

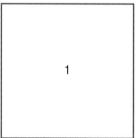

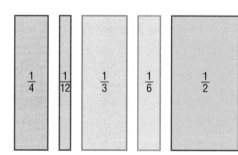

## Example

One $\frac{1}{2}$ piece and three $\frac{1}{6}$ pieces cover the square. Represent this with the addition equation.

Removing three $\frac{1}{6}$ pieces leaves a $\frac{1}{2}$ piece. Represent this with the subtraction equation.

$$\frac{1}{2} + \frac{3}{6} = 1$$

$$1 - \frac{3}{6} = \frac{1}{2}$$

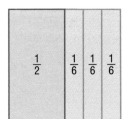

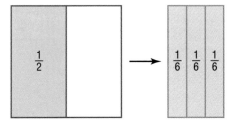

## ✅ Develop & Understand: A

1. Choose fraction pieces with two different denominators.

    a. Find as many ways as you can to cover the whole square with those two types of fraction pieces. Rearranging the same pieces in different positions does *not* count as a new way to cover the square.

    For each combination you find, make a sketch and write an equation. Each equation should be a sum of two fractions equal to 1. For example, the equation in the example above is $\frac{1}{2} + \frac{3}{6} = 1$.

    b. Choose another pair of denominators. Look for ways to cover the square with those two types of fraction pieces. Record a sketch and an equation for each combination.

    c. Continue this process until you think you have found all the ways to cover the square with two types of fraction pieces.

2. For each addition equation, you can write two related subtraction equations. In the example above, the sixths pieces were removed to get $1 - \frac{3}{6} = \frac{1}{2}$. You could instead remove the $\frac{1}{2}$ piece to get $1 - \frac{1}{2} = \frac{3}{6}$.

    Choose three of the addition equations you wrote in Exercise 1. Write two related subtraction equations for each.

3. Now find as many ways as you can to cover the square with *three* different types of fraction pieces. Record a sketch and an equation for each combination.

All of the sums in Exercises 1–3 on page 199 equal 1. You can also use your fraction pieces to form sums greater than or less than 1.

## Example

Use fraction pieces to find $\frac{2}{4} + \frac{1}{6}$.

Choose the appropriate pieces.
Arrange them on your fraction mat.

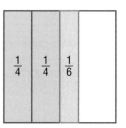

Look for a set of identical fraction pieces that cover the same area.
You could use two $\frac{1}{3}$ pieces.

$$\text{So, } \frac{2}{4} + \frac{1}{6} = \frac{2}{3}.$$

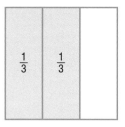

The addition equation above has two related subtraction equations.

$$\frac{2}{3} - \frac{2}{4} = \frac{1}{6} \qquad \frac{2}{3} - \frac{1}{6} = \frac{2}{4}$$

## ✔ Develop & Understand: B

4. Form sums less than 1 by combining two different types of fraction pieces. Find at least six different combinations.

   a. Record each combination by drawing a sketch and writing an addition equation.

   b. For each addition equation you wrote in Part a, write two related subtraction equations.

5. Now combine two different types of fraction pieces to form sums greater than 1. Find at least six different combinations.

   a. Record a sketch and an addition equation for each combination.

   b. For each addition equation you wrote in Part a, write two related subtraction equations.

**6.** Look for a pair of fractions with sum $\frac{7}{12}$.

   **a.** Use the $\frac{1}{12}$ pieces to cover $\frac{7}{12}$ of the square.

   **b.** Look for a combination of two types of fraction pieces that cover $\frac{7}{12}$ of the square. Record the result with a sketch and an addition equation.

   **c.** Write two related subtraction equations for the addition equation.

## Share & Summarize

Consider the equation $\frac{2}{3} + \frac{1}{4} = \frac{10}{12}$.

**1.** Explain how you could use fraction pieces to determine whether the equation is true.

**2.** Is the equation true? If it is not, change it to a true equation by replacing one of the fractions in it.

# Investigation 2 Common Denominators

## Materials
- fraction cards
- 3-by-3 grid

You have used fraction pieces to add and subtract fractions with unlike denominators. What do you do if you do not have your fraction pieces or if you want to add fractions with denominators other than 2, 3, 4, 6, or 12?

### Explore

Try to find this sum without using fraction pieces or writing anything.

$$\frac{1}{6} + \frac{3}{4}$$

Were you able to calculate the sum in your head? If so, what strategy did you use? If not, why did you find the problem difficult?

The sum would be much easier to find if the fractions had the same denominator.

Now use your fraction pieces or another method to find two fractions with the same denominator. The fractions should be equivalent to $\frac{1}{6}$ and $\frac{3}{4}$. Then add the two fractions in your head.

## ✅ Develop & Understand: A

In Exercises 1–4, use your fraction pieces or any other method you know to help rewrite the fractions.

1. Choose two of the addition equations that you wrote for Exercise 1 on page 199. Rewrite each equation so the fractions being added have a common denominator. Check to be sure each sum is equal to 1. An example follows.

$$\frac{1}{3} + \frac{8}{12} = 1 \longrightarrow \frac{4}{12} + \frac{8}{12} = \frac{12}{12} = 1$$

2. Choose two of the addition equations that you wrote for Exercise 3 on page 199. Rewrite each equation so the fractions being added have a common denominator. Check to be sure that each sum is equal to 1.

3. Choose two of the addition equations that you wrote for Exercise 4 on page 200. Rewrite each equation and the two related subtraction equations using common denominators. Check each equation to make sure that it is correct.

4. Choose two of the addition equations you wrote for Exercise 5 on page 200. Rewrite each equation and the two related subtraction equations using common denominators. Check that each equation is correct.

Many of the following expressions would be difficult or impossible to solve using fraction pieces. Use what you know about common denominators and equivalent fractions to rewrite the sums and differences.

## ✅ Develop & Understand: B

**Math Link**

A fraction is in lowest terms if the only common factor of its numerator and denominator is 1.

Rewrite each expression using a common denominator. Then find the sum or difference. Give your answers in lowest terms. If an answer is greater than 1, write it as a mixed number.

5. $\frac{1}{2} + \frac{7}{8}$       6. $\frac{2}{5} + \frac{2}{3}$       7. $\frac{1}{9} + \frac{5}{6}$

8. $\frac{4}{5} - \frac{1}{4}$       9. $\frac{7}{8} - \frac{2}{3}$       10. $\frac{5}{9} - \frac{1}{3}$

11. Daniela has $\frac{1}{2}$ of a bag of potting soil. She uses $\frac{1}{3}$ of a bag to plant some flower seeds. How much of the bag remains?

## Example

Jahmal, Marcus, and Caroline discuss some strategies that they use to find common denominators.

As you work the following exercises, you might try some of the methods described in the above example.

## ✓ Develop & Understand: C

Find each sum or difference and show each step of your work. Give your answers in lowest terms. If an answer is greater than 1, write it as a mixed number.

12. $\dfrac{1}{3} + \dfrac{3}{7}$

13. $\dfrac{1}{4} + \dfrac{3}{9}$

14. $\dfrac{4}{5} - \dfrac{3}{8}$

15. $\dfrac{13}{27} - \dfrac{2}{9}$

16. $\dfrac{7}{15} + \dfrac{1}{3} + \dfrac{1}{5}$

17. $\dfrac{11}{12} + \dfrac{3}{8}$

18. $1 - \dfrac{3}{4} - \dfrac{1}{5}$

19. $1\dfrac{31}{42} + \dfrac{17}{21}$

20. $1\dfrac{3}{4} - \dfrac{5}{8}$

21. $\dfrac{32}{75} + \dfrac{32}{50}$

A *magic square* is a square grid of numbers in which every row, column, and diagonal has the same sum. This magic square has a sum of 15.

| 8 | 1 | 6 |
|---|---|---|
| 3 | 5 | 7 |
| 4 | 9 | 2 |

## ✓ Develop & Understand: D

You can create your own magic squares using a set of fraction cards and a grid.

**22.** Arrange the following numbers into a magic square with a sum of 1. Record your grid.

$$\frac{1}{15} \quad \frac{2}{15} \quad \frac{4}{15} \quad \frac{7}{15} \quad \frac{8}{15} \quad \frac{1}{5} \quad \frac{2}{5} \quad \frac{3}{5} \quad \frac{1}{3}$$

**23.** Arrange the following numbers into a magic square with a sum of $\frac{1}{2}$. Record your grid.

$$\frac{5}{36} \quad \frac{7}{36} \quad \frac{1}{18} \quad \frac{5}{18} \quad \frac{1}{12} \quad \frac{1}{9} \quad \frac{2}{9} \quad \frac{1}{6} \quad \frac{1}{4}$$

### Share & Summarize

**1.** Use an example to help explain the steps you follow to add or subtract two fractions with different denominators.

**2.** Mandy wrote the following statement on her homework paper.

$$\frac{2}{7} + \frac{9}{4} + \frac{11}{11} = 1$$

Is Mandy correct? Explain your reasoning.

# Investigation 3 Add and Subtract Mixed Numbers

In this investigation, you will apply what you have learned about adding and subtracting fractions to solve exercises involving mixed numbers.

## Explore

Rosita is making a box to hold her pencils. She needs one piece of wood for the bottom and four pieces of wood for the sides. The top of the box will be open.

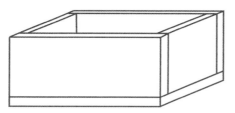

Rosita plans to cut the five pieces from a long wooden board that is 4 inches wide and $\frac{5}{8}$ inch thick. New pencils are about $7\frac{1}{2}$ inches long. Rosita wants to make the inside of the box $\frac{3}{4}$ inch longer, so it is easy to take out the pencils.

She made this sketch of the top view of the box. All measurements are in inches.

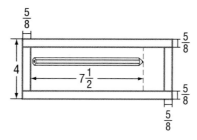

- What lengths will Rosita need to cut for the bottom, ends, and sides of her box? Give your answers as fractions or mixed numbers. Show how you found your answers.
- What will be the height of the box?

## ✓ Develop & Understand: A

1. Suppose a piece of wood $6\frac{7}{8}$ inches long is cut from a $36\frac{1}{2}$-inch board. How much of the board is left? Explain how you found your answer.

2. Juana needs two red ribbons for a costume she is making. One ribbon must be $2\frac{1}{3}$ feet long, and the other must be $3\frac{1}{2}$ feet long. What total length of red ribbon does she need?

3. Juana has a piece of blue ribbon $6\frac{1}{2}$ feet long. She cuts off a piece $3\frac{3}{4}$ feet long. How much blue ribbon will be left?

4. Juana has a length of green ribbon that is $3\frac{2}{3}$ yards long. Suppose she cuts off a piece $1\frac{3}{4}$ yards long. How many yards of green ribbon will remain?

Althea and Jing are comparing how they thought about Exercise 4.

### Real-World Link

Sewing originated more than 20,000 years ago. The earliest sewing was done using animal tendons for thread and needles made from bones or horns.

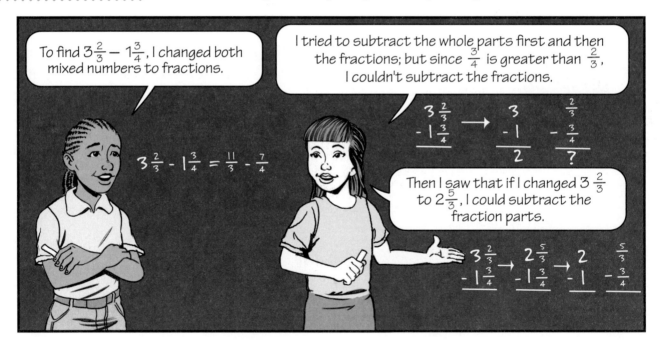

## Think & Discuss

Althea changed both mixed numbers to fractions. Finish her solution.

Jing changed $3\frac{2}{3}$ to $2\frac{5}{3}$. Explain why these two numbers are equal. How does this step help Jing find the solution?

Finish Jing's solution.

# ✅ Develop & Understand: B

**Use any method you like to solve these exercises.**

5. Jahmal organized a foot-long sub sandwich sale as a student council fundraiser. At lunch, students could purchase whole sandwiches or portion of sandwiches. The amounts sold during the four lunch periods are listed below.

| Lunch A | Lunch B | Lunch C | Lunch D |
|---------|---------|---------|---------|
| $5\frac{3}{4}$ ft | $7\frac{1}{8}$ ft | $7\frac{1}{2}$ ft | $5\frac{2}{3}$ ft |

  a. During which lunch periods did students purchase the least and the most number of sandwiches?

  b. What was the total amount of sandwiches purchased?

6. Miguel's brother, Carlos, is $69\frac{1}{2}$ inches tall. Last year, Carlos was $63\frac{3}{4}$ inches tall. How much did he grow in the year?

**Estimate each sum or difference by rounding mixed numbers to whole numbers. Then find each sum or difference. Give your answers in lowest terms.**

7. $3\frac{7}{8} + 2\frac{1}{4}$

8. $3\frac{7}{8} - 2\frac{1}{4}$

9. $6\frac{1}{3} - 5\frac{3}{4}$

10. $13\frac{3}{4} + 8\frac{19}{20}$

11. $22\frac{7}{10} - 13\frac{3}{4}$

12. $9\frac{1}{2} + 3\frac{7}{8}$

---

## Share & Summarize

Look back at Althea's and Jing's methods for subtracting mixed numbers.

1. For which types of exercises do you prefer Althea's method? Give an example.

2. For which types of exercises do you prefer Jing's method? Give an example.

# *Inquiry*

## Investigation ④ Use a Calculator

### Materials

- calculator with fraction capabilities keys
- two decks of Fraction Match cards

In this lab, you will use a calculator to add and subtract fractions and mixed numbers.

### Learning the Basics

To enter a fraction on your calculator:

- Enter the numerator.
- Press ÷ .
- Enter the denominator.

To enter a mixed number:

- Enter the whole-number part.
- Press UNIT .
- Enter the numerator of the fraction part.
- Press ÷ .
- Enter the denominator of the fraction part.

**1.** Use your calculator to find $\frac{1}{4} + 1\frac{2}{3}$.

If you enter a fraction that is not in lowest terms, or if a calculation results in a fraction that is not in lowest terms, the calculator may display something like N/D → n/d. Use the following steps to put the fraction in lowest terms.

- Press SIMP .
- Enter a common factor of the numerator and denominator.
- Press ENTER .

The calculator will divide the numerator and denominator by the factor you specify and display the result. If the fraction is still not in lowest terms, the calculator will continue to display N/D → n/d. In that case, repeat the above steps to divide by another common factor.

**2.** Estimate the value of $\frac{180}{210}$. Then use your calculator to help you write $\frac{180}{210}$ in lowest terms. Compare your answer to your estimate.

Use the following steps to change a fraction to a mixed number or to change a mixed number to a fraction.

- Press 2nd [A$^b$/$_c$ ◄► $^d$/$_e$].
- Press ENTER .

**3.** Use your calculator to change $3\frac{5}{7}$ to a fraction and to change $\frac{43}{21}$ to a mixed number.

## Playing *Fraction Match*

*Fraction Match* is a memory game for two players. Here are the rules.

- Choose Deck 1 or Deck 2.
- Shuffle the cards. Place them face down in five rows of six cards each.

- The first player turns over two cards. If needed, he or she uses a calculator to determine whether the values on the cards are the same.
- If the cards have the same value, the player keeps them and takes another turn. If they have different values, the player turns them back over and his or her turn ends.
- Play continues until all the cards have been taken. The player with the most cards at the end of the game wins.

When you finish the game, play again with the other deck.

## What Did You Learn?

**4.** Describe step by step how to use a calculator to find $2\frac{11}{14} - \frac{2}{7}$.

**5.** Design your own deck of *Fraction Match* cards. Your deck should have at least 16 cards. Test your deck by playing *Fraction Match* with a friend or classmate.

Solve each equation in your head.

**1.** $\frac{3}{4} - \frac{2}{4} =$ _____

**2.** $\frac{12}{17} - \frac{4}{17} =$ _____

**3.** _____ $- \frac{3}{8} = \frac{1}{8}$

**4.** $\frac{5}{12} -$ _____ $= \frac{1}{12}$

**5.** _____ $- \frac{8}{9} = \frac{1}{9}$

**6.** _____ $- \frac{7}{11} = \frac{5}{11}$

Use your fraction pieces or another method to help fill in each blank.

**7.** $\frac{1}{4} +$ _____ $= 1$

**8.** $\frac{1}{2} + \frac{1}{3} +$ _____ $= 1$

**9.** $\frac{1}{12} + \frac{2}{6} +$ _____ $+ \frac{1}{4} = 1$

**10.** $\frac{1}{6} + \frac{5}{12} +$ _____ $+ \frac{1}{3} = 1$

**11.** $\frac{1}{2} + \frac{1}{3} +$ _____ $= 1\frac{1}{6}$

**12.** $\frac{2}{3} + \frac{11}{12} -$ _____ $= \frac{5}{6}$

**13.** $\frac{5}{6} + \frac{1}{3} + \frac{1}{2} +$ _____ $= 2$

**14.** $\frac{1}{6} + \frac{1}{2} +$ _____ $= \frac{3}{4}$

Use your fraction pieces or another method to write each sum or difference with a common denominator. Then find the sum or difference. Give your answers in lowest terms. If an answer is greater than 1, write it as a mixed number.

**15.** $\frac{5}{6} - \frac{1}{2}$

**16.** $\frac{11}{12} - \frac{3}{4}$

**17.** $\frac{7}{6} - \frac{1}{3}$

**18.** $\frac{13}{12} + \frac{3}{4}$

Find each sum or difference. Give your answers in lowest terms. If an answer is greater than 1, write it as a mixed number.

**19.** $\frac{3}{8} + \frac{2}{3}$

**20.** $\frac{9}{22} - \frac{3}{10}$

**21.** $\frac{25}{32} + \frac{7}{24}$

**22.** $\frac{8}{26} + \frac{9}{39}$

**23.** On a sheet of paper, create a magic square with a sum of 2 using the numbers $\frac{11}{12}, \frac{3}{4}, \frac{5}{6}, \frac{1}{2}, \frac{2}{3}, \frac{5}{12}, \frac{1}{3}$, 1, and $\frac{7}{12}$.

**24.** On a sheet of paper, create a magic square with a sum of $1\frac{1}{2}$ using the numbers $\frac{1}{2}, \frac{1}{4}, \frac{2}{3}, \frac{1}{6}, \frac{5}{12}, \frac{5}{6}, \frac{1}{3}, \frac{3}{4}$, and $\frac{7}{12}$.

**Estimate each sum or difference. Then find each sum or difference, showing each step of your work. Give your answers in lowest terms. If an answer is greater than 1, write it as a mixed number.**

**25.** $2\frac{1}{2} - \frac{7}{9}$

**26.** $1\frac{8}{15} - \frac{3}{5}$

**27.** $10\frac{2}{5} - 4\frac{1}{3}$

**28.** $3\frac{5}{6} + \frac{6}{7}$

**29.** $3\frac{1}{4} + 1\frac{1}{3}$

**30.** $4\frac{1}{3} - 2\frac{3}{8}$

**31.** For the second year, a scout troop has participated in community efforts to clear litter from local roads. Last year, the scouts cleared $8\frac{3}{4}$ miles of roadways. So far this year, the scouts have cleared the following lengths of roadways.

$$\frac{7}{8} \text{ mi} \qquad 1\frac{1}{2} \text{ mi} \qquad 1\frac{1}{4} \text{ mi} \qquad \frac{3}{4} \text{ mi}$$

**a.** This year, what are the least and greatest lengths of roadways cleared by the scouts?

**b.** Since the beginning of last year, what is the total length of roadway cleared by the scouts?

**Connect & Extend**

**32. Number Sense** Cover your whole square in as many different ways as you can using *four* different types of fraction pieces.

**a.** Write an equation for each combination you find.

**b.** Describe the strategy you used to find all the possible equations.

**Give the rule for finding each term in the sequence from the previous term. Then use your rule to find the missing terms. Use the last term to help check your answers.**

**33.** $0, \frac{1}{4}, \frac{2}{4}, \frac{3}{4}, \underline{\qquad}, \underline{\qquad}, \underline{\qquad}, \underline{\qquad}, \frac{8}{4}$

**34.** $0, \frac{2}{3}, \frac{4}{3}, \frac{6}{3}, \underline{\qquad}, \underline{\qquad}, \underline{\qquad}, \underline{\qquad}, \frac{16}{3}$

**35.** $\frac{24}{4}, \frac{21}{4}, \frac{18}{4}, \frac{15}{4}, \underline{\qquad}, \underline{\qquad}, \underline{\qquad}, \underline{\qquad}, \frac{0}{4}$

**Rolling Fractions Game Card**

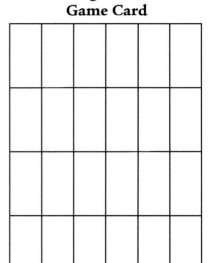

*Rolling Fractions* is played with a number cube with faces labeled $\frac{1}{2}, \frac{1}{3}, \frac{1}{4}, \frac{1}{6}, \frac{1}{8}$, and $\frac{1}{12}$. Each player has a game card divided into 24 equal rectangles. Players take turns rolling the cube and shading that fraction of the card. For example, if a player rolls $\frac{1}{2}$, he would shade 12 of the 24 rectangles.

If the fraction rolled is greater than the unshaded fraction of the card, the player shades no rectangles for that turn. The first player to shade the card completely is the winner.

**36.** In her first two turns, Caroline rolled $\frac{1}{2}$ and $\frac{1}{6}$. To win the game on her next turn, which fraction would she need to roll?

**37.** In his first two turns, Miguel rolled $\frac{1}{3}$ and $\frac{1}{12}$. To win the game in *two* more turns, which two fractions would he need to roll?

**38.** In their first three turns, Conor rolled $\frac{1}{8}, \frac{1}{6}$, and $\frac{1}{3}$, and Jahmal rolled $\frac{1}{12}, \frac{1}{2}$, and $\frac{1}{3}$. Who is more likely to win on his next roll? Explain why.

**39.** In her first two turns, Rosita rolled $\frac{1}{2}$ and $\frac{1}{3}$.

    **a.** To win on her next turn, which fraction would she need to roll?

    **b.** To win in *two* more turns, which fractions would she need to roll?

    **c.** Which fractions would give Rosita a sum greater than 1 on her third turn?

**40.** What is the fewest number of turns it could take to win this game? Which fractions would a player have to roll in that number of turns?

**41.** Luke rolled $\frac{1}{2}, \frac{1}{3}$, and $\frac{1}{8}$. He says he might as well quit because there is no way for him to win. Do you agree? Explain.

**42.** Arrange the numbers $\frac{3}{4}, \frac{2}{3}, \frac{1}{4}, \frac{11}{12}, \frac{5}{12}, \frac{1}{3}, \frac{7}{12}, \frac{1}{2}$, and $\frac{5}{6}$ into a magic square. What is the sum for your magic square?

**43.** Arrange the numbers $1, \frac{2}{3}, \frac{5}{6}, \frac{1}{3}, \frac{1}{2}, \frac{7}{6}, \frac{3}{2}, \frac{4}{3}$, and $\frac{5}{3}$ into a magic square. What is the sum for your magic square?

**44.** Arrange 1, 2, 3, 4 in the boxes to create the least possible sum. Use each number exactly once.

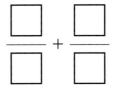

**45.** Arrange 1, 2, 3, 4 in the boxes to create the least possible positive difference. Use each number exactly once.

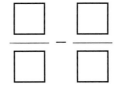

**46.** Arrange 2, 3, 4 and 12 in the boxes to create the least possible sum. Use each number exactly once.

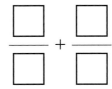

**47.** Arrange 2, 3, 4, and 12 in the boxes to create the least possible positive difference. Use each number exactly once.

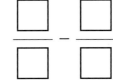

**48. Challenge** Create a magic square with a sum of 1 in which the denominators of the fractions are factors of 24.

**Patterns** Give the rule for finding each term in the sequence from the previous term. Then use your rule to find the missing terms. Use the last term to check your answers.

**49.** $1\frac{2}{5}$, $2\frac{4}{5}$, $4\frac{1}{5}$, $5\frac{3}{5}$, _____, _____, _____, _____, $12\frac{3}{5}$

**50.** $3\frac{1}{2}$, $6\frac{3}{4}$, $10$, $13\frac{1}{4}$, _____, _____, _____, _____, $29\frac{1}{2}$

**51.** $15$, $14\frac{5}{8}$, $14\frac{1}{4}$, $13\frac{7}{8}$, _____, _____, _____, _____, $12$

**52. Measurement** Below is part of a ruler.

a. What fraction of an inch does the smallest division of this ruler represent? How do you know?

**b.** How could you use this ruler to find $1\frac{1}{8} + \frac{5}{16}$? Find the sum.

**c.** How could you use this ruler to find $2\frac{1}{4} - \frac{5}{16}$? Find the difference.

**d.** Could you use this ruler to find $2\frac{1}{2} - \frac{5}{12}$? Explain.

**e.** Could you use this ruler to find $\frac{3}{8} + 1\frac{7}{16}$? Explain.

**53. In Your Own Words** Explain why you do not add fractions by just adding the numerators and adding the denominators.

*Mixed Review*

**Geometry**   The measure of two angles of a triangle are given. Find the measure of the third angle.

**54.** 23° and 47°

**55.** 60° and 60°

**56.** 10° and 120°

**57.** 94° and 20°

**Identify the property illustrated by each equation.**

**58.** $5 + 7 = 7 + 5$

**59.** $19 + (81 + 25) = (19 + 81) + 25$

**60.** $8(3 + 2) = 8 \cdot 3 + 8 \cdot 2$

**61.** $a \cdot (b \cdot c) = (a \cdot b) \cdot c$

**62.** $x \cdot y = y \cdot x$

**Find each sum or difference without using a calculator.**

**63.** $165.7 + 47.5$

**64.** $3.179 - 0.238$

**65.** $976,556 + 0.002$

**66.** $87.78 + 94.76$

**67.** $10.0101 + 1.101$

**68.** $9.02 - 7.34$

# LESSON 4.2

# Multiply and Divide Fractions

In this lesson, you will learn how to multiply and divide with fractions. As you work the exercises, it may be helpful to think about what multiplication and division mean and about how these operations work with whole numbers.

**Explore**

Work with your group to answer these questions. Try to find more than one way to answer each question.

- You want to serve lemonade to 20 people. Each glass holds $\frac{3}{4}$ cup. How many cups of lemonade do you need?

- You grew 12 pounds of peas. You give some away, keeping $\frac{2}{3}$ of the peas for yourself. How many pounds do you have left?

The questions you just answered can be represented with multiplication equations. Write a multiplication equation to represent each question. Explain why the equation fits the situation.

# Investigation 1 Multiply Fractions and Whole Numbers

In this investigation, you will explore more exercises involving multiplication with fractions. As you work the exercises, you might try some of the strategies that you and your classmates used in Explore.

**Real-World Link**

The vanilla extract used in recipes is derived from the vanilla planifolia orchid. This plant is native to Mexico, where it is pollinated by bees and hummingbirds.

## ✅ *Develop & Understand: A*

1. Suppose you want to bake a fraction cake.

   a. If you want enough cake to serve 12 people, how much of each ingredient do you need?

   b. For each ingredient, write a multiplication equation to represent the work you did in Part a.

   > **Fraction Cake**
   > $\frac{3}{4}$ cup sugar
   > $2\frac{1}{2}$ cups flour
   > $\frac{1}{2}$ teaspoon salt
   > $1\frac{1}{2}$ teaspoons vanilla
   > 2 eggs
   > Serves four people

2. Find each product using any method you like.

   a. $\frac{1}{4} \cdot 20$    b. $20 \cdot \frac{1}{2}$    c. $20 \cdot \frac{3}{4}$    d. $\frac{3}{2} \cdot 20$

3. Describe the strategies you used to find the products in Exercise 2.

### *Think & Discuss*

Hannah and Jahmal have different strategies for multiplying a whole number by a fraction.

The following is Hannah's strategy.

*I multiply the whole number by the numerator of the fraction. Then I divide the result by the denominator.*

The following is Jahmal's strategy.

*I do just the opposite. I divide the whole number by the denominator of the fraction. Then I multiply the result by the numerator.*

Try both methods on Parts a–d of Exercise 2 above. Do they both work?

You will probably find that some multiplication expressions are easier to simplify with Hannah's method and others are easier with Jahmal's method. For each expression below, tell whose method you think would be easier to use.

$$7 \cdot \frac{5}{14} \qquad \frac{3}{4} \cdot 8 \qquad 10 \cdot \frac{7}{36} \qquad \frac{9}{10} \cdot 5$$

The methods described above also work for multiplying a fraction by a decimal. The exercises on page 218 provide an opportunity to practice Hannah's and Jahmal's methods.

**Use Hannah's or Jahmal's method, or one of your own, to solve these exercises. Show how you find your answers.**

4. A cocoa recipe for one person calls for $\frac{3}{4}$ cup milk. Tell how much milk is needed to make the recipe for the following numbers of people.

   a. 3 people   b. 5 people   c. 6 people   d. 8 people

5. Fudge costs $6 per pound. Use the fact that 16 ounces equals 1 pound to help find the cost of each amount of fudge.

   a. $\frac{3}{4}$ pound   b. 8 ounces   c. 4 ounces   d. $1\frac{1}{2}$ pounds

6. At Fiona's Fabrics, plaid ribbon costs $2 per yard. Give the cost of each length of ribbon.

   a. $\frac{2}{3}$ yard   b. $1\frac{1}{2}$ feet   c. 8 inches   d. $1\frac{3}{4}$ yards

### Real-World Link

Milk chocolate was first created in Switzerland in 1876. The country consumes more than 20 pounds per person every year, more than anywhere else in the world.

✓ *Develop & Understand: C*

**Complete each multiplication table.**

7.

| ×             | 24 | 120 | 60 | 72 |
| ------------- | -- | --- | -- | -- |
| $\frac{1}{2}$ | 12 |     |    |    |
| $\frac{2}{3}$ |    |     |    |    |
| $\frac{1}{4}$ | 6  |     | 15 |    |
| $\frac{3}{4}$ |    |     |    |    |

8.

| ×             | 180 | ?   | 60 | 120 |
| ------------- | --- | --- | -- | --- |
| $\frac{1}{5}$ | 36  |     |    |     |
| $\frac{1}{2}$ |     |     | 30 |     |
| $\frac{2}{3}$ |     | 200 |    |     |
| $\frac{1}{4}$ |     |     |    | 30  |

9.

| ×             | ?  | ?  | ?  | 120 |
| ------------- | -- | -- | -- | --- |
| $\frac{3}{2}$ |    |    | 36 |     |
| ?             | 64 | 96 |    |     |
| $\frac{3}{8}$ |    | 54 |    |     |
| ?             |    |    | 20 | 100 |

**Share & Summarize**

1. Create a real-world situation that can be solved by multiplying a whole number and a fraction. Explain how to solve it.

2. What calculation do you need to do to find $\frac{3}{4}$ of 6 inches? What is $\frac{3}{4}$ of 6 inches?

# Investigation 2 Model Fraction Multiplication

You can visualize the product of two whole numbers by drawing a rectangle with side lengths equal to the numbers. The area of the rectangle represents the product. This rectangle represents 4 · 6.

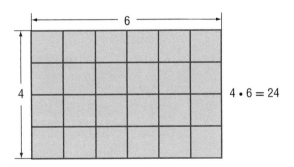

$4 \cdot 6 = 24$

### Math Link
The area of a rectangle is its length times its width.

You can represent the product of a whole number and a fraction in the same way. The shaded portion of this diagram represents the product $\frac{1}{2} \cdot 6$.

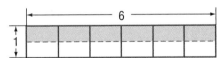

### Think & Discuss

What is the area of each small rectangle in the above diagram? ☐

How many small rectangles are shaded?

Use your answers to the previous questions to find the total shaded area. Explain why this area is equal to $\frac{1}{2} \cdot 6$.

### ✅ *Develop & Understand: A*

1. Consider this diagram.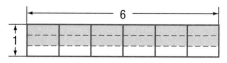

    a. The shaded region shows the product of what two numbers?

    b. What is the area of each small rectangle? How many small rectangles are shaded?

    c. Use your answers from Part b to find the total shaded area.

    d. Write a multiplication equation to represent the product of the numbers from Part a.

2. You can use similar diagrams to represent the product of two fractions. The shaded portion of this diagram represents the product $\frac{2}{3} \cdot \frac{3}{4}$.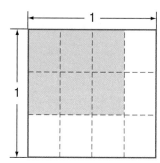

    a. What are the dimensions of the entire figure?

    b. Look at the entire shaded region. What is the height of this region? What is the width of this region?

    c. Use your answers from Part b to explain why the area of the shaded region is $\frac{2}{3} \cdot \frac{3}{4}$.

    d. What is the area of each small rectangle? How many small rectangles are shaded?

    e. Use your answers from Part d to find $\frac{2}{3} \cdot \frac{3}{4}$.

Draw a diagram like the one in Exercise 2 to represent each product. Then use your diagram to find the product. Give your answer as a multiplication equation, for example, $\frac{2}{3} \cdot \frac{3}{4} = \frac{6}{12}$.

3. $\frac{1}{2} \cdot \frac{1}{3}$

4. $\frac{3}{5} \cdot \frac{3}{4}$

5. $\frac{1}{2} \cdot \frac{5}{6}$

6. Look at your diagrams and equations from Exercises 3–5. Can you see a shortcut for multiplying two fractions *without* making a diagram? If so, use your shortcut to find $\frac{4}{7} \cdot \frac{2}{3}$. Then draw a diagram to see if your shortcut worked.

You may have noticed the following shortcut for multiplying two fractions.

*The product of two fractions is the product of the numerators over the product of the denominators.*

Use one of your rectangle diagrams from Exercises 3–5 on page 220 to explain why this shortcut works.

## ✅ Develop & Understand: B

**Use the shortcut described above to find each product. Express your answers in both the original form and in lowest terms.**

7. $\frac{3}{5} \cdot \frac{1}{4}$

8. $\frac{1}{6} \cdot \frac{2}{3}$

9. $\frac{4}{5} \cdot \frac{5}{8}$

10. $\frac{2}{3} \cdot \frac{3}{7}$

11. $\frac{2}{3} \cdot \frac{1}{8}$

12. $\frac{2}{3} \cdot \frac{3}{8}$

13. Rob wants to create a small herb garden in his backyard. The space he marked off is a square, $\frac{7}{8}$ of a meter on each side. What will be the area of his herb garden?

**Share & Summarize**

Draw a diagram to represent the product of two fractions you have not yet multiplied together in this investigation. Explain how the diagram shows the product.

# Investigation 3 Multiply with Fractions

In the last investigation, you found a shortcut for multiplying fractions. Now you will look at products involving mixed numbers.

---**Example**

This diagram illustrates $1\frac{1}{2} \cdot 2$. Each shaded section is labeled with its area. The total area is $1 + 1 + \frac{1}{2} + \frac{1}{2} = 3$. So,

$$1\frac{1}{2} \cdot 2 = 3$$

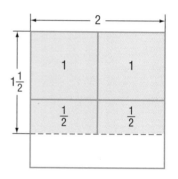

## ✅ Develop & Understand: A

**Draw a diagram to illustrate and find each product. Give your answer as a multiplication equation.**

1. $1\frac{1}{2} \cdot \frac{1}{3}$

2. $1\frac{1}{2} \cdot 2\frac{1}{2}$

3. Marcus suggested this shortcut for multiplying mixed numbers.

   *Multiply the whole number parts, multiply the fraction parts, and add the two results.*

   Try Marcus' method to find the products in Exercises 1 and 2. Does it work?

4. Miguel suggested this shortcut for multiplying mixed numbers.

   *Change the mixed numbers to fractions and multiply.*

   Try Miguel's method on the products in Exercises 1 and 2. Does it work?

There are several calculations involved in multiplying two mixed numbers. It is a good idea to estimate the product before you multiply.

## Think & Discuss

Consider $1\frac{1}{3} \cdot 5\frac{2}{3}$.

Before multiplying, make an estimate of the product. Explain how you found your answer.

Now change both mixed numbers to fractions and multiply.

How does your result compare to your estimate?

## ✅ Develop & Understand: B

In Exercises 5–10, complete Parts a and b.

a. Estimate the product.

b. Find the product, showing all of your steps. Give your result as a mixed number. If your answer is far from your estimate, check your calculations.

5. $1\frac{3}{8} \cdot 2\frac{1}{2}$

6. $3\frac{1}{3} \cdot \frac{8}{5}$

7. $\frac{1}{4} \cdot 8\frac{3}{5}$

8. $3\frac{1}{2} \cdot 1\frac{2}{3}$

9. $9\frac{2}{3} \cdot 1\frac{1}{2}$

10. $2\frac{1}{4} \cdot \frac{7}{8}$

11. Wei-Ling wants to hang wallpaper on two walls in her kitchen. One wall measures $11\frac{1}{2}$ feet by $8\frac{1}{2}$ feet, and the other measures $15\frac{2}{3}$ feet by $8\frac{1}{2}$ feet.

About how many square feet of wallpaper will she need? Estimate first and then calculate.

Suppose one roll of wallpaper contains 55 square feet. How many rolls of wallpaper will Wei-Ling need?

After you multiply fractions or mixed numbers, you often have to put the product in lowest terms. Sometimes, it is easier to simplify *before* you multiply. Check whether the numerator of each fraction shares a common factor with the denominator of the other fraction.

## Example

Find $\frac{1}{3} \cdot \frac{3}{4}$.

Notice that **3** is a factor of the denominator of $\frac{1}{3}$ and the numerator of $\frac{3}{4}$.

$$\frac{1}{3} \cdot \frac{3}{4} = \frac{1 \cdot 3}{3 \cdot 4}$$ Rewrite as the product of numerators over the product of denominators.

$$= \frac{3}{3} \cdot \frac{1}{4}$$ Group the common factors to form a fraction equal to 1.

$$= 1 \cdot \frac{1}{4}$$ Simplify.

$$= \frac{1}{4}$$

Find $\frac{2}{3} \cdot \frac{9}{16}$.

Notice that **2** is a factor of 2 and 16, and **3** is a factor of 3 and 9. As in the previous example, group these common factors to form a fraction equal to 1.

$$\frac{2}{3} \cdot \frac{9}{16} = \frac{2 \cdot 9}{3 \cdot 16}$$ Rewrite as the product of numerators over the product of denominators.

$$= \frac{2 \cdot 3 \cdot 3}{3 \cdot 2 \cdot 8}$$ Rewrite 9 as $3 \cdot 3$ and 16 as $2 \cdot 8$.

$$= \frac{2 \cdot 3}{2 \cdot 3} \cdot \frac{3}{8}$$ Group the common factors to form a fraction equal to 1

$$= 1 \cdot \frac{3}{8}$$ Simplify.

$$= \frac{3}{8}$$

## ✅ *Develop & Understand: C*

In Exercises 12–17, find the product in two ways.

- Multiply the fractions. Write the product in lowest terms.
- Simplify before finding the product.

**Show all of your steps.**

12. $\dfrac{3}{5} \cdot \dfrac{15}{6}$

13. $\dfrac{5}{6} \cdot \dfrac{3}{10}$

14. $\dfrac{1}{8} \cdot \dfrac{2}{3}$

15. $\dfrac{7}{12} \cdot \dfrac{3}{5}$

16. $\dfrac{3}{10} \cdot \dfrac{2}{3}$

17. $\dfrac{1}{2} \cdot 2\dfrac{4}{5}$

**Find each product.**

18. $\dfrac{3}{8} \cdot \dfrac{16}{7}$

19. $\dfrac{2}{3} \cdot \dfrac{7}{8}$

20. $\dfrac{2}{5} \cdot \dfrac{5}{6}$

21. $\dfrac{4}{5} \cdot \dfrac{3}{4}$

22. $\dfrac{1}{5} \cdot 1\dfrac{1}{2}$

23. $\dfrac{4}{5} \cdot 5\dfrac{3}{16}$

24. Mai wants to make enough fraction pasta for six servings. Rewrite the recipe for her.

Fraction Pasta

$1\frac{1}{2}$ cups vegetable stock

$\frac{2}{3}$ cup diced carrots

$\frac{3}{4}$ cup asparagus tips

$\frac{5}{6}$ cup peas

$1\frac{1}{3}$ pounds pasta

2 tabelspoons olive oil

$1\frac{1}{6}$ cups Parmesan cheese

Makes five servings

## Share & Summarize

1. Explain how to multiply two mixed numbers. Tell how you can use estimation to determine whether your answer is reasonable.

2. Explain how you could find the product $\dfrac{3}{4} \cdot \dfrac{8}{9}$ by simplifying before you multiply.

# Investigation (4) Divide Whole Numbers by Fractions

**Real-World Link**

Apples belong to the rose family. More than 7,500 varieties of apples are grown worldwide, including 2,500 varieties grown in the United States.

You know how to add, subtract, and multiply fractions. Now you will learn how to divide a whole number by a fraction.

## Explore

Work with your group to answer the following questions. Try to find more than one way to answer each question.

- Suppose you have five apples to share with your friends. If you divide each apple in half, how many halves will you have to share?
- There are 10 cups of punch left in the punchbowl. Each glass holds $\frac{2}{3}$ of a cup. How many glasses can you fill?

The questions above can be represented by division equations. Write an equation to represent each question. Explain why the equation fits the situation.

Caroline and Marcus solved the second Explore question in different ways.

## ✅ Develop & Understand: A

**Use Caroline's or Marcus' method to find each quotient. Try each method at least once. Show your work.**

1. $5 \div \frac{1}{3}$  2. $6 \div \frac{1}{6}$

3. $4 \div \frac{2}{3}$  4. $8 \div \frac{4}{5}$

5. $3 \div \frac{3}{6}$  6. $5 \div \frac{5}{6}$

Every multiplication equation has two related division equations. Here are two examples.

| Multiplication Equations | Related Division Equations | |
|---|---|---|
| $2 \cdot 10 = 20$ | $20 \div 10 = 2$ | $20 \div 2 = 10$ |
| $\frac{1}{2} \cdot 40 = 20$ | $20 \div 40 = \frac{1}{2}$ | $20 \div \frac{1}{2} = 40$ |

You can use this idea to perform divisions involving fractions.

### Example

Find $20 \div \frac{1}{4}$.

The *quotient* is the number that goes in the blank in this division equation.

$$20 \div \frac{1}{4} = \underline{\hspace{1cm}}$$

You can find the quotient by thinking about the related multiplication equation.

$$\frac{1}{4} \cdot \underline{\hspace{1cm}} = 20$$

Now just think, "One fourth of what number equals 20?" The answer is 80. So, $20 \div \frac{1}{4} = 80$.

## ✅ Develop & Understand: B

**Fill in the blanks in each pair of related equations.**

7. $15 \div \frac{1}{2} = \underline{\hspace{1cm}}$    $\frac{1}{2} \cdot \underline{\hspace{1cm}} = 15$

8. $20 \div \frac{2}{3} = \underline{\hspace{1cm}}$    $\frac{2}{3} \cdot \underline{\hspace{1cm}} = 20$

**Find each quotient by writing and solving a related multiplication equation.**

9. $20 \div \frac{1}{5} =$ _____

10. $14 \div \frac{2}{3} =$ _____

11. $15 \div \frac{3}{5} =$ _____

12. $12 \div \frac{3}{4} =$ _____

All the quotients you have found so far are whole numbers. Of course, this is not always the case. Find each quotient below using any method you like. Explain how you found your answer.

13. $3 \div \frac{2}{5} =$ _____

14. $7 \div \frac{3}{4} =$ _____

## ✅ Develop & Understand: C

Use any methods you like to complete each division table. Each entry is the result of dividing the first number in that row by the top number in that column. As you work, look for patterns that might help you complete the table without computing every quotient.

15.

| ÷ | $\frac{1}{3}$ | $\frac{2}{3}$ | $\frac{3}{3}$ |
|---|---|---|---|
| 6 | | | |
| 4 | | | |
| 2 | | | |

16.

| ÷ | $\frac{1}{4}$ | $\frac{2}{4}$ | $\frac{3}{4}$ | $\frac{4}{4}$ |
|---|---|---|---|---|
| 12 | | | | |
| 18 | | | | |
| 24 | | | | |

17.

| ÷ | $\frac{1}{5}$ | $\frac{2}{5}$ | $\frac{3}{5}$ | $\frac{4}{5}$ | $\frac{5}{5}$ |
|---|---|---|---|---|---|
| 24 | | | | | |
| 18 | | | | | |
| 9 | | | | | |

## ✓ Develop & Understand: A

**Use the method described in the example to find each quotient. Show all your steps.**

1. $\dfrac{7}{8} \div \dfrac{2}{3}$

2. $\dfrac{5}{6} \div \dfrac{1}{4}$

3. $5 \div \dfrac{3}{8}$

4. $\dfrac{1}{4} \div \dfrac{3}{4}$

5. $\dfrac{2}{5} \div 4$

6. $1\dfrac{1}{3} \div \dfrac{4}{5}$

**Fill in each blank with *greater than*, *less than*, or *equal to*. Give an example to illustrate each completed sentence.**

7. When you divide a fraction by a greater fraction, the quotient is

   _____ 1.

8. When you divide a fraction by a lesser fraction, the quotient is

   _____ 1.

Many people use a shortcut when dividing with fractions. The patterns you find in Think & Discuss will help you understand the shortcut and why it works.

---

### Think & Discuss

Find each product in your head. Describe any patterns you see in the expressions and the answers.

$$\dfrac{3}{4} \cdot \dfrac{4}{3} \qquad \dfrac{2}{15} \cdot \dfrac{15}{2}$$

$$\dfrac{5}{8} \cdot \dfrac{8}{5} \qquad \dfrac{25}{100} \cdot \dfrac{100}{25}$$

Now find these products. How are these expressions similar to those above?

$$\dfrac{1}{4} \cdot 4 \qquad \dfrac{1}{6} \cdot 6$$

$$\dfrac{1}{20} \cdot 20 \qquad \dfrac{1}{100} \cdot 100$$

Two numbers with a product of 1 are **reciprocals** of one another. Every number except 0 has a reciprocal. You can find the reciprocal of a fraction by switching its numerator and denominator.

---

**Math Link**

A reciprocal is also called a multiplicative inverse. The inverse property of multiplication states the product of a number and its inverse is 1.

---

The shortcut for dividing fractions follows.

*To divide a fraction by a fraction, multiply the first fraction by the reciprocal of the second fraction.*

## Example

Find $\frac{5}{7} \div \frac{10}{12}$.

To find $\frac{5}{7} \div \frac{10}{12}$, multiply $\frac{5}{7}$ by the reciprocal of $\frac{10}{12}$.

$$\frac{5}{7} \div \frac{10}{12} = \frac{5}{7} \cdot \frac{12}{10} = \frac{60}{70} = \frac{6}{7}$$

To see why the shortcut works, rewrite the division expression as a fraction. Multiply *both* the numerator and denominator by the reciprocal of the denominator. The denominator becomes 1.

$$\frac{5}{7} \div \frac{10}{12} = \frac{\frac{5}{7}}{\frac{10}{12}} = \frac{\frac{5}{7} \cdot \frac{12}{10}}{\frac{10}{12} \cdot \frac{12}{10}} = \frac{\frac{5}{7} \cdot \frac{12}{10}}{1} = \frac{5}{7} \cdot \frac{12}{10}$$

So, $\frac{5}{7} \div \frac{10}{12} = \frac{5}{7} \cdot \frac{12}{10}$.

## ✅ Develop & Understand: B

**Find each quotient using any method you like.**

9. $\frac{3}{2} \div \frac{9}{6}$

10. $\frac{2}{5} \div \frac{5}{2}$

11. $\frac{1}{8} \div \frac{1}{9}$

**Estimate whether each quotient will be greater than, less than, or equal to 1. Then find the quotient.**

12. $\frac{3}{5} \div \frac{3}{4}$

13. $2 \div \frac{3}{5}$

14. $\frac{2}{3} \div \frac{5}{6}$

15. $1 \div 3\frac{1}{2}$

16. $3\frac{1}{2} \div \frac{2}{7}$

17. $4\frac{1}{2} \div 2\frac{1}{4}$

## Share & Summarize

1. Describe two methods for dividing a fraction by a fraction.

2. Write two fraction division expressions, one with a quotient greater than 1 and the other with a quotient less than 1.

**Practice & Apply**

**Real-World Link**

The first practical sewing machines were built in the mid 1800s and could sew only straight seams. Modern sewing machines utilize computer technology. For instance, scanners can take an image and reproduce it in an embroidered version on cloth.

. . . . . . . . . . . . . . . . . . . . . . . .

**1.** Caroline is sewing a costume with 42 small ribbons on it. Each ribbon is $\frac{1}{3}$ of a yard long. How many yards of ribbon does she need?

**2.** Caroline is sewing a costume from a pattern that calls for 4 yards of material. Because the costume is for a small child, she is reducing all the lengths to $\frac{2}{3}$ of the lengths given by the pattern. How many yards of material should she buy?

**3.** Consider this table of products.

   **a.** Copy the table. Write the result of each multiplication expression in the second column.

   **b.** What patterns do you see in the expressions and the results?

   **c.** What would be the next two expressions and results?

| Expression | Result |
|---|---|
| $\frac{1}{4} \cdot 10$ | |
| $\frac{2}{4} \cdot 10$ | |
| $\frac{3}{4} \cdot 10$ | |
| $\frac{4}{4} \cdot 10$ | |
| $\frac{5}{4} \cdot 10$ | |
| $\frac{6}{4} \cdot 10$ | |
| $\frac{7}{4} \cdot 10$ | |
| $\frac{8}{4} \cdot 10$ | |

**4.** Consider this table of products.

   **a.** Copy the table. Write the result of each multiplication expression in the second column.

   **b.** What relationships do you see between the fraction in each expression and the result?

   **c.** If you changed the expressions so the numerator of each fraction was 2 instead of 1, how would the products change?

| Expression | Result |
|---|---|
| $\frac{1}{30} \cdot 60$ | |
| $\frac{1}{20} \cdot 60$ | |
| $\frac{1}{15} \cdot 60$ | |
| $\frac{1}{12} \cdot 60$ | |
| $\frac{1}{6} \cdot 60$ | |
| $\frac{1}{5} \cdot 60$ | |
| $\frac{1}{4} \cdot 60$ | |
| $\frac{1}{3} \cdot 60$ | |
| $\frac{1}{2} \cdot 60$ | |

**5.** Consider the product $\frac{3}{4} \cdot 8$.

    **a.** Draw a rectangle diagram to represent this product.

    **b.** What is the area of each small rectangle in your diagram? How many small rectangles are shaded?

    **c.** What does $\frac{3}{4} \cdot 8$ equal?

**6.** Consider the product $\frac{2}{3} \cdot \frac{4}{5}$.

    **a.** Draw a rectangle diagram to represent this product.

    **b.** What is the area of each small rectangle in your diagram? How many small rectangles are shaded?

    **c.** What does $\frac{2}{3} \cdot \frac{4}{5}$ equal?

**Find each product. Give your answers in lowest terms.**

**7.** $\frac{7}{8} \cdot \frac{2}{5}$                       **8.** $\frac{2}{3} \cdot \frac{7}{12}$

**9.** $\frac{6}{11} \cdot \frac{2}{3}$                   **10.** $\frac{30}{50} \cdot \frac{15}{20}$

**11.** $\frac{4}{7} \cdot \frac{7}{9}$                     **12.** $\frac{13}{15} \cdot \frac{5}{6}$

**13.** Copy and complete this multiplication table. Give each answer as both a fraction and a mixed number. The first answer is provided.

| $\times$ | $1\frac{1}{2}$ | $2\frac{2}{3}$ | $3\frac{3}{4}$ |
|---|---|---|---|
| $1\frac{1}{2}$ | $\frac{9}{4} = 2\frac{1}{4}$ | | |
| $2\frac{1}{2}$ | | | |
| $3\frac{1}{2}$ | | | |

**Find each product by simplifying before you multiply.**

**14.** $\frac{3}{4} \cdot \frac{8}{9}$                   **15.** $\frac{2}{3} \cdot \frac{3}{8}$

**16.** $\frac{3}{5} \cdot \frac{4}{9}$                   **17.** $\frac{2}{5} \cdot \frac{5}{9}$

**18.** Consider the following multiplication table.

| ×           | $\frac{2}{3}$ | $1\frac{1}{3}$ | $2$ | $2\frac{2}{3}$ | $3\frac{1}{3}$ |
|-------------|---------------|----------------|-----|----------------|----------------|
| $\frac{1}{2}$ |               |                |     |                |                |
| $\frac{1}{4}$ |               |                |     |                |                |
| $\frac{1}{8}$ |               |                |     |                |                |

**a.** Copy and complete the table. Give your answers in lowest terms. Express answers greater than 1 as mixed numbers.

**b.** What patterns and relationships do you see in the table?

**Fill in each blank.**

**19.** $\frac{2}{3} \cdot$ _____ $= 20$    **20.** $\frac{2}{3} \cdot$ _____ $= 15$    **21.** $\frac{2}{3} \cdot$ _____ $= 10$

**22.** In Parts a–f, use any method you like to find the answer.

**a.** How many $\frac{1}{4}$s are in 16?    **b.** $\frac{1}{4}$ of what number is 16?

**c.** How many $\frac{2}{3}$s are in 16?    **d.** $\frac{2}{3}$ of what number is 16?

**e.** How many $\frac{4}{3}$s are in 16?    **f.** $\frac{4}{3}$ of what number is 16?

**g.** How are the answers to Parts a and b related? Parts c and d? Parts e and f? Explain why this makes sense.

**23.** Find the first quotient. Use your result to predict the second quotient. Check your answers by multiplying.

**a.** $14 \div \frac{1}{3}$      $14 \div \frac{2}{3}$

**b.** $5 \div \frac{5}{8}$      $15 \div \frac{5}{8}$

**c.** $6 \div \frac{1}{5}$      $6 \div \frac{3}{5}$

**d.** $6 \div \frac{3}{7}$      $24 \div \frac{3}{7}$

**Fill in each blank.**

**24.** $\frac{2}{3} \cdot$ _____ $= 1$    **25.** _____ $\cdot \frac{1}{8} = 1$    **26.** $1\frac{1}{2} \cdot \frac{2}{3} =$ _____

**27.** _____ $\cdot 1\frac{1}{8} = 1$    **28.** $\frac{5}{3} \cdot$ _____ $= 1$    **29.** $2\frac{1}{2} \cdot \frac{4}{5} =$ _____

**30.** In Parts a–f, estimate whether the quotient will be greater than, less than, or equal to 1. Then find the quotient. Leave answers greater than 1 in fraction form.

   **a.** $\frac{1}{3} \div \frac{2}{5}$                          **b.** $\frac{2}{5} \div \frac{1}{3}$

   **c.** $\frac{7}{8} \div \frac{3}{4}$                          **d.** $\frac{3}{4} \div \frac{7}{8}$

   **e.** $\frac{1}{5} \div \frac{1}{7}$                          **f.** $\frac{1}{7} \div \frac{1}{5}$

   **g.** Look at the expressions and answers for Parts a–f. How are the answers to Parts a and b related? Parts c and d? Parts e and f? Explain why this makes sense.

**Find each quotient. Give all answers in lowest terms. If an answer is greater than 1, write it as a mixed number.**

**31.** $7\frac{1}{2} \div 1\frac{1}{2}$         **32.** $3\frac{1}{3} \div \frac{3}{4}$         **33.** $2\frac{2}{3} \div \frac{8}{5}$

**34.** $\frac{9}{8} \div \frac{2}{3}$         **35.** $\frac{8}{10} \div \frac{1}{100}$         **36.** $5\frac{1}{3} \div 2\frac{2}{3}$

**37.** $2\frac{1}{3} \div 3\frac{1}{3}$         **38.** $18 \div 4\frac{1}{2}$         **39.** $7\frac{1}{2} \div 4\frac{1}{2}$

**Preview** Find each missing factor.

**40.** $\frac{1}{3} \cdot$ _____ $= 60$         **41.** $\frac{2}{3} \cdot$ _____ $= 60$

**42.** $\frac{1}{5} \cdot$ _____ $= 60$         **43.** $\frac{2}{5} \cdot$ _____ $= 60$

**44.** $\frac{3}{5} \cdot$ _____ $= 60$         **45.** $\frac{4}{5} \cdot$ _____ $= 60$

**46. Economics** Rosita paid $8 for a CD marked "$\frac{1}{2}$ off." How much did she save?

**47. Life Science** The St. Louis Zoo has about 700 species of animals.

   **a.** The Detroit Zoological Park has $\frac{2}{5}$ as many species as the St. Louis Zoo. About how many species does the Detroit park have?

   **b.** The Toronto Zoo has $\frac{23}{70}$ *fewer* species than the St. Louis Zoo. About how many species does the Toronto Zoo have?

**Real-World Link**

The first modern zoo was the Imperial Managerie at the Schönbrunn Palace in Vienna, which opened to the public in 1765. The National Zoological Park in Washington, D.C., was the first zoo created to preserve endangered species.

**Tell whether each statement is true or false.**

**48.** $\frac{2}{3}$ of 300 = $\frac{3}{4}$ of 200

**49.** $\frac{1}{3}$ of 150 = $\frac{1}{2}$ of 100

**50.** $\frac{2}{3}$ of 300 = $\frac{1}{2}$ of 400

**51.** $\frac{2}{3}$ of 100 = $\frac{1}{3}$ of 200

**52. Economics** Last week, Sal's Shoe Emporium held its semiannual clearance sale. All winter boots were marked down to $\frac{4}{5}$ of the original price. This week, the sale prices of the remaining boots were cut in half. What fraction of the original price is the new sale price?

**53.** Using what you have learned about multiplying fractions, explain why multiplying the numerator and denominator of a fraction by the same number does not change the fraction's value. Give examples if they help you to explain.

**54. Measurement** The Danson's horse ranch is a rectangular shape measuring $\frac{3}{5}$ mile by $\frac{4}{7}$ mile.

**a.** What is the area of the ranch?

**b.** There are 640 acres in 1 square mile. What is the area of the Danson's ranch in acres?

**55. Measurement** The left column of the table shows the ingredient list for a spice cake that serves 12 people. Complete the table to show the amount of each ingredient needed for the given numbers of people.

| | 12 People | 10 People | 8 People | 6 People | 4 People | 2 People |
|---|---|---|---|---|---|---|
| **Flour** | $2\frac{1}{4}$ c | | | | | |
| **Sugar** | $1\frac{1}{3}$ c | | | | | |
| **Salt** | $\frac{3}{4}$ tsp | | | | | |
| **Butter** | $1\frac{1}{2}$ sticks | | | | | |
| **Ginger** | $\frac{1}{2}$ tsp | | | | | |
| **Raisins** | $\frac{2}{3}$ c | | | | | |

**56. Number Sense** In Parts a and b, use the numbers 1, 2, 3, 4, 5, and 6.

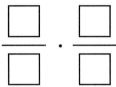

a. Fill in each square with one of the given numbers to create the greatest possible product. Use each number only once.

b. Fill in each square with one of the given numbers to create the least possible product. Use each number only once.

c. Choose four different whole numbers. Repeat Parts a and b using your four numbers.

**57.** Consider this multiplication table.

| ×             | $4\frac{1}{2}$ | $2\frac{1}{4}$ | $1\frac{1}{8}$ | $\frac{9}{16}$ | $\frac{9}{32}$ |
|---------------|----------------|----------------|----------------|----------------|----------------|
| $\frac{1}{2}$  | $2\frac{1}{4}$ |                |                |                |                |
| $1\frac{1}{2}$ |                |                |                |                |                |
| $2\frac{1}{2}$ |                |                |                |                |                |
| $3\frac{1}{2}$ |                |                |                |                |                |

a. Copy the table. Use your knowledge of fraction multiplication and number patterns to help complete it. Try to do as few paper-and-pencil calculations as possible.

b. Describe the pattern in the table as you read across the rows from left to right.

c. Describe the pattern in the table as you read down the columns.

**58.** Rosita bought six yards of ribbon for a sewing project.

   **a.** She wants to cut the ribbon into pieces $\frac{1}{2}$ foot long. How many pieces can she cut from her six-yard ribbon?

   **b.** What calculations did you do to solve Part a?

   **c.** Rosita decides instead to cut pieces $\frac{2}{3}$ foot long. How many pieces can she cut?

   **d.** What calculations did you do to solve Part c?

**59.** Hannah made up riddles about the ages of some people in her family. See if you can solve them.

   **a.** My brother Tim is $\frac{3}{4}$ my cousin Janice's age. Tim is 15 years old. How old is Janice?

   **b.** To find my grandpa Henry's age, divide my aunt Carol's age by $\frac{4}{5}$ and then add 10. Aunt Carol is 40. How old is Grandpa Henry?

   **c.** Today is Uncle Mike's 42nd birthday. He is now $\frac{2}{3}$ of his father's age and $\frac{7}{3}$ of his daughter's age. How old are Uncle Mike's father and daughter?

**60. Measurement** In this exercise, use the following facts.

   1 cup = 8 ounces     1 quart = 4 cups     1 gallon = 4 quarts

   **a.** Conor has 3 gallons of lemonade to serve at his party. If he pours $\frac{3}{4}$-cup servings, how many servings can he pour?

   **b.** What calculations did you do to solve Part a?

   **c.** If Conor pours 4-ounce servings instead, how many servings can he pour?

   **d.** What calculations did you do to solve Part c?

**61.** Consider this division table.

| ÷ | $\frac{1}{2}$ | $\frac{1}{4}$ | $\frac{1}{8}$ | $\frac{1}{16}$ | $\frac{1}{32}$ | $\frac{1}{64}$ |
|---|---|---|---|---|---|---|
| $\frac{1}{2}$ | | | | | | |
| $\frac{1}{4}$ | | | | | | |
| $\frac{1}{8}$ | | | | | | |
| $\frac{3}{2}$ | | | | | | |
| $\frac{3}{4}$ | | | | | | |
| $\frac{3}{8}$ | | | | | | |

   **a.** Copy the table. Use your knowledge of fraction division and number patterns to help complete it. Try to do as few paper-and-pencil calculations as possible.

   **b.** Describe the number pattern in the table as you read across the rows from left to right. Explain why this pattern occurs.

   **c.** Describe the number pattern in the table as you read down the columns. Explain why this pattern occurs.

**Tell whether each statement is true or false. If a statement is false, give the correct quotient.**

**62.** $\frac{9}{12} \div \frac{1}{4} = 3$  **63.** $\frac{1}{4} \div \frac{9}{12} = \frac{2}{3}$

**64.** $3\frac{1}{2} \div \frac{7}{4} = 2$  **65.** $\frac{7}{4} \div 3\frac{1}{2} = \frac{1}{2}$

**66.** $\frac{10}{4} \div \frac{4}{10} = 1$  **67.** $\frac{4}{10} \div \frac{10}{4} = 1$

**Evaluate each expression.**

**68.** $\frac{1}{2} \cdot \frac{3}{4} + \frac{1}{2} \div \frac{5}{6}$

**69.** $\left(\frac{3}{5} + \frac{7}{8}\right) \cdot \frac{1}{5} \div \frac{2}{15}$

**70.** In Parts a and b, use the numbers 2, 4, 6, 8, 10, and 12.

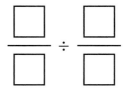

 a. Fill in each blank with one of the given numbers to create the greatest possible quotient. Use each number only once.

 b. Fill in each blank with one of the given numbers to create the least possible quotient. Use each number only once.

 c. Choose four different whole numbers. Repeat Parts a and b using your four numbers.

**71.** Explain how you could solve the problem below without doing any calculations. Give the result, and check your answer by performing the calculations.

$$\left(\frac{5}{6} \div \frac{4}{7}\right) \cdot \left(\frac{4}{7} \div \frac{5}{6}\right)$$

**72. In Your Own Words** Which methods do you prefer for multiplying fractions? Which methods do you prefer for dividing fractions? Explain why you chose each method.

*Mixed Review*

**Write a rule for each relationship between $p$ and $q$.**

**73.**

| p | 2 | 3 | 4 | 5 | 6 | 7 |
|---|---|---|---|---|---|---|
| q | 4 | 6 | 8 | 10 | 12 | 14 |

**74.**

| p | 14 | 6 | 8.35 | 8 | 12 | 9 |
|---|---|---|---|---|---|---|
| q | 9 | 1 | 3.35 | 3 | 7 | 4 |

**75.**

| p | 2 | $\frac{2}{3}$ | 7 | 3 | 10 | 1 |
|---|---|---|---|---|---|---|
| q | 4 | 0 | 19 | 7 | 28 | 1 |

**76.**

| p | $\frac{1}{2}$ | 0 | 1 | 2 | 3 | $3\frac{1}{2}$ |
|---|---|---|---|---|---|---|
| q | 2 | 1 | 3 | 5 | 7 | 8 |

**Write each expression as repeated multiplication. Then find the value of each expression.**

**77.** $5^2$

**78.** $10^2$

**79.** $4^3$

**80.** $2^4$

**81.** $5^4$

**82.** $2^6$

# Multiply and Divide Decimals

Figuring out the tip on a restaurant bill, converting measurements from one unit to another, finding lengths for a scale model, and exchanging money for different currencies are just a few activities that involve multiplying and dividing with decimals.

In this lesson, you will learn how to multiply and divide decimals and to use estimation to determine whether your results are reasonable.

## Think & Discuss

Fill in the blank with a whole number or a decimal so the product is:

$16 \cdot \underline{\hspace{2cm}}$

- greater than 100.

- greater than 32 but less than 100.

- at least 17 but less than 32.

- equal to 16.

- greater than 8 but less than 16.

- less than 8 but greater than 0.

## Investigation 1 Multiply Whole Numbers and Decimals

Although you may not be able to calculate 172 • 97 in your head, you know the product is greater than both 172 and 97. In fact, when you multiply any two numbers greater than 1, the product must be greater than either of the numbers.

In the next exercises, you will explore what happens when one of the numbers is less than 1.

## ✅ *Develop & Understand: A*

**Real-World Link**

Cats purr at the same frequency as an idling diesel engine, about 26 cycles per second.

1. Luke considered two brands of food for his cats. Kitty Kans cost $0.32 per can, and Purrfectly Delicious cost $0.37 per can. Find the cost of each of the following.

    a. 10 cans of Kitty Kans

    b. 10 cans of Purrfectly Delicious

    c. 12 cans of Kitty Kans

    d. 12 cans of Purrfectly Delicious

    e. Look at your answers to Parts a–d. In each case, was the cost more or less than the number of cans? Explain why this makes sense.

2. After buying cat food, Luke went to CD-Rama and bought a CD for $14.00 plus 4% sales tax. To calculate a 4% sales tax, multiply the cost of the purchase by 0.04.

    a. The cashier tried to charge Luke $19.60, $14.00 for the CD and $5.60 for sales tax. Without actually calculating the tax, how do you know that the cashier made a mistake?

    b. What is the correct amount of sales tax on Luke's $14 purchase?

3. Multiply each number by 0.01. If you see a shortcut for finding the answer without doing any calculations, use it.

    a. 1,776

    b. 28,000

    c. 29,520

    d. 365.1

4. If you used a shortcut for Exercise 3, explain what you did. If you did not use a shortcut, compare the four original numbers with the answers. What pattern do you see?

You can multiply with decimals by first ignoring the decimal points and multiplying whole numbers. Then you can use your estimation skills to help determine where to put the decimal point in the answer.

## ✅ Develop & Understand: B

Jahmal's calculator is broken. It calculates correctly but no longer displays decimal points. The result 452.07 is displayed as 45207. Unfortunately, 4.5207 is displayed the same way.

5. List all other numbers that would be displayed as 45207.

6. Jahmal's calculator displays 45207 when he enters 5.023 • 9. Without using a calculator, estimate the product. Use your estimate to figure out where to place the decimal point in the result.

7. Jahmal's calculator also displays 45207 when 5 • 904.14 is entered. Use estimation to figure out the correct result of this calculation. Explain how you found your answer.

8. When Jahmal enters 279 • 0.41, his calculator displays 11439.

   a. Is the correct result greater than or less than 279? Explain.

   b. Jahmal estimated the result by multiplying 300 by $\frac{40}{100}$. Will this calculation give a good estimate of the actual answer? Explain.

   c. What estimate will Jahmal get for the product?

   d. What is the exact answer to 279 • 0.41? Explain how you know.

9. Below are some calculations that Jahmal entered and the results that his calculator displayed. Estimate each product. Explain how you made your estimate. Then give the actual product by placing the decimal point in the correct place in the display number.

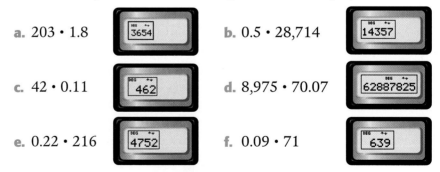

   a. 203 • 1.8    [DEG] 3654
   
   b. 0.5 • 28,714    [DEG] 14357

   c. 42 • 0.11    [DEG] 462
   
   d. 8,975 • 70.07    [DEG] 62887825

   e. 0.22 • 216    [DEG] 4752
   
   f. 0.09 • 71    [DEG] 639

10. Look back at your results for Exercise 9. For each part, tell whether the product is greater than or less than the whole number. Explain why.

11. Jahmal's calculator displayed a result as 007. What might be the actual result? Give all the possibilities.

## Share & Summarize

1. Suppose you are multiplying a whole number by a decimal.

   a. When do you get a result less than the whole number? Give some examples.

   b. When do you get a result greater than the whole number? Give some examples.

2. Explain how using estimation is helpful when multiplying with decimals.

# Investigation 2 Multiply Decimals as Fractions

One strategy for multiplying decimals is to write them as fractions, multiply the fractions, and then write the answer as a decimal. For example, this strategy is used to calculate $0.7 \cdot 0.02$.

$$0.7 \cdot 0.02 = \frac{7}{10} \cdot \frac{2}{100} = \frac{14}{1,000} = 0.014$$

## Think & Discuss

Althea said the $\frac{2}{100}$ in the calculation above should have been written in lowest terms before multiplying. Rosita disagreed. She said it is easier to change the answer to a decimal if the fractions are not written in lowest terms before multiplying.

What do you think? Defend your answer.

## Develop & Understand: A

**Math Link**

A decimal greater than 1 can be represented by a fraction or a mixed number. For example, 5.12 can be written as $\frac{512}{100}$ or $5\frac{12}{100}$.

**Write the decimals as fractions and then multiply. Give each product as a decimal. Show all of your steps.**

1. $0.6 \cdot 0.7$

2. $1.06 \cdot 0.07$

3. $5.12 \cdot 0.2$

4. $0.002 \cdot 0.003$

**5.** Conor calculated 0.625 • 0.016 on his calculator and got 0.01. Jing told him, "You must have made a mistake. When you write the expression with fractions, you get $\frac{625}{1,000} \cdot \frac{16}{1,000}$. When you multiply these fractions, you get a denominator of 1,000,000. Your answer is equal to $\frac{1}{100}$, which has a denominator of 100."

Find 0.625 • 0.016 using your calculator. Did Conor make a mistake? If so, explain it. If not, explain the mistake in Jing's reasoning.

**6.** In a particular state, the sales tax is 5%. To calculate the sales tax on an item, multiply the cost by 0.05. Assume the sales tax is always rounded up to the next whole cent.

**a.** Calculate the sales tax on an $0.89 item. Show your steps.

**b.** Find a price less than $1 so that the sales-tax calculation results in a whole number of cents without rounding.

**Real-World Link**

The first battery-operated, hand-held calculators were introduced in 1970 and sold for about $400.

Multiplying decimals is just like multiplying whole numbers, except you need to determine where to place the decimal point in the product. In the last investigation, you used estimation to locate the decimal point. The idea of writing decimals as fractions before multiplying leads to another method for multiplying decimals.

**Think & Discuss**

Write each number as a fraction.

> 10.7          2.43          0.073          13.0601

How does the number of digits to the right of the decimal point compare to the number of zeros in the denominator of the fraction?

Find each product.

> 10 • 10,000          100 • 100          1,000,000 • 10,000

How is the number of zeros in each product related to the numbers of zeros in the numbers being multiplied?

Now consider the product 0.76 • 0.041.

If you wrote 0.76 and 0.041 as fractions, how many zeros would be in each denominator? How many zeros would be in the denominator of the product?

Use the fact that 76 • 41 = 3,116 and your answers to the above questions to find the product 0.76 • 0.041. Explain how you found your answer.

The following rule was used to multiply decimals in Think & Discuss.

- Ignore the decimal points. Multiply the numbers as if they were whole numbers.
- Place the decimal point so that the number of digits to its right is equal to the total number of digits to the right of the decimal points in the numbers being multiplied.

## ✓ Develop & Understand: B

7. Consider the product 0.1123 · 0.92.

   a. Use the rule above to find the product. Explain each step.

   b. Check your answer to Part a by changing the numbers to fractions and multiplying.

   c. Without multiplying, how could you know that the correct answer could not be 1.03316 or 0.0103316?

8. Use the rule above to recalculate the answers to Exercises 1–4 on page 245. Do you get the same results you found by writing the decimals as fractions and then multiplying?

9. Jing said, "The rule doesn't work for the product 0.625 · 0.016. When I find the product on my calculator, I get 0.01, which has only two digits to the right of the decimal point. According to the rule, there should be six digits to the right of the decimal point. What's going on?" Explain why this is happening.

**Find each product without using a calculator.**

10. 43.3 · 2.05

11. 0.0005 · 10.5

12. 2.667 · 0.11

### Share & Summarize

1. Explain how you can use fractions to determine where to place the decimal point in the product of two decimals. Give an example to illustrate your answer.

2. The total number of digits to the right of the decimal points in 0.25 and 0.012 is five. Explain why the product 0.25 · 0.012 has only three decimal places.

# Investigation ③ Multiply Decimals in Real Life

Decimals are easy to understand and compare. They are easier to use than fractions when doing computations with calculators and computers. For these reasons, numerical information is often given in decimal form. In this lesson, you will explore some real situations involving calculations with decimals.

## Think & Discuss

You can predict approximately how tall a four-year-old will be at age 12 by multiplying his or her height by 1.5.

Hannah's four-year-old brother, Jeremy, likes to play with her calculator. She let him press the buttons as she predicted his height at age 12. He is now 101.5 cm tall, so she told him to enter 1.5 · 101.5. The calculator's display read 1522.5.

Hannah realized right away that Jeremy had pressed the wrong keys. How did she know?

## ✅ Develop & Understand: A

1. On her fourth birthday, Shanise was exactly as tall as a meterstick. Predict her height at age 12.

2. Ynez is a four-year-old who likes to use very precise numbers. She claims she is exactly 98.75 cm tall.

   a. Use this value to predict Ynez's height at age 12.

   b. Rosita says it does not make sense to use the exact decimal answer as a prediction for Ynez's height. Do you agree? Explain.

   c. What is a reasonable prediction for Ynez's height at age 12?

**Real-World Link**

The practice of tipping began in the 18th century in British inns. Patrons gave waiters money before a meal with a note indicating the money was "to insure promptness." The word tip is an abbreviation of this message.

The most common decimal calculations are probably those that involve money. You will now explore some situations involving money.

## ✅ Develop & Understand: B

3. To figure out the cost of dinner at a restaurant, Mr. Rivera multiplies the total of the prices for the items by 1.25. The result includes the cost of the items ordered, the 5% sales tax, and a 20% tip.

   Mr. Rivera and his friend went to dinner. The cost of the items that they ordered was $42. When Mr. Rivera tried to calculate the cost with tax and tip, he got $85. Without doing any calculations, explain why this result cannot be correct.

Since you cannot pay for something in units smaller than a penny, most prices are given in dollars and whole numbers of cents. Gasoline prices, however, are often given to tenths of a cent, which is thousandths of a dollar, and displayed in a form that combines decimals and fractions. For example, a price of $3.579 is given as $3.57$\frac{9}{10}$. The fraction $\frac{9}{10}$ represents $\frac{9}{10}$ of a cent.

4. Imagine you are buying exactly 10 gallons of gas priced at $3.57$\frac{9}{10}$ per gallon.

   a. Without using a calculator, figure out exactly how much the gas will cost.

   b. If the price were rounded to $3.58 per gallon, how much would you pay?

   c. How much difference is there in your answers to Parts a and b? Why do you think gas prices are not just rounded up to, for example, $3.58 per gallon?

5. Ms. Kenichi filled her sports utility vehicle with 38.4 gallons of gas. The gas cost $3.57$\frac{9}{10}$ per gallon.

   a. Estimate the total cost for the gas.

   b. Now calculate the exact price, rounding off to the nearest penny at the end of your calculation.

When visiting other countries, travelers exchange the currency of their home country for the currency used in the country they are visiting. Converting from one unit of currency to another involves operations with decimals.

## ✅ Develop & Understand: C

The unit of currency in Japan is the yen. On October 30, 2007, one yen was worth $0.008715, slightly less than one U.S. penny. In the following exercises, round your answers to the nearest cent.

6. On October 30, 2007, what was the value of 100 yen in U.S. dollars?

7. Dr. Kuno was traveling in Japan on October 30, 2007. On that day, she purchased a digital camera priced at 14,100 yen for her nephew. What was the equivalent price in U.S. dollars?

8. Copy and complete the table to show the dollar equivalents for the given numbers of yen.

| Yen | Dollars |
|---|---|
| 10 | |
| 50 | |
| 100 | |
| 150 | |
| 200 | |
| 300 | |
| 1,000 | |
| 2,000 | |
| 3,000 | |
| 10,000 | |
| 1,000,000 | |

9. Use your table to help estimate the dollar equivalent of 6,000 yen. Explain how you made your estimate.

10. Use your table to help estimate the yen equivalent of $100. Explain how you made your estimate.

## Share & Summarize

On her return from Japan, Dr. Kuno traded her yen for dollars. She handed the teller 23,500 yen. The teller gave her a $20 bill, a $1 bill, and some change. Use estimation to explain how you know the teller gave Dr. Kuno the incorrect amount.

# Investigation (4) Divide Decimals

In this investigation, you will divide decimals. First, you will examine patterns relating multiplication and division.

## ✅ Develop & Understand: A

**Math Link**
The ÷ symbol was used by editors of early manuscripts to indicate text to be cut. The symbol was used to indicate division as early as 1650.

1. Copy and complete this table.

| | |
|---|---|
| 3.912 • 0.1 = | 3.912 ÷ 0.1 = |
| 3.912 • 0.01 = | 3.912 ÷ 0.01 = |
| 3.912 • 0.001 = | 3.912 ÷ 0.001 = |
| 4,125.9 • 0.1 = | 4,125.9 ÷ 0.1 = |
| 4,125.9 • 0.01 = | 4,125.9 ÷ 0.01 = |
| 4,125.9 • 0.001 = | 4,125.9 ÷ 0.001 = |

2. Look for patterns in your completed table.

   a. What happens when a number is multiplied by 0.1, 0.01, and 0.001?

   b. What happens when a number is divided by 0.1, 0.01, and 0.001?

3. Copy and complete this table.

| | |
|---|---|
| 3.912 ÷ 10 = | 3.912 • 10 = |
| 3.912 ÷ 100 = | 3.912 • 100 = |
| 3.912 ÷ 1,000 = | 3.912 • 1,000 = |
| 4,125.9 ÷ 10 = | 4,125.9 • 10 = |
| 4,125.9 ÷ 100 = | 4,125.9 • 100 = |
| 4,125.9 ÷ 1,000 = | 4,125.9 • 1,000 = |

4. Compare your results for Exercises 1 and 3. Then complete these statements.

   a. Multiplying a number by 0.1 is the same as dividing it by _____.

   b. Multiplying a number by 0.01 is the same as dividing it by _____.

   c. Multiplying a number by 0.001 is the same as dividing it by _____.

   d. Dividing a number by 0.1 is the same as multiplying it by _____.

   e. Dividing a number by 0.01 is the same as multiplying it by _____.

   f. Dividing a number by 0.001 is the same as multiplying it by _____.

Now you can probably divide by such decimals as 0.1, 0.01, and 0.001 mentally. For other decimal divisors, it is not easy to find an exact answer in your head. However, the next example shows a method for estimating the quotient of two decimals.

## Example

Estimate $0.0351 \div 0.074$.

Think of the division expression as a fraction.

$$\frac{0.0351}{0.074}$$

Multiply both the numerator and denominator by 10,000 to get an equivalent fraction involving whole numbers.

$$\frac{0.0351 \cdot 10{,}000}{0.074 \cdot 10{,}000} = \frac{351}{740}$$

$\frac{351}{740}$ is close to $\frac{1}{2}$, or 0.5. So, $0.0351 \div 0.074$ is about 0.5.

**Real-World Link**
Currency exchange rates change continually. Internet sites provide currency conversion tools to convert between any two units of currency.

## ✅ Develop & Understand: B

**In Exercises 5–7, estimate each quotient. Then use your calculator to find the quotient to the nearest thousandth.**

**5.** $25.27 \div 0.59$     **6.** $32.47 \div 81.5$     **7.** $0.4205 \div 0.07$

**8.** Find $10 \div 0.01$ without using a calculator.

**9.** Suppose one Japanese yen is worth 0.008715 U.S. dollar. You have $10 to exchange.

   **a.** Which of the following calculations will determine how many yen you will receive? Explain your answer.

       $0.008715 \cdot 10$      $0.008715 \div 10$      $10 \div 0.008715$

   **b.** How many yen will you receive for $10?

   **c.** Your answer to Part b should be fairly close to your answer for Exercise 8. Explain why.

   **d.** Caroline calculated that she could get about 22 yen in exchange for $20. Is Caroline's answer reasonable? Explain.

When you do not have a calculator to divide decimals, you can use the method shown in the next example to change a decimal-division expression into a division expression involving whole numbers.

## Example

Calculate $5.472 \div 1.44$.

Write the division expression as a fraction. Then multiply the numerator and denominator by 1,000 to get a fraction with a whole-number numerator and denominator.

$$\frac{5.472}{1.44} = \frac{5.472 \cdot 1,000}{1.44 \cdot 1,000} = \frac{5,472}{1,440}$$

Now just divide the whole numbers.

$$
\begin{array}{r}
3.8 \\
1440{\overline{\smash{\big)}\,5472\phantom{0}}} \\
\underline{4320\phantom{0}} \\
1152\,0 \\
\underline{1152\,0} \\
0
\end{array}
$$

So, $5.472 \div 1.44 = 3.8$.

## ✅ Develop & Understand: C

**Solve these exercises without using a calculator.**

For Exercises 10–12, find each quotient.

**10.** $10.5 \div 0.42$   **11.** $37.5 \div 1.25$   **12.** $0.00045 \div 0.06$

**13.** There are 2.54 centimeters in 1 inch.

  **a.** Estimate the number of inches in 25 cm.

  **b.** Use your answer to Part a to estimate the number of inches in 100 cm.

  **c.** Find the actual number of inches in 25 cm and in 100 cm.

  **d.** The average height of a 12-year-old girl in the United States is 153.5 cm. Convert this height to inches. Round to the nearest tenth of an inch.

## Math Link

Mass measures the material an object contains. Weight measures the force with which it is attracted toward the planet. For practical purposes, the distinction is not important.

### *Share & Summarize*

1. Describe a quick way to divide a number by 0.01. Use your knowledge of fraction division to explain why your method works.

2. Luke and Rosita are arguing about whose cat is heavier. Luke says his cat Tom is huge, almost 23 pounds. Rosita says her cat Spike is even heavier, close to 11 kilograms. There are about 2.2 pounds in 1 kilogram. Describe a calculation you could do to figure out whose cat is heavier.

# Investigation 5 Multiply or Divide

Your calculator multiplies and divides decimals accurately, as long as you do not make any mistakes when pressing the keys. However, it cannot tell you *whether* to multiply or divide. In this investigation, you will decide whether to multiply or divide in specific situations.

### *Think & Discuss*

For each question, choose the correct calculation. Explain your selection. Although you do not have to do the calculation, you may want to make an estimate to check that the answer is reasonable.

- A package of fifty recordable DVDs costs $15.95. Which calculation could you do to find the cost per DVD?

$$15.95 \cdot 50 \text{ or } 15.95 \div 50$$

- There are 2.54 centimeters in one inch. A sheet of paper is $8\frac{1}{2}$ inches wide. Which calculation could you do to find the paper's width in centimeters?

$$2.54 \cdot 8.5 \text{ or } 2.54 \div 8.5$$

- Marcus is building a model-railroad layout in HO scale, in which 0.138 inch in the model represents one foot in the real world. He wants to include a model of the Sears Tower in Chicago, which is about 1,450 feet tall. Which calculation could he do to find the height of the model tower in inches?

$$1,450 \cdot 0.138 \text{ or } 1,450 \div 0.138$$

In Exercises 1–4, you will need to think carefully about whether to multiply or divide to find each answer.

## ✓ Develop & Understand: A

1. On January 1, 1999, the *euro* was introduced as the common currency of 11 European nations. Currency markets opened on January 4. On that day, one euro was worth 1.1874 U.S. dollars.

   a. On January 4, 1999, what was the value of 32 euros in U.S. dollars? Explain how you decided whether to multiply or divide.

   b. On January 4, 1999, what was the value of $32.64 in euros? Explain how you decided whether to multiply or divide.

2. A kilometer is equal to approximately 0.62 mile.

   a. Molly ran a 42-km race. Find how far she ran in miles. Explain how you decided whether to multiply or divide.

   b. The speed limit on Duncan Road is 55 miles per hour. Convert this speed to kilometers per hour. Round to the nearest whole number. Explain how you decided whether to multiply or divide.

## ✓ Develop & Understand: B

3. Miguel's father's car can travel an average of 18.4 miles on one gallon of gasoline. Gas at the local station costs $3.53\frac{9}{10}$ per gallon.

   a. Miguel's father filled the car with 12.8 gallons of gas. How much did he pay?

   b. Miguel's brother took the car to the gas station and handed the cashier two $20 bills. How much gas could he buy with $40? Round your answer to the nearest hundredth of a gallon.

   c. How far could Miguel's brother drive on the amount of gas he bought? Round your answer to the nearest mile.

   d. Over spring break, Miguel's family drove the car on a 500-mile trip. About how much gas was used? Round your answer to the nearest gallon.

**Real-World Link**

The euro is used by countries in the European Union's Eurozone. Having a common currency makes buying and selling products among these nations easier.

4. Althea is building a model-railroad layout using the Z scale, the smallest scale for model railroads. In the Z scale, 0.055 inch on a model represents one foot in the real world.

  a. The caboose in Althea's model is one inch long. How long is the real caboose?

  b. Althea is making model people for her layout. She wants to make a model of her favorite basketball player, who is 6 feet 9 inches tall. If she could make her model exactly to scale, how tall would it be?

  c. Althea is using an architect's ruler, which measures to the nearest tenth of an inch. Using this ruler, it is impossible for her to build her basketball-player model to the exact scale height that you calculated in Part b. How tall do you think she should make the model?

### Share & Summarize

In the United States, people measure land area in acres. People in many other countries use hectares, the metric unit for land area. There is approximately 0.405 hectare in one acre.

1. To convert 3.5 acres to hectares, do you multiply or divide 3.5 by 0.405? Explain how you know.

2. To convert 3.5 hectares to acres, do you multiply or divide 3.5 by 0.405? Explain how you know.

**Practice & Apply**

1. Dae Ho made a spreadsheet showing products of decimals and whole numbers. The toner in his printer is running low. When he printed his spreadsheet, the decimal points were nearly invisible. Copy the spreadsheet. Insert decimal points in the first and third columns to make the products correct.

| Decimal | Whole Number | Product |
|---------|--------------|---------|
| 4 8 3 9 8 | 3 0 6 | 1 4 8 0 9 7 8 8 |
| 3 6 4 | 9 6 7 | 3 5 1 9 8 8 |
| 1 7 0 6 | 6 9 8 | 1 1 9 0 7 8 8 |
| 1 6 7 9 3 5 | 5 3 4 | 8 9 6 7 7 2 9 |
| 7 5 0 7 2 | 9 7 6 | 7 3 2 7 0 2 7 2 |
| 9 3 | 1 6 0 | 1 4 8 8 0 |

2. **Economics** Gloria, Wilton, and Alex have a band that plays for dances and parties. Gloria does most of the song writing, and Wilton acts as the manager. The band divides their earnings as shown below.

   - Gloria gets 0.5 times the band's profit.
   - Wilton gets 0.3 times the band's profit.
   - Alex gets 0.2 times the band's profit.

   a. This month, the band earned $210. How much money should each member receive?

   b. A few months later, the band members changed how they share their profit. Now, Gloria gets 0.42 times the profit, Wilton gets 0.3 times the profit, and Alex gets 0.28 times the profit. Alex said his share of $210 would now be $66. Explain why this estimate could not be correct. Calculate the correct amount.

   c. If the band makes $2,000 profit over the next several months, how much more will Wilton earn than Alex? Try finding the answer without calculating how much money Wilton earns.

**Measurement** Use the fact that 1 m = 0.001 km to convert each distance to kilometers without using a calculator.

3. 283 m

4. 314,159 m

5. 2,000,000 m

6. 1,776 m

7. 7 m

8. 0.12 m

**Write the decimals as fractions and then multiply. Give the product as a decimal.**

**9.** $0.17 \cdot 0.003$

**10.** $0.0005 \cdot 0.8$

**11.** $0.00012 \cdot 12.34$

**12.** $0.001 \cdot 0.2 \cdot 0.3$

**Find each product without using a calculator.**

**13.** $0.023 \cdot 17.51$

**14.** $0.15 \cdot 1.75$

**15.** $0.34 \cdot 0.0072$

**16.** $3.02 \cdot 100.25$

**17.** $0.079 \cdot 0.970$

**18.** $0.0354 \cdot 97.3$

**Calculate each product mentally.**

**19.** $0.0002 \cdot 2.5$

**20.** $7 \cdot 0.006$

**21.** $0.03 \cdot 0.05$

**22.** $0.4 \cdot 0.0105$

**23. Economics** The unit of currency in Guatemala is the *quetzal*. On October 31, 2007, one quetzal was worth 0.1300 U.S. dollar.

**a.** How much was a 100-quetzal note worth in U.S. dollars?

**b.** On this same day, a small rug in a Guatemalan market was priced at 52 quetzals. Convert this amount to U.S. dollars and cents.

**24. Economics** At Sakai's Sweet Shop, gummy worms cost $.28 per ounce and chocolate-covered raisins cost $.37 per ounce.

**a.** How much do six ounces of gummy worms cost?

**b.** Use the fact that 16 ounces is equal to one pound to calculate the cost of a pound of chocolate-covered raisins.

**c.** In the state where Sakai's is located, sales tax on candy is computed by multiplying the price by 0.1 and rounding up to the next penny. Without using a calculator, find the tax on six ounces of gummy worms. Also, find the tax on one pound of chocolate-covered raisins.

***Real-World Link***

The quetzal is a bird that is found in the rain forests of Central America. Today the male quetzal appears on Guatemalan currency and on the Guatemalan flag.

**25. Economics** When she eats at a restaurant, Viviana likes to leave a 15% tip. She multiplies the price of the meal by 0.15. Franklin usually leaves a 20% tip. He multiplies the price by 0.20. They both round up to the nearest 5¢.

a. How much tip would Viviana leave for a meal costing $24.85?

b. How much tip would Franklin leave for a meal costing $24.85?

c. The price $24.85 does not include tax. Viviana and Franklin live in a state where the meal tax is 6%, and fractions of a cent are rounded to the nearest penny. Figure out the tax on the $24.85 meal by multiplying by 0.06.

d. Calculate the tips Viviana and Franklin would leave if they tipped based on the cost of the meal plus tax.

**26. Measurement** In Investigation 3, you learned that you can multiply a four-year-old's height by 1.5 to predict his or her height at age 12. You can work backward to estimate what a 12-year-old's height might have been at age 4.

a. Nicky is 127.5 cm tall at age 12. Estimate her height at age 4 to the nearest centimeter.

b. Javon is 152.4 cm tall at age 12. Estimate his height at age 4 to the nearest centimeter.

***Real-World Link***

The HO scale is the most popular size for model railroads. Using this scale, models are $\frac{1}{87}$ the size of real trains.

**27. Measurement** To determine a length in inches on an HO-scale model of an object, divide the actual length in feet by 7.25. For example, a 15-foot-high building would have a height of about 2 inches in the model.

a. Without using a calculator, estimate how long a model train would be if the real train is 700 feet long. Explain how you made your estimate.

b. Calculate the length of the model train to the nearest quarter of an inch.

c. Estimate the length of the model train to the nearest foot. Is the length of the model shorter or longer than your estimate?

**28. Economics** Last summer, Rosita, Miguel, Luke, and Marcus shared a paper route. At the end of the summer, they divided their $482.50 profit according to how much each had worked.

• To get Rosita's share, the profit was divided by 2.5.

• To get Miguel's share, the profit was divided by 4.0.

• To get Luke's share, the profit was divided by 6.25.

• To get Marcus' share, the profit was divided by 10.

**a.** Without using a calculator, estimate how much each friend earned. Explain how you made your estimates.

**b.** Now calculate the exact amount each friend received.

**c.** Do their shares add to $482.50? If not, change one person's share so they *do* add to $482.50.

**Evaluate each expression without using a calculator.**

**29.** $0.1 \cdot 17 + 15 \cdot 0.001$

**30.** $8.82 \div 0.63 \div 0.7$

**31.** $2.75 - 0.05 \cdot 10$

**32. Economics** Sasha wants to buy a gallon of orange juice. He is considering two brands. Sunny Skies costs $.77 per quart, and Granger's Grove costs $2.99 per gallon. Sasha likes the taste of both brands. Which brand is a better deal? Explain.

**33. Economics** The unit of currency in Vanuatu is the *vatu*. In March 2003, there were 125.37 vatus to one U.S. dollar.

**a.** What is the value of 853.25 vatus in dollars?

**b.** What is the value of $853.25 in vatus? Round your answer to the nearest hundredth of a vatu.

**c.** In January 1988, there were 124.56 vatus to one U.S. dollar. Did the value of a vatu go up or down between January 1988 and March 2003? Justify your answer.

**34. Measurement** In this exercise, use the following facts.

1 quart = 0.947 liter     1 quart = 32 ounces

**a.** VineFresh grape juice is $1.25 a quart. Groovy Grape is $1.35 a liter. If you like both brands, which is the better buy? Explain.

**b.** How many quarts are in 0.5 liter? In 1 liter? In 1.5 liters? In 2.0 liters? Express your answers to the nearest hundredth.

**c.** How many liters are in 10 ounces? In 12 ounces? In 20 ounces? In 2 quarts? Express your answers to the nearest thousandth.

*Connect & Extend*   **35. Number Sense** Try to get as close to 262 as you can, without going over, by multiplying 210 by a number. You can use a calculator, but the only operation you may use is multiplication.

**a.** Should you multiply 210 by a number greater than 1 or less than 1?

**b.** Get as close as you can to 262 by multiplying 210 by a number with only one decimal place. By what number did you multiply 210? What is the product?

**c.** Get as close as you can to 262 by multiplying 210 by a number with two decimal places. By what number did you multiply 210? What is the product?

**d.** Now get as close as you can using a number with three decimal places. Give the number by which you multiplied and the product.

**e.** Try multiplying 210 by numbers with up to nine decimal places to get as close to 262 as possible. Give each number by which you multiplied and the product. Describe any strategies you develop for choosing numbers to try.

**36. Measurement** In this exercise, you will figure out how much water it would take to fill a room in your home.

**a.** Choose a room in your home that has a rectangular floor. Find the length, width, and height of the room to the nearest foot. If you do not have a yardstick or tape measure, just estimate. Multiply the three measurements to find the volume, or number of cubic feet, in the room.

**b.** A cubic foot holds 7.48 gallons of water. How many gallons will it take to fill the room?

**c.** Suppose 748 gallons of water cost $164. How much would it cost to fill the room? Explain how you found your answer.

**Real-World Link**

The common light bulb is an incandescent bulb, thin glass filled with a mixture of nitrogen and argon gas and a tungsten wire filament. Fluorescent and halogen bulbs are usually shaped like tubes or spotlights.

37. **Physical Science** A light bulb's wattage indicates how much energy the bulb uses in one hour. For example, a 75-watt bulb uses 75 watt-hours of energy per hour. Electric companies charge by the kilowatt-hour.

 a. One watt-hour equals 0.001 kilowatt-hour. How many kilowatt-hours does a 75-watt bulb use in an hour? In 24 hours?

 b. Suppose your electric company charges $0.21 per kilowatt-hour. How much would it cost to leave a 75-watt bulb on for 24 hours?

 c. Figure out how much it would cost to leave on all of the light bulbs in your home for 24 hours. You will need to count the bulbs and note the wattage of each bulb. Do not look directly at a light bulb. If it is not possible to find the wattage of some bulbs, assume each is a 75-watt bulb. Count only incandescent bulbs, not fluorescent or halogen bulbs.

38. **Economics** The imaginary country of Glock uses *utils* for its currency. One util has the same value as five U.S. dollars. In other words, there are $5 per util.

 a. What is the value of $1 in utils? That is, how many utils are there per dollar? Express your answer as both a fraction and a decimal.

 b. In Part a, you used the number of dollars per util to find the number of utils per dollar. What mathematical operation did you use to find your answer?

 c. Use the same process to find the number of yen per dollar if one yen is worth $0.008715. Round your answer to the nearest whole yen.

 d. If you are given the value of one unit of a foreign currency in dollars, describe a rule you could use to find the value of $1 in that foreign currency.

 e. If a unit of foreign currency is worth more than a dollar, what can you say about how much $1 is worth in that currency?

 f. If a unit of foreign currency is worth less than a dollar, what can you say about how much $1 is worth in that currency?

**39. Architecture** Miguel built a scale model of the Great Pyramid of Giza as part of his history project. He used a scale factor of 0.009 for his model. This means he multiplied each length on the actual pyramid by 0.009 to find the length for his model.

Some of the measurements of Miguel's model are below. Find the measurements of the actual pyramid. Round your answers to the nearest meter.

**a.** Height of pyramid: 1.23 m

**b.** Length of each side of pyramid's base: 2.07 m

**c.** Height of king's chamber: 0.05 m

**d.** Length of king's chamber: 0.04 m

**e.** Width of king's chamber: 0.09 m

**40.** Imagine you are playing a game involving multiplying and dividing decimals. The goal is to score as close to 100 as possible without going over. Each player starts with ten points. The following are directions for what to do on each turn.

- Draw two cards with decimals on them.

- Using estimation, choose to multiply or divide your current score by one of the decimals.

- Once you have made your decision, compute your new score by doing the calculation and rounding to the nearest whole number. If the result is over 100, you lose.

- You may decide to stop at the end of any turn. If you do, your opponent gets one more turn to try to score closer to 100 than your score.

**a.** On one turn, you start with a score of 50 and draw 0.2 and 1.75. Tell what your new score will be if you do each of the following.

**i.** divide by 0.2      **ii.** multiply by 0.2

**iii.** divide by 1.75      **iv.** multiply by 1.75

**b.** Suppose your score is 88, and you draw 1.3 and 0.6. Use estimation to figure out your best move. Then calculate your new score.

**c.** Jahmal and Hannah are playing against each other. On his last turn, Jahmal's score was 57. He drew 0.8 and 1.8. On her last turn, Hannah's score was 89 and she drew 0.7 and 1.2. If each player made the best move, who has the greater score now? Explain.

***Real-World Link***

The Great Pyramid of Giza was built around 2560 B.C. The pyramid was constructed from about two million blocks of stone, each weighing more than 2 tons.

**41. In Your Own Words** Describe a realistic situation in which you would have to multiply two decimal numbers. Then describe another situation in which you would have to divide by a decimal.

**Mixed Review**

**Geometry** Identify each geometric figure.

**42.** a quadrilateral with four equal sides

**43.** a five-sided polygon

**44.** a polygon whose interior angles add to 180°

**45.** two rays with the same endpoint

**46.** a six-sided polygon with six equal sides

**47.** a segment from the center of a circle to a point on the circle

**48. Geometry** Choose which of these terms describe each polygon, quadrilateral, pentagon, hexagon, concave, symmetric, regular. List all that apply.

a.    b.    c.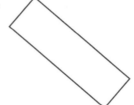

**Find each sum or difference.**

**49.** $\dfrac{7}{10} - \dfrac{3}{15}$

**50.** $3\dfrac{4}{7} + 2\dfrac{18}{21}$

**51.** $\dfrac{7}{8} - \dfrac{5}{12}$

**52.** $\dfrac{37}{13} - 2\dfrac{1}{2}$

**53.** $\dfrac{3}{8} + 10\dfrac{4}{5}$

**54.** $\dfrac{1}{15} - \dfrac{1}{40}$

# LESSON 4.4

# What Is Typical?

To help people understand a set of data, it is useful to give them an idea of what is *typical*, or average, about the data. In this lesson, you will work with data sets involving fractions and decimals, and you will learn three ways to describe the typical value in a data set. You will also learn about a type of graph that is useful for showing the distribution of values in a set of data.

## Think & Discuss

Have each student in your class estimate how many minutes he or she spent doing homework yesterday. Your teacher should record the data on the board. How would you describe to someone who is not in your class what is typical about your class' data?

## Investigation 1 — Mode and Median

### Vocabulary

line plot

median

mode

range

The Jump Shot shoe store sells basketball shoes to college players. The tables show the brand and size of each pair the store sold one Saturday.

| Brand | Size |
| --- | --- |
| Swish | 13 |
| Dunkers | 14.5 |
| Hang Time | 8.5 |
| Swish | 13 |
| Hang Time | 13 |
| Airborne! | 15 |
| Swish | 12.5 |
| Swish | 8 |
| Dunkers | 13 |
| Big J | 13.5 |
| Hang Time | 14 |
| Big J | 14.5 |
| Dunkers | 14 |

| Brand | Size |
| --- | --- |
| Swish | 14 |
| Hang Time | 10 |
| Airborne! | 11 |
| Dunkers | 12 |
| Airborne! | 14 |
| Big J | 12.5 |
| Swish | 14 |
| Swish | 14.5 |
| Hang Time | 10.5 |
| Swish | 10.5 |
| Swish | 14 |
| Hang Time | 13.5 |

In the following exercises, you will learn ways to summarize the shoe store's data.

## ✅ *Develop & Understand: A*

1. You can create a line plot to show the sizes of the shoes sold on Saturday. A **line plot** is a number line with X's indicating the number of times each data value occurs.

   a. To make the line plot, copy the number line below. Mark an X above a shoe size each time it appears in the data set. For example, 14.5 appears three times, so put three X's above 14.5.

   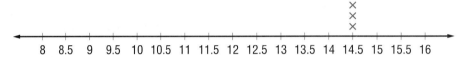

   b. Describe the shape of your line plot. Tell what the shape indicates about the distribution of shoe sizes.

2. When you describe a data set, it is helpful to give the *minimum,* or least, and *maximum,* or greatest, values.

   a. Give the minimum and maximum shoe sizes in the data set.

   b. How can you find the minimum and maximum values by looking at a line plot?

3. The **range** of a data set is the difference between the minimum and maximum values. Give the range of the shoe-size data.

4. The **mode** of a data set is the value that occurs most often. A data set may have no mode or several modes.

   a. Give the mode of the shoe-size data.

   b. How can you find the mode of a data set by looking at a line plot?

5. The **median** is the middle value when all the values in a data set are ordered from least to greatest.

   a. List the shoe-size data in order from least to greatest and then find the median size.

   b. How can you find the median of a data set by looking at a line plot?

6. Suppose the store discovered five Saturday sales that were not recorded. Add 14.5, 14.5, 15, 14.5, and 16 to your line plot.

   a. What is the range of the data now?

   b. What is the mode now?

   c. When a data set has an even number of values, there is no single middle value. In such cases, the median is the number halfway between the two middle values. Find the median of the new data set.

In Exercises 1–6, you looked at ways to summarize *numerical data,* that is, data that are numbers. You will now look at the brand-name data, which are not numbers. Non-numerical data are sometimes called *categorical data* because they can be thought of as names of *categories,* or groups.

## ✅ Develop & Understand: B

7. Is it possible to make a line plot to show the distribution of the brand-name data? Explain.

8. Can you find the range of the brand-name data? If so, find it. If not, explain why it is not possible to find the range.

9. Do the brand-name data have a mode? If so, find it. If not, explain why it is not possible to find the mode.

10. Do the brand-name data have a median? If so, find it. If not, explain why it is not possible to find the median.

11. What are some other ways that you might summarize the brand-name data?

The mode and the median are two measures of the typical, or average, value of a data set. In some cases, one of these measures describes the data better than the other.

## ✅ Develop & Understand: C

12. Ms. Washington gave her class a ten-point quiz. Her students' scores follow.

    7  9  10  5  5  8  6  10  6  7  10  2
    7  5  8  8  4  9  10  4  10  7  6

    a. Find the range, mode, and median of the quiz scores.

    b. Do you think the mode or the median is a better measure of what is typical in this data set? Explain.

**13.** Hannah asked nine of her classmates how many pets they have. The results follow.

$$0 \quad 4 \quad 1 \quad 0 \quad 0 \quad 4 \quad 4 \quad 0 \quad 4$$

**a.** Find the range, mode, and median of these data.

**b.** Do you think the mode or the median is a better measure of what is typical in this data set? Explain.

**14.** During one afternoon practice, an athlete threw a javelin 13 times. The distances are listed here.

257.3   210.5   210   255.2   210   220.8
275.7   253   210   253.6   250.1   252.4   200

**a.** Find the range, mode, and median of the data.

**b.** If you had to use only one type of average, the mode or the median, to summarize this athlete's performance, which would you choose? Give reasons for your choice.

**15.** During one week in January, the following precipitation amounts were reported. Data were reported in inches.

| M | T | W | Th | F | S | S |
|---|---|---|---|---|---|---|
| 0 | $\frac{1}{10}$ | $7\frac{1}{2}$ | 3 | $6\frac{1}{4}$ | $\frac{1}{10}$ | 5 |

**a.** Find the range, mode, and median of the data.

**b.** Which would you choose to represent the week's precipitation, the mode or the median?

**16.** When you are given summary information about a set of data, you can sometimes get an overall picture of how the values are distributed. Suppose you know the following facts about a data set.
- It has 15 values.
- The minimum value is 50.5, and the maximum value is 99.5.
- The mode is 55.
- The median is 57.

**a.** What do you know about how the data values are distributed?

**b.** Create a data set that fits this description.

***Real-World Link***

The sport of javelin throwing evolved from ancient spear-throwing contests introduced as part of the pentathlon in the Olympics in 708 B.C.

## Share & Summarize

**1.** Describe what the range, mode, and median tell you about a set of numerical data.

**2.** Which measure, the range, the mode, or the median, can be used to describe a set of categorical data? Explain.

# Investigation ② The Meaning of *Mean*

## Vocabulary

mean

**Real-World Link**

Strawberries were originally known as "strewberries" because they appeared to be strewn among the leaves of the strawberry plant.

The median and the mode are two ways to describe what is typical, or average, about a set of data. These values are sometimes referred to as *measures of central tendency,* or simply *measures of center,* because they give an idea of where the data values are centered. In this investigation, you will explore a third measure of center, the *mean.*

## ✓ Develop & Understand: A

1. Althea's scouting troop went strawberry picking. They decided to divide the strawberries they picked equally, so each girl would take home the same amount. The illustration shows how many quarts each girl picked.

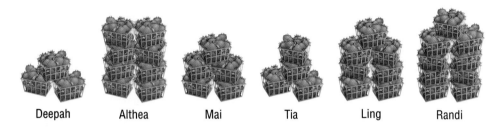

Deepah    Althea    Mai    Tia    Ling    Randi

How many quarts did each girl take home? How did you find your answer?

2. Another group of friends went strawberry picking. They divided their berries equally. How many quarts did each friend receive?

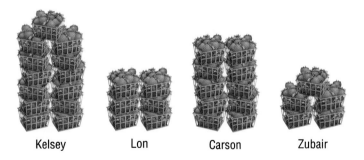

Kelsey    Lon    Carson    Zubair

3. A group of ten friends picked the following numbers of quarts of strawberries.

$$5 \quad 10 \quad 4 \quad 5 \quad 7 \quad 9 \quad 9 \quad 6 \quad 8 \quad 7$$

Suppose the friends divided the strawberries equally. How many quarts did each friend get?

In each situation in Exercises 1–3, you redistributed the quarts to give each person the same number. The result was the *mean* of the number of quarts picked. The **mean** of a set of values is the number you get by evenly distributing the total among the members of the data set. You can compute the mean by adding the values and dividing the total by the number of values.

The mean is another measure of the typical, or average, value of a data set. In everyday language, the word *average* is often used for *mean*. However, it is important to remember that the mean, median, and mode are *all* types of averages.

## ✓ Develop & Understand: B

4. The astronomy club is selling calendars to raise money to purchase a telescope. The ten club members sold the following numbers of calendars.

    3   5   7   10   5   3   4   6   9   8

   a. Find the mean, median, and mode of the numbers of calendars the club members sold.

   b. Suppose two very motivated students join the club. One sells 20 calendars and the other sells 22. Find the new mean, median, and mode.

   c. In Part b, suppose that instead of 22 calendars, the 12th club member had sold 100 calendars. What would be the new mean, median, and mode?

   d. How does the median in Part c compare to the median in Part b? How do the two means compare? Explain why your answers make sense.

5. Luke asked 12 students in his class how many books they had read, other than school books, in the past six months. The following responses were given by 11 of the students.

    3   5   7   10   5   3   4   6   9   8   20

   a. Suppose you know that the mean number of books read by the 12 students is 10. Is it possible to find the number of books the 12th student read? If so, explain how. If not, explain why not.

   b. Suppose you know that the median number of books read by the 12 students is 5.5. Is it possible to find the number of books the 12th student read? If so, explain how. If not, explain why not.

## Develop & Understand: C

6. Ines calls her grandmother each week. The lengths, in minutes, of the last nine phone calls are listed below.

22.7   23.9   25.5   28   32.1   32.1   35   37.4   37.8

a. Find the mean and median of the data set.

b. Add two values to the data set so the median remains the same but the mean decreases. Give the new mean.

c. Start with the original data set. Add two values to the set so the median remains the same but the mean increases. Give the new mean.

d. Start with the original data set. Add two values to the set so the mean remains the same but the median changes. Give the new median.

**For each description below, create a data set with ten values.**

7. The minimum is 45, the maximum is 55, and the median and mean are both 50.

8. The minimum is 10, the maximum is 90, and the median and mean are both 50.

9. The range is 85, the mean is 50, and the median is 40.

10. The range is 55, the mean is 40, and the median is 50.

## Share & Summarize

1. Jing said that the students in her class have an average of three pets each.

   a. If Jing is referring to the mode, explain what her statement means.

   b. If Jing is referring to the median, explain what her statement means.

   c. If Jing is referring to the mean, explain what her statement means.

2. Suppose you have a data set for which the mean and median are the same. If you add a value to the set that is much greater than the other values in the set, would you expect the median or the mean to change more? Explain.

# Investigation 3 Mean or Median?

**Vocabulary**

outlier

This investigation will help you better understand what the mean and median reveal about a set of data.

## ✓ Develop & Understand: A

1. Lee and Arturo collected the heights of the students in their math classes. They found that the median height of students in Arturo's class is greater than the median height of students in Lee's class.

   Tell whether each statement below is *definitely true,* is *definitely false,* or *could be true or false* depending on the data. In each case, explain why your answer is correct. (Hint: It may help to create data sets for two small classes with three or four students each.)

   a. The tallest person is in Arturo's class.

   b. Lee's class must have the shortest person.

   c. If you line up the students in each class from shortest to tallest, each person in Arturo's class will be taller than the corresponding person in Lee's class. Assume the classes have the same number of students.

   d. If you line up the students in each class from shortest to tallest, the middle person in Arturo's class would be taller than the middle person in Lee's class. Assume the classes have the same odd number of students.

2. Marta and Grace collected height data for their math classes. They found that Marta's class has a greater mean height than Grace's class.

   Tell whether each statement below is definitely true, is definitely false, or could be true or false depending on the data. In each case, explain how you know your answer is correct.

   a. The tallest person is in Marta's class.

   b. Grace's class must have the shortest person.

   c. If you line up the students in each class from shortest to tallest, each person in Marta's class will be taller than the corresponding person in Grace's class. Assume the classes have the same number of students.

   d. If you line up the students in each class from shortest to tallest, the middle person in Marta's class would be taller than the middle person in Grace's class. Assume the classes have the same odd number of students.

Books, news reports, and advertisements often mention average values.

You have learned about three types of average. They are the mode, the median, and the mean. The average reported in a particular situation depends on many factors. Sometimes, one measure is "more typical" than the others. Other times, a measure is selected to give a particular impression or to support a particular opinion.

## ✅ Develop & Understand: B

Career Connections is a small company that helps college graduates find jobs. The company is creating a brochure to attract new clients and would like to include the average starting salary of recent clients. Career Connections has asked Data, Inc. to help it determine which type of average to use.

Listed below are the starting salaries of the clients that Career Connections has helped in the past three months.

| | | | | |
|---|---|---|---|---|
| $30,000 | $25,000 | $60,000 | $40,000 | $25,000 |
| $50,000 | $70,000 | $50,000 | $25,000 | $60,000 |
| $25,000 | $1,000,000 | $60,000 | $25,000 | $40,000 |
| $50,000 | $25,000 | $50,000 | $25,000 | $25,000 |

3. You know how to compute three types of averages, which are the mode, the median, and the mean.

   a. Find the mode, median, and mean for these data.

   b. Which average do you think best describes a typical value in this data set? Explain.

**4.** One of the salaries, $1,000,000, is much greater than the rest. A value that is much greater than or much less than most of the other values in a data set is called an **outlier**. A Data, Inc. analyst suggested that this outlier should not be included when determining the average salary.

   **a.** Remove $1,000,000 from the data set. Recompute the mode, median, and mean.

   **b.** How does removing the outlier affect the three measures of center?

**5.** Write a brief letter to Career Connections telling the company what value you recommend it reports as the average starting salary of its clients. Give reasons for your choice. Consider all the averages you have computed for the salary data, both including and not including the $1,000,000 salary.

In the following exercises, you will use a single set of data to support two very different points of view.

## ✅ Develop & Understand: C

The Hillsdale School District will hold its annual girls' basketball banquet. The head of athletics will present an award to the best scorer in the two high schools. The points per game scored by each school's best offensive player are shown below.

Points per game for Westside Wolves' best offensive player:

$$30, 61, 10, 0, 28, 48, 55, 12, 23, 55, 6, 25,$$
$$39, 18, 55, 31, 30$$

Points per game for Eastside Eagles' best offensive player:

$$22, 35, 12, 37, 19, 36, 39, 13, 13, 36, 11,$$
$$37, 13, 38, 21, 37, 35$$

Each coach wants to be able to argue that her player deserves the award. Both coaches have come to Data, Inc. for help.

**6.** Use your knowledge of statistics to argue that the Westside Wolves' player deserves the award.

**7.** Use your knowledge of statistics to argue that the Eastside Eagles' player deserves the award.

***Real-World Link***

The first women's intercollegiate basketball game was played in 1896 in San Francisco. The game pitted Stanford University against the University of California at Berkeley. Stanford won the game by a score of 2 to 1. Male spectators were not allowed at the game.

The next exercise set will help you better understand what the mean and median tell you about the distribution of a data set.

## ✓ Develop & Understand: D

8. Create a data set with eight values from 1 to 20 that fits each description. Give the median and mean of each data set you create.

   a. The median is greater than the mean.

   b. The mean is greater than the median.

   c. The mean and median are equal.

   d. The mean is 3 more than the median.

   e. The median is 3 more than the mean.

9. The data set 1, 2, 3, 4, 5 has a mean of 3. Change two values so the new data set has a mean of 4.

### Share & Summarize

1. Reports of the typical income of a city, state, or country often use the median rather than the mean. Why do you think this is so?

2. What might cause the mean of a data set to be much greater than the median?

3. What might cause the median of a data set to be much greater than the mean?

4. What might cause the median of a data set to be equal to the mean?

**Practice & Apply**

1. The table shows the style and size of all the hats sold at the Put a Lid on It! hat shop last Thursday.

| Style | Size | Style | Size | Style | Size |
|---|---|---|---|---|---|
| Cap | $6\frac{5}{8}$ | Cap | $7\frac{1}{4}$ | Fedora | $7\frac{1}{4}$ |
| Beret | $7\frac{3}{8}$ | Beret | $6\frac{7}{8}$ | Chef's hat | $7\frac{1}{8}$ |
| Fedora | $7\frac{1}{4}$ | Panama hat | $7\frac{5}{8}$ | Beret | 7 |
| Sombrero | 7 | Fedora | $6\frac{7}{8}$ | Derby | $7\frac{1}{8}$ |
| Cap | $7\frac{1}{4}$ | Cap | $7\frac{1}{4}$ | Beret | $7\frac{1}{8}$ |
| Cap | $7\frac{3}{8}$ | Sombrero | $7\frac{1}{2}$ | Top hat | $7\frac{3}{4}$ |
| Fedora | $7\frac{3}{8}$ | Fedora | $7\frac{1}{2}$ | Panama hat | $7\frac{3}{8}$ |

   a. Make a line plot of the hat-size data.

   b. Describe the shape of your line plot. Tell what the shape indicates about the distribution of the hat sizes sold.

   c. Find the range, mode, and median of the hat-size data.

   d. Find the mode of the hat-style data.

   e. Is it possible to find the median and range of the hat-style data? If so, find them. If not, explain why it is not possible.

2. This list shows the number of hits Jing got in each softball game this season.

   0 3 2 7 4 2 3 0 4 0 6 5 5 2 4 0

   a. Find the mode and median of these data.

   b. Do you think the mode or the median is a better measure of what is typical in this data set? Explain.

3. Create a data set with 13 values that has a minimum value of 3, a maximum value of 13, a mode of 4, and a median of 8.

4. A scientist in a science fiction story finds the mass of alien creatures from four areas of the planet Xenon.

| Area of Xenon | Mass (kilograms) |
|---|---|
| Alpha | 6, 21, 12, 36, 15, 12, 27, 12 |
| Beta | 18, 36, 36, 27, 21, 48, 36, 33, 21 |
| Gamma | 12, 18, 12, 21, 18, 12, 21, 12 |
| Delta | 30, 36, 30, 39, 36, 39, 36 |

a. Find the range, mean, median, and mode of the masses for each area of Xenon. Round to the nearest tenth.

b. The scientist realizes he made a mistake. One of the 39 kilogram creatures in the Delta area actually has a mass of 93 kilograms. Compute the new mean and median for the Delta area.

c. Compare the original Delta mean and median to the mean and median you computed in Part b. Which average changed more? Explain why this makes sense.

d. The scientist realized that one value for the Alpha area is missing from the table. He does not remember what the value is, but he remembers that the complete data set has a mean of 19. What value is missing?

5. Listed below are the number of unusual birds spotted by each member of a bird-watching club on a weekend excursion.

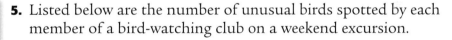

4   4   6   10   11   11   11   14   19

a. Find the mean and median of the data.

b. Add two values to the data set so the median remains the same but the mean increases. Give the new mean.

c. Add two values to the original data set so the median decreases but the mean remains the same. Give the new median.

d. Add two values to the original data set so both the mean and median stay the same.

**6.** Lonnie tutors students in algebra. The following are the scores his students received on their most recent algebra tests.

0   60   78   79   90   95   95

**a.** Lonnie claims that the students he tutored received an average score of 95 on their tests. Which measure of center is he using? Do you think 95 is a good measure of what is typical about these tests scores? Explain.

**b.** Find the mean and median of the test scores.

**c.** Lonnie said the score of 0 should not be counted when finding the average because the student did not show up for the test. Delete the 0, and find the new mean and median.

**d.** Which of the averages computed in Parts b and c do you think best represents the typical test scores for the students Lonnie tutored?

**7.** Elsa received the scores of 81, 79, 90, and 70 on her first four math tests this semester. There is one more test left. Elsa's teacher has told her she may choose to use her mean or her median test score as her final grade, but she must decide *before* she takes the final test.

**a.** Calculate Elsa's current mean and median test scores.

**b.** If Elsa is not confident she will do well on the final test, should she choose the mean or the median? Explain.

**c.** If Elsa is confident she will do well on the final test, should she choose the mean or the median? Explain.

*Real-World Link*

The first public swimming pool in the United States was built in Brookline, Massachusetts, in 1887. There are now more than 200,000 public swimming pools in the United States.

8. Alano and Kate are swimming instructors at the local recreation center. One day, both instructors asked their students to swim as many laps as they could. The results are shown in the table.

   a. Find the range, mean, median, and mode for all 12 swimmers.

| Student | Instructor | Laps Swum |
|---------|-----------|-----------|
| Lucinda | Kate | 7 |
| Jay | Alano | 15 |
| Guto | Kate | 9 |
| Carson | Alano | 11 |
| Ebony | Kate | 6 |
| Darius | Kate | 7 |
| Curtis | Alano | 9 |
| Carmen | Alano | 4.5 |
| Avi | Kate | 8 |
| Theo | Kate | 7 |
| Gil | Alano | 4 |
| Louis | Alano | 4 |

   b. Find the range, mean, median, and mode for each instructor's students.

   c. Alano said his students were stronger swimmers. Kate argued that her students were stronger. Use your knowledge of statistics to write two arguments, one to support Alano's position and one to support Kate's position.

**Connect & Extend**

9. **Data Analysis** Zeke's class made a line plot showing the number of people in each student's family.

   a. What is the total number of people in all of the students' families?

   b. Zeke said, "The plot can't be right! My family has 8 people. If I have the largest family, why is the stack of X's over the 8 the shortest one on the graph?" Answer Zeke's question.

**Real-World Link**

*Double Dutch* jump roping involves skipping two ropes as they are swung in opposite directions. Since 1973, the American Double Dutch League has held an annual rope-jumping competition.

10. **Sports** The students in Consuela's gym class recorded how many times they could jump rope without missing. The results are shown in the table.

| Group 1 | | | Group 2 | | |
|---|---|---|---|---|---|
| **Name** | **Gender** | **Jumps** | **Name** | **Gender** | **Jumps** |
| Jorge | male | 1 | Lucas | male | 4 |
| Felise | female | 1 | Colin | male | 4 |
| Lana | female | 5 | Olivia | female | 4 |
| Sean | male | 7 | Trent | male | 23 |
| Matt | male | 8 | Lawana | female | 35 |
| David | male | 11 | Tyrone | male | 48 |
| Aaron | male | 16 | Francisca | female | 68 |
| Kara | female | 26 | Enola | female | 83 |
| Brandon | male | 26 | Shari | female | 89 |
| Enrique | male | 26 | Kiran | male | 96 |
| Emma | female | 40 | Meela | female | 110 |
| Nicholas | male | 50 | Consuela | female | 138 |
| Shondra | female | 95 | Tariq | male | 151 |
| Selena | female | 300 | | | |

a. Group 1 claims it did better. Use what you have learned about statistics, along with any other information you think is useful, to write an argument Group 1 could use to support its claim.

b. Group 2 says it did better. Use what you have learned about statistics, along with any other information you think is useful, to write an argument Group 2 could use to support its claim.

11. If you have a data set that includes only whole numbers, which measures of center, mode, median, or mean, will *definitely be* whole numbers? Which measures of center *may* or *may not be* whole numbers? Explain your answers.

12. Suppose a data set has a mean and a median that are equal.

a. What must be true about the distribution of the data values?

b. Suppose one value is added to the set, and the new mean is much greater than the median. What must be true about the new value? Explain.

c. Now suppose you start with a new data set in which the mean and median are equal. You add one value to the set, and the new median is much greater than the mean. What must be true about the new value?

**13.** Emelia asked her friends to rate three movies on a scale from 1 to 5, with 5 being terrific and 1 being terrible.

**Movie Ratings**

| Friend | Star Wars | The Sound of Music | The Wizard of Oz |
|--------|-----------|--------------------|------------------|
| Adam | 2 | 1 | 4 |
| Ashley | 5 | 1 | 5 |
| Corey | 2 | 4 | 3 |
| Emelia | 4 | 4 | 2 |
| Eric | 4 | 4 | 4 |
| Hector | 3 | 5 | 5 |
| Ilene | 2 | 5 | 3 |
| Jay | 3 | 2 | 1 |
| Jose | 5 | 1 | 5 |
| Kareem | 3 | 3 | 2 |
| Kara | 2 | 5 | 3 |
| Lawana | 4 | 5 | 4 |
| Letonya | 3 | 5 | 4 |
| Lynn | 1 | 2 | 2 |
| Mark | 4 | 3 | 3 |
| Maria | 3 | 2 | 1 |
| Marcos | 3 | 1 | 1 |
| Peter | 1 | 1 | 2 |

**Real-World Link**

It is difficult to think of *The Wizard of Oz* without picturing Dorothy's ruby slippers. However, in the book on which the movie was based, Dorothy wore silver shoes, not ruby slippers.

**a.** For each movie, make a line plot of the friends' ratings.

**b.** Compute the mean and median rating for each movie.

**c.** Do you think reporting the means and medians is a good way to summarize the ratings for the three movies? Explain.

**d.** How would you summarize these data if you wanted to emphasize the differences in the ratings among the three movies? Explain why you would summarize the data this way.

**14. In Your Own Words** How are mean, median, and mode alike? How are they different?.

**Mixed Review**

**Simplify each expression.**

**15.** $\dfrac{1}{3} \div \dfrac{2}{3}$

**16.** $\dfrac{7}{5} \div \dfrac{5}{7}$

**17.** $\dfrac{7}{5} \div \dfrac{7}{5}$

**18.** $\dfrac{1}{8} \div \dfrac{1}{4}$

**Find three fractions equivalent to each given fraction.**

**19.** $\dfrac{7}{9}$

**20.** $\dfrac{12}{54}$

**21.** $\dfrac{6}{13}$

**22.** $\dfrac{14}{5}$

**23.** Order these fractions from least to greatest.

$$\dfrac{1}{3} \qquad \dfrac{12}{30} \qquad \dfrac{9}{28} \qquad \dfrac{11}{30} \qquad \dfrac{12}{29}$$

**24.** Hannah and Rosita each wrote rules for the number of toothpicks in each stage of this sequence.

Stage 1     Stage 2     Stage 3     Stage 4

Both girls used $t$ to represent the number of toothpicks and $n$ to represent the stage number. Use words or diagrams to explain why each rule is correct.

**a.** Hannah's rule: $t = 2 \cdot n + 2 \cdot n$

**b.** Rosita's rule: $t = 4 + 4 \cdot (n - 1)$

# Review & Self-Assessment

## Vocabulary

line plot

mean

median

mode

outlier

range

reciprocal

## Chapter Summary

In this chapter, you learned how to do calculations with fractions and decimals. You used fraction pieces to add and subtract fractions with different denominators. You found that by rewriting the fractions with a common denominator, you could add and subtract without fraction pieces.

You then used what you know about multiplying whole numbers to figure out how to multiply a whole number by a fraction. You used rectangle diagrams to discover a method for multiplying two fractions.

You learned how to divide a whole number by a fraction by using a model and by writing a related multiplication problem. Then you learned two methods for dividing two fractions.

You also turned your attention to operations with decimals. You learned that you could use estimation, along with what you already know about multiplying and dividing whole numbers, to multiply and divide decimals.

Finally, you were also introduced to some statistics used to summarize a data set. The *range* of a data set is the difference between the minimum and maximum values. The *mode* is the value that occurs most often. The *median* is the middle value. The *mean* is the value found by dividing the sum of the data values equally among the data items. The mean, median, and mode are all measures of what is typical, or average, about a set of data.

## Strategies and Applications

The questions in this section will help you review and apply the important ideas and strategies developed in this chapter.

### Adding and subtracting fractions and mixed numbers

1. Explain the steps you would follow to add two fractions with different denominators. Give an example to illustrate your steps.

2. Describe two methods for subtracting one mixed number from another. Use one of the methods to find $7\frac{1}{3} - 4\frac{5}{6}$. Show your work.

### Multiplying fractions and mixed numbers

3. Find $\frac{2}{3} \cdot \frac{4}{5}$ by making a rectangle diagram. Show how this method of finding the product is related to finding the product of the numerators over the product of the denominators.

4. Describe how you would multiply two mixed numbers. Give an example to illustrate your method.

### Dividing fractions and mixed numbers

**5.** Describe two ways to divide a whole number by a fraction. Illustrate both methods by finding $4 \div \frac{2}{3}$.

**6.** Use $\frac{5}{6} \div \frac{4}{9}$ to illustrate a method for dividing fractions.

**7.** Without dividing, how can you tell whether a quotient will be greater than 1? Less than 1?

### Multiplying and dividing decimals

**8.** Describe how to multiply $9.475 \cdot 0.0012$ without using a calculator. Be sure to explain how to decide where to put the decimal point.

**9.** Describe how you can use what you know about dividing whole numbers to divide two decimals without using a calculator. Illustrate your method by finding $15.665 \div 0.65$.

**10.** Suppose one U.S. dollar is equivalent to 10.3678 Moroccan dirham.

    **a.** To convert $100 to dirham, what calculation would you do? Explain how you know this calculation is correct.

    **b.** To convert 100 dirham to dollars, what calculation would you do? Explain how you know this calculation is correct.

    **c.** Caroline said that $750 is equal to about 7.5 dirham. Is Caroline's estimate reasonable? Explain.

***Real-World Link***

Morocco is located on the westernmost tip of north Africa. Although the climate in most of the country is quite warm, parts of the mountains enjoy snow for most of the year.

## Demonstrating Skills

**Find each sum or difference. Give your answers in lowest terms. If an answer is greater than 1, write it as a mixed number.**

**11.** $\frac{4}{5} + \frac{1}{10}$                 **12.** $\frac{3}{7} + \frac{5}{9}$

**13.** $\frac{13}{15} + \frac{1}{3}$             **14.** $\frac{5}{8} - \frac{1}{12}$

**15.** $\frac{11}{24} - \frac{3}{9}$             **16.** $\frac{4}{5} - \frac{9}{25}$

**17.** $3\frac{1}{3} + 2\frac{3}{5}$             **18.** $1\frac{3}{8} - \frac{3}{4}$

**19.** $5\frac{1}{7} - \frac{11}{3}$             **20.** $6\frac{4}{9} - 2\frac{5}{6}$

**Find each product. Give your answers in lowest terms. If an answer is greater than 1, write it as a mixed number.**

**21.** $5 \cdot \dfrac{3}{10}$

**22.** $3\dfrac{1}{2} \cdot 6$

**23.** $\dfrac{2}{3} \cdot 12$

**24.** $\dfrac{5}{8} \cdot 14$

**25.** $1\dfrac{2}{3} \cdot \dfrac{7}{8}$

**26.** $\dfrac{6}{11} \cdot \dfrac{5}{3}$

**27.** $\dfrac{123}{12} \cdot \dfrac{12}{123}$

**28.** $\dfrac{1}{3} \cdot \dfrac{75}{100}$

**29.** $\dfrac{9}{14} \cdot \dfrac{16}{21}$

**30.** $\dfrac{18}{35} \cdot \dfrac{14}{17}$

**Find each quotient. Give your answers in lowest terms. If an answer is greater than 1, write it as a mixed number.**

**31.** $4 \div \dfrac{1}{8}$

**32.** $7 \div \dfrac{1}{5}$

**33.** $45 \div \dfrac{3}{5}$

**34.** $20 \div \dfrac{2}{3}$

**35.** $\dfrac{15}{21} \div \dfrac{5}{7}$

**36.** $\dfrac{7}{9} \div \dfrac{14}{27}$

**37.** $3\dfrac{2}{7} \div 2\dfrac{3}{7}$

**38.** $8\dfrac{1}{3} \div 1\dfrac{5}{9}$

**39.** $\dfrac{3}{7} \div \dfrac{3}{7}$

**40.** $\dfrac{343}{425} \div \dfrac{343}{425}$

**Use the fact that 652 • 25 = 16,300 to find each product without using a calculator.**

**41.** $65.2 \cdot 2.5$

**42.** $0.625 \cdot 25$

**43.** $6.52 \cdot 0.00025$

**44.** $6.52 \cdot 2{,}500$

**45.** $0.00652 \cdot 0.25$

**46.** $65.2 \cdot 0.25$

**Find each product or quotient.**

**47.** 0.25 · 400

**48.** 64 ÷ 0.8

**49.** 32.07 · 0.001

**50.** 32.07 ÷ 0.001

**51.** 7.75 · 12.4

**52.** 0.009 · 1.2

**53.** 0.144 ÷ 0.6

**54.** 87.003 · 5.5

**55.** 21 ÷ 0.0025

**56.** 19 ÷ 0.00038

**57.** Kristin determined the ages of all the children playing at a fast food play area. Here are her results.

> 2   2   3   4   2   3   2   5   3   2   8

**a.** Find the range, mode, and median of these data.

**b.** Which average best describes the data? Explain your choice

**58.** The students in Mrs. Foley's class are counting the number of pretzels they get in a small bag. Here are the numbers of pretzels.

> 10   11   12   10   9   13   12   11

**a.** Find the mean and median of the data set.

**b.** Add two values to the data set so that the median remains the same, but the mean decreases. What is the new mean?

**c.** Start with the original set of data. Add two values to the set so the median remains the same, but the mean increases. What is the new mean?

**59.** Students at Central High School are donating money to a charity. The administration wants to know which average should be used to show how much the students donated.

Listed below are the amounts donated by individual students.

| | | | | | | |
|----|----|----|----|----|----|----|
| 10 | 5  | 12 | 2  | 6  | 3  | 11 |
| 8  | 9  | 11 | 2  | 5  | 5  | 6  |
| 7  | 1  | 3  | 3  | 7  | 8  | 9  |
| 20 | 5  | 2  | 3  | 7  | 10 | 2  |

**a.** Find the mode, median, and mean for these data.

**b.** Which average best describes a typical value in this set of data?

# Test-Taking Practice

*SHORT RESPONSE*

**1** The data set shows the amounts Mya earned working on different days. Find the mean, median, and mode of the amounts.

$35, $28, $54, $29, $47, $35, $42

*Show your work.*

*Answer* _____

*MULTIPLE CHOICE*

**2** Lea needs $2\frac{2}{3}$ cups of flour for one recipe and $3\frac{1}{4}$ cups of flour for a second recipe. How much flour does Lea need for both recipes?

A $5\frac{1}{4}$

B $5\frac{3}{7}$

C $5\frac{11}{12}$

D $6\frac{11}{12}$

**3** Shannon rode her bicycle $\frac{5}{6}$ mile Friday afternoon and $3\frac{1}{6}$ miles Saturday afternoon. How much farther did she ride her bicycle Saturday than Friday?

F $2\frac{1}{3}$ miles

G $2\frac{2}{3}$ miles

H $3\frac{1}{3}$ miles

J $3\frac{2}{3}$ miles

**4** Find the product of 7.3 and 0.6.

A 4.38

B 7.9

C $12.1\overline{6}$

D 43.8

**5** Lauren cut a $\frac{2}{3}$-yard piece of material into $\frac{1}{6}$-yard pieces. How many pieces did she cut?

F 2

G 3

H 4

J 6

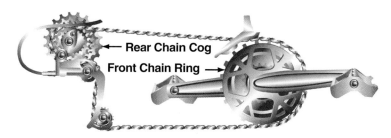

# CHAPTER 5

# Rate, Ratio, and Proportion

## Real-Life Math

**Gearing Up** Understanding *gear ratios* can help you ride your bike more efficiently.

A bike chain goes around a front chain ring, which is connected to the pedal, and a rear chain cog, which turns the back wheel. Changing gears moves the chain to a different rear cog or a different front ring.

← **Rear Chain Cog**
**Front Chain Ring** →

The gear ratio for a particular gear tells you how many times the rear wheel rotates each time you rotate the pedals once. You find the gear ratio by counting teeth.

$$\text{gear ratio} = \frac{\text{number of teeth on front chain ring}}{\text{number of teeth on rear chain cog}}$$

For example, if the chain is on a front ring with 54 teeth and a rear cog with 27 teeth, the gear ratio is $\frac{54}{27}$, or $\frac{2}{1}$. This means that the back wheel rotates twice every time the pedals rotate once.

**Think About It** If you want to travel as far as possible with the least amount of pedaling, would you want to be in a gear with a high gear ratio or a low gear ratio?

**Math Online**
Take the **Chapter Readiness Quiz** at glencoe.com.

# Dear Family,

Chapter 5 is about ratio and proportion.

## Key Concept—Ratios

The class will begin with the idea of mixing different strengths of dye to explore ratios. Using this model makes the idea of ratio and proportion easy to understand. For example, Mixture A will be darker because the ratio of dye to water is 3:1. It is only 2:1 in Mixture B.

**Mixture A**          **Mixture B**

The class will also learn how to scale ratios. To make a larger batch of Mixture A, keep the ratio the same but increase the number of cans. This can be done by multiplying both parts of the ratio by the same number.

Students will use proportions to find missing quantities and to estimate large quantities that would be difficult to count. For example, the total number of people affected by a flu epidemic can be etimated by counting the number in a small sample and using the proportion to estimate the total.

## Chapter Vocabulary

| | |
|---|---|
| congruent | equivalent ratios |
| corresponding angles | ratio |
| corresponding sides | similar |
| counterexample | unit rate |

## Home Activities

• Encourage your student to point out different instances where ratios are used in his or her life, such as finding the cost of five cans of beans if two cans cost 70 cents.

# Ratios and Rates

TastySnacks, Inc. is introducing Lite Crunchers, a reduced-fat version of its best-selling Crunchers popcorn.

The tables list nutrition information for both products.

| Original Crunchers Serving size: 35 g | | |
|---|---|---|
| **Total Fat** | 6 | g |
| Saturated fat | 5 | g |
| **Cholesterol** | 0 | mg |
| **Sodium** | 200 | mg |
| **Total Carbohydrate** | 15 | g |
| Dietary fiber | 1 | g |
| Sugars | 0 | g |
| **Protein** | 2 | g |

| Lite Crunchers Serving size: 35 g | | |
|---|---|---|
| **Total Fat** | 3 | g |
| Saturated fat | 2 | g |
| **Cholesterol** | 0 | mg |
| **Sodium** | 160 | mg |
| **Total Carbohydrate** | 25 | g |
| Dietary fiber | 2 | g |
| Sugars | 0 | g |
| **Protein** | 3 | g |

Many comparison statements can be made from this information. Below are some comparisons involving differences, ratios, and rates.

### Difference Comparisons

- A serving of Lite Crunchers has 22 more grams of carbohydrate than grams of protein.
- Original Crunchers has 40 more milligrams of sodium per serving than Lite Crunchers.

### Rate Comparisons

- Lite Crunchers contains 3 g of protein per serving.
- Original Crunchers contains 200 mg of sodium per serving.

### Ratio Comparisons

- The ratio of saturated fat grams to total fat grams in Original Crunchers is 5 to 6.
- The ratio of fiber grams in Lite Crunchers to fiber grams in Original Crunchers is 2:1.

Work with a partner to write more comparison statements about the nutrition data. Try to write a difference comparison, a ratio comparison, and a rate comparison.

Members of the TastySnacks advertising department are designing a bag for the new popcorn. They want the bag to include a statement comparing the amount of fat in Lite Crunchers to the amount in Original Crunchers. Write a statement that you think would entice people to try Lite Crunchers.

# Investigation 1 Ratios

## Vocabulary

ratio

A **ratio** is a way to compare two numbers. When one segment is twice as long as another, the ratio of the length of the longer segment to the length of the shorter segment is "two to one."

One way to write "two to one" is 2:1. That means that for every two units of length on the longer segment, there is one unit of length on the shorter segment.

Ratios may be written in several ways. Three ways to express the ratio "two to one" are 2 to 1, 2:1, $\frac{2}{1}$.

## ✓ Develop & Understand: A

1. Consider this keyboard.

a. What is the ratio of white keys to black keys on the keyboard?

b. This pattern of keys is repeated on larger keyboards. How many black keys would you expect to find on a keyboard with 42 white keys?

c. What is the ratio of black keys to *all* keys on this keyboard?

d. How many black keys would you expect to find on a keyboard with 72 keys in all?

**2.** The square tiles on Efrain's kitchen floor are laid in this pattern.

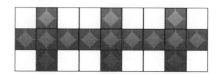

  **a.** What is the ratio of white tiles to purple tiles in this pattern?

  **b.** The entire kitchen floor contains 1,000 purple tiles. How many white tiles does it have?

  **c.** What is the ratio of white tiles to all tiles in this pattern?

  **d.** A floor with this pattern has 2,880 tiles in all. How many white tiles does it have?

**3.** Mercedes made this bead necklace at summer camp.

  **a.** What is the ratio of spherical beads to cube-shaped beads on the necklace?

  **b.** Mercedes wants to make a longer necklace with beads in the same pattern. She plans to use 20 spherical beads. How many cube-shaped beads will she need?

  **c.** What is the ratio of cube-shaped beads to all of the beads on this necklace?

  **d.** Mercedes wants to make a bracelet in the same pattern. She uses ten beads in all. How many cube-shaped beads will she need?

4. The two numbers on each card in this set are in the same ratio.

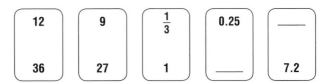

a. Find the missing numbers. Explain how you found your answers.

b. What ratio expresses the relationship between the top number and the bottom number on each card?

c. Draw three more cards that belong in this set.

5. The two numbers on each card in this set are in the same ratio.

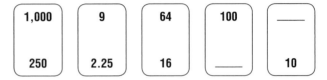

a. Fill in the missing numbers.

b. Draw three more cards that belong in this set.

## ✓ Develop & Understand: B

The lists show the ten most popular first names in the United States given to children born during the 2000s.

| Boys | Girls |
|------|-------|
| 1. Jacob | 1. Emily |
| 2. Michael | 2. Madison |
| 3. Joshua | 3. Emma |
| 4. Matthew | 4. Hannah |
| 5. Andrew | 5. Abigail |
| 6. Christopher | 6. Olivia |
| 7. Daniel | 7. Ashley |
| 8. Joseph | 8. Samantha |
| 9. Ethan | 9. Alexis |
| 10. Nicholas | 10. Sarah |

**Source:** Social Security Administration

6. What is the ratio of names that start with *J* to names that do not?

7. What is the ratio of students in your class with one of these first names to students whose names are not on these lists?

8. What is the ratio of students in your class whose first *or* middle names are on these lists to the total number of students?

9. Now you will use the lists to make some other ratio comparisons.

a. Write two ratio statements based on the information on these lists. For example, one possible statement is "The ratio of girls' names that have only the letter *a* as a vowel to all the girls' names is 3 to 10."

b. Try to write two different comparisons that involve the same ratio.

### Share & Summarize

1. Work with a partner to write at least five ratio statements about this quilt with white, red, and pink squares.

2. Write a ratio statement about the quilt that involves each given ratio.

   a. 1:4

   b. 1:3

   c. 3:5

# Investigation 2 Compare and Scale Ratios

You have practiced writing ratios to express comparisons between quantities. Reasoning about ratios will help you solve the exercises in this investigation.

# ✅ *Develop & Understand: A*

Researchers at First-Rate Rags are developing a shade of green for a new line of shirts. They are experimenting with various shades by mixing containers of blue and yellow dye.

1. Below are two mixtures the researchers tested. Each blue can represents a container of blue dye. Each yellow can represents a container of yellow dye. All containers contain the same amount of liquid. Which mixture is darker green? Explain how you decided.

**Mixture A**          **Mixture B**

In Exercises 2–4, tell which mixture is darker green.

2.          **Mixture C**          **Mixture D**

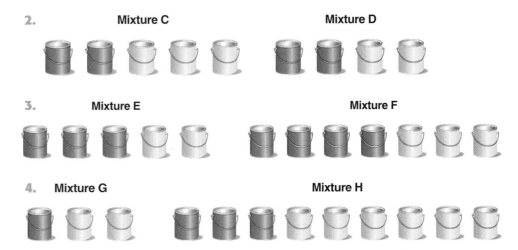

3.          **Mixture E**          **Mixture F**

4.          **Mixture G**          **Mixture H**

## *Think & Discuss*

Shaunda and Simon found different answers for Exercise 4.

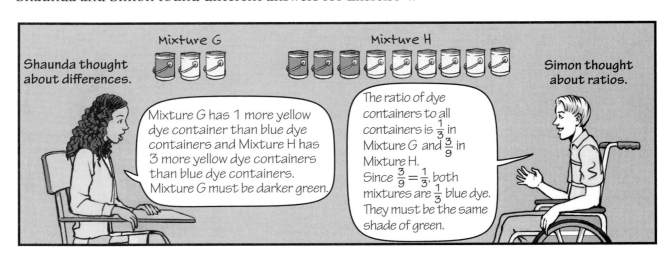

Mixture G          Mixture H

**Shaunda thought about differences.**          **Simon thought about ratios.**

Mixture G has 1 more yellow dye container than blue dye containers and Mixture H has 3 more yellow dye containers than blue dye containers. Mixture G must be darker green.

The ratio of dye containers to all containers is $\frac{1}{3}$ in Mixture G and $\frac{3}{9}$ in Mixture H. Since $\frac{3}{9} = \frac{1}{3}$, both mixtures are $\frac{1}{3}$ blue dye. They must be the same shade of green.

Whose reasoning is correct? How would you use the correct student's method to solve Exercise 5?

## ✅ *Develop & Understand: B*

The researchers at First-Rate Rags have found some shades that they like. Now they want to make larger batches of dye.

**5.** The researchers call this mixture Grassy Green. Draw a picture to show how Grassy Green can be created using 12 containers in all.

**6.** This mixture is called Grazing Green. Draw a set of containers that could be used to create a larger batch of Grazing Green.

**7.** This mixture is called Turtle Green. Draw a set of containers that could be used to create a larger batch of Turtle Green.

**8.** Order the three mixtures above from darkest to lightest. Use ratios to explain your ordering.

---

### *Share* & *Summarize*

**1.** Describe a strategy for determining which of two shades of green is darker. Use your strategy to determine which mixture is darker.

| Mixture X | Mixture Y |
|:---:|:---:|

**2.** Draw a set of containers that would create a larger batch of dye the same shade as Mixture X. Explain how you know the shade would be the same.

# Investigation 3 Ratio Tables

In the last investigation, you figured out how to make larger batches of dye that would be the same shade as a given mixture. There are many ways to think about exercises like these.

The mixture at the right is Grazing Green. Jin Lee, Zach, and Maya tried to make a batch of Grazing Green using nine containers.

Jin Lee: I'll draw copies of the 3 containers until I have 9 in all. First, I draw the mixture, that's 3 containers. Then I draw it again to get 6 containers. I draw it once more to get 9 containers, and I'm done.

Zach: The original mixture has 1 more container of blue than yellow. The new mixture must also have 1 more container of blue than yellow. So, I'll draw 9 containers, 5 blue and 4 yellow.

Maya: In the original mixture, $\frac{2}{3}$ of the containers are blue. If I use 9 containers, $\frac{2}{3}$ of them must be blue. $\frac{2}{3}$ of 9 is 6, so 6 containers must be blue and 3 must be yellow.

## Think & Discuss

Did all the students reason correctly? Explain any mistakes that they made.

Did any of the students use reasoning similar to your own?

When Jin Lee answered the question above, she made copies of the original mixture until she had the correct number of containers. Jin Lee could have used a ratio table to record her work. A *ratio table,* a tool for recording many equal ratios, can help you think about how to find equal ratios.

This ratio table shows Jin Lee's thinking.

**Math Link**

The terms *rate* and *ratio* come from a Latin word that means "to calculate."

| Blue Containers | 2 | 4 | 6 |
|---|---|---|---|
| Total Containers | 3 | 6 | 9 |

All columns of a ratio table contain numbers in the same ratio. The equivalent ratios in this table are $\frac{2}{3}$, $\frac{4}{6}$, and $\frac{6}{9}$.

## ✅ *Develop & Understand: A*

1. Complete this ratio table to show the number of blue containers and the total number of containers for various batches of this shade.

| Blue  |   | 3 | 4 | 8 | 25 |    | 200 |
|-------|---|---|---|---|----|----|-----|
| Total | 2 |   | 8 |   |    | 74 |     |

2. The ratio table in Exercise 1 compares the number of blue containers to the total number of containers. Complete the next ratio table to compare the number of blue containers to the number of yellow containers in this mixture.

| Blue   | 2 |     | 6 | 10 |    | 90 | $n$ |
|--------|---|-----|---|----|----|----|-----|
| Yellow |   | 1.5 | 3 |    | 15 |    |     |

3. The school band is holding a car wash to raise money for new uniforms. Ms. Chang, the band director, wants to order pizza for everyone. After the car wash last year, 20 people ate eight pizzas.

   a. Complete this ratio table based on last year's information.

| People |   | 10 | 15 | 20 |   |   |
|--------|---|----|----|----|---|---|
| Pizzas | 2 |    |    | 8  |   |   |

   b. How many people will two pizzas feed?

   c. Ms. Chang is planning to feed 25 people. How many pizzas will she need?

4. Jayvyn and Rosario are planning a party. They want to figure out how many pints of juice to buy. At Toya's party the month before, 16 people drank 12 pints of juice. Jayvyn and Rosario want the same ratio of juice to people at their party.

   a. Complete this ratio table based on the information about Toya's party.

| People |   |   | 12 | 16 | 20 | 24 |
|--------|---|---|----|----|----|----|
| Pints  |   |   |    | 12 |    |    |

**b.** How many people will 1.5 pints of juice serve?

**c.** Jayvyn said that if they extend the table, it will show that 21 pints are needed to serve 28 people. Is Jayvyn correct? How do you know?

## Share & Summarize

This mixture is called Sea Green.

**1.** Describe how you could find a mixture of Sea Green that uses 9 containers in all.

**2.** Make a ratio table to show the number of blue containers and the total number of containers that you would need to make different-sized batches of Sea Green.

# Investigation 4 Comparison Shopping

### Vocabulary

unit rate

In the blue-dye exercises, you can *compare* ratios to find the darkest mixture, and you can *scale* ratios to make larger and smaller batches of a given shade. Comparing and scaling ratios and rates is useful in many real-life situations.

In this investigation, you will see how these skills can help you get the most for your money.

## ✓ Develop & Understand: A

**1.** Alexi is inviting several friends to camp in her backyard. She wants to serve bagels for breakfast. She knows that Bagel Barn charges $3 for half a dozen bagels, and Ben's Bagels charges $1 for three bagels. At which store will Alexi get more for her money? Explain how you found your answer.

**2.** Alexi is considering serving muffins instead of bagels. Mollie's Muffins charges $6.25 per dozen. The East Side Bakery advertises "Two muffins for 99¢." Where will Alexi get more for her money? Explain.

Here is how Simon thought about solving Exercise 1.

Shaunda thought about scaling ratios.

### Think & Discuss

Did you solve the bagel exercise using a method similar to either Simon's or Shaunda's?

Solve Exercise 2 using Simon's or Shaunda's method. Use a different method from the one you used to solve it the first time.

Simon's method involves finding *unit rates*. In a **unit rate**, one quantity is compared to one unit of another quantity. Simon found the prices for one bagel and compared them. Here are some other examples of unit rates.

| | | |
|---|---|---|
| $1.99 per lb | 65 miles per hour | $15 for each CD |
| 24 students per teacher | 3 tsp in a tbsp | 4 quarts in 1 gallon |

Unit rates that involve prices, such as 50¢ per bagel and $1.99 per pound, are sometimes called *unit prices*. Supermarket shelves often have tags displaying unit prices. These tags can help consumers make more informed decisions about their purchases.

## ✓ Develop & Understand: B

Use unit rates to help you answer Exercises 3–5.

3. Camisha has a long-distance plan that charges the same amount for each minute of a call. The rate depends on where she calls. A 12-minute call to Honolulu costs $4.44. A 17-minute call to Hong Kong costs $7.14. Is the rate to Honolulu or the rate to Hong Kong higher? Explain how you found your answer.

4. At FreshStuff Produce, Antoine paid $3.52 for four mangoes. Manuela bought six mangoes at FruitMart for $5.52. Which store offers the better price for mangoes? Explain how you found your answer.

5. Blank CDs are sold in four packages at Xavier's Music Store.

   Which package costs the least per CD? Explain how you found your answer.

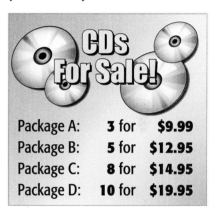

6. Coach Thomas found several ads for soccer balls in the sports section of the local paper. Use the technique of scaling ratios to find which store offers the best buy. Explain how you made your decision.

### Share & Summarize

1. Explain how unit prices can help you make wise purchasing decisions. Give an example to explain your thinking.

2. Suppose a package of eight pencils costs 92¢. A package of 12 costs $1.45. Which method would you prefer to use to find which package costs the least per pencil? Explain.

**Practice & Apply**

**Real-World Link**

The highest mountain on the North American Continent, Mount McKinley, was named for William McKinley, the 25th U.S. president (1897–1901).

· · · · · · · · · · · · · · · · · · · ·

**1.** Mount McKinley is located in Denali National Park, Alaska. Between 1980 and 1992, 10,470 climbers attempted to reach its summit, a height of 20,320 feet. Of these climbers, 5,271 were successful. The five comparisons below are based on this information.

   **i.** More than half the people who attempted to reach the summit were successful.

   **ii.** The ratio of climbers who successfully reached the summit to those who failed is 5,271 to 5,199.

   **iii.** Of the climbers who attempted to reach the summit, 72 more succeeded than failed.

   **iv.** The ratio of the total number of climbers to the number of successful climbers is about 2 to 1.

   **a.** Explain why each statement is true.

   **b.** Suppose that Expert Expeditions arranges group climbing trips on Mount McKinley. Its guides help less-experienced climbers reach the summit. The company is designing an advertising brochure and would like to include one of the above statements. Which do you think should be used? Explain your reasoning.

**2.** Eduardo is laying out the pages of a newspaper. He divides one of the pages into 24 sections and decides which sections will be devoted to headlines, to text, to pictures, and to advertisements.

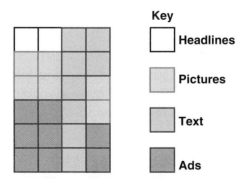

Describe something in the layout that has each given ratio.

**a.** 1 to 3

**b.** 1:4

**c.** 3:4

**3.** First-Rate Rags wants to create a new shade of green using blue and yellow dye. They create two mixtures to compare. Decide whether Mixture A or Mixture B is a darker shade of green. Explain why.

**Mixture A**                **Mixture B**

**4.** First-Rate Rags has finally agreed on a shade of green for its shirts. A sample of the dye is shown below.

How many containers of blue dye and yellow dye are needed for a batch of green dye that has a total of 75 containers?

For each pair of mixtures, decide which is a *lighter* shade of green. Then draw or describe a larger batch that will make the same shade as the lighter mixture's shade.

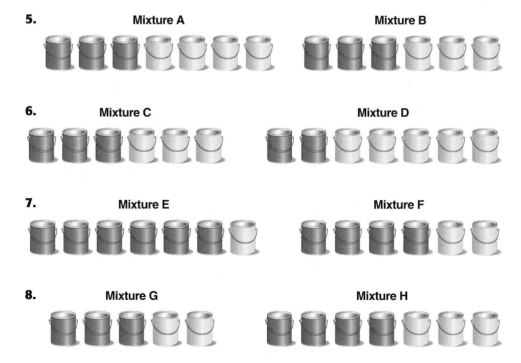

**5.**     **Mixture A**                          **Mixture B**

**6.**     **Mixture C**                          **Mixture D**

**7.**     **Mixture E**                          **Mixture F**

**8.**     **Mixture G**                          **Mixture H**

**Complete the ratio table for each mixture.**

9.

| Blue | 1 | 3 | 6 | 10 | 50 | 100 | $n$ |
|---|---|---|---|---|---|---|---|
| Yellow | | | | | | | |

10.

| Blue | 1 | 2 | 5 | 10 | 50 | 100 | $n$ |
|---|---|---|---|---|---|---|---|
| Total | | | | | | | |

In Exercises 11–13, is the given table a ratio table? If so, tell the ratio. If not, explain why not.

11.

| 3 | 6 | 9 | 12 | 15 |
|---|---|---|---|---|
| 5 | 10 | 15 | 20 | 25 |

12.

| 1 | 4 | 7 | 10 | 13 |
|---|---|---|---|---|
| 2 | 5 | 8 | 11 | 14 |

13.

| 3 | 30 | 300 | 3,000 | 30,000 |
|---|---|---|---|---|
| 9 | 90 | 900 | 9,000 | 90,000 |

In Exercises 14–17, tell whether the given rate is a unit rate or a non-unit rate.

14. 25 heartbeats in 20 seconds

15. 72 heartbeats per minute

16. 30 mph

17. 200 miles per five-hour period

**18.** To serve seven people, Natasha needed two pizzas. What was the per-person rate?

**19.** At basketball camp, Belicia made 13 baskets for every 25 shots attempted. What was Belicia's success rate per attempt?

**20.** Jogging burns about 500 calories every five miles. What is the rate of calorie consumption per mile?

In Exercises 21–23, determine the unit price for each offer. Then tell which offer is best.

**21.** Two pens for $3, ten pens for $16, or five pens for $7

**22.** Five lb of potatoes for $2.99, ten lb of potatoes for $4.99, or 15 lb of potatoes for $6.99

**23.** Three toy racing cars for $7, five toy racing cars for $11, two toy racing cars for $4.50

**Connect & Extend**

**24. Architecture** In ancient Greece, artists and architects believed there was a particular rectangular shape that looked very pleasing to the eye. For rectangles of this shape, the ratio of the long side to the short side is roughly 1.6:1. This ratio is very close to what is known as the *Golden Ratio*.

**a.** Try drawing three or four rectangles of different sizes that look like the most "ideal" rectangles to you. For each rectangle, measure the side lengths and find the ratio of the long side to the short side.

**b.** Which of your rectangles has the ratio closest to the Golden Ratio?

**Real-World Link**

The famous Greek temple, the Parthenon, made entirely of white marble in the fifth century B.C., was built according to the Golden Ratio.

**25.** Consider this mixture.

    **a.** Draw a mixture that is darker green than this one but uses fewer containers. Do not make *all* the containers blue.

    **b.** Explain why you think your mixture is darker.

**26. Measurement** Trevor always forgets how to convert miles to kilometers and back again. However, he remembers that his car's speedometer shows both miles and kilometers. Trevor knows that traveling 50 miles per hour is the same as traveling 80 kilometers per hour. In one hour, he would travel 50 miles, or 80 kilometers.

    **a.** From this information, find the ratio of miles to kilometers in simplest terms.

    **b.** To cover 100 km in an hour, how fast would Trevor have to go in miles per hour? Explain how you found your answer.

**27.** Denay's class sold posters to raise money. Denay wanted to create a ratio table to find how much money her class would make for different numbers of posters sold. She knew that $25 would be raised for every 60 posters sold.

    **a.** Describe how Denay can use that one piece of information to make a ratio table. Assume the relationship is proportional.

    **b.** How much money would Denay's class make for selling 105 posters?

    **c.** Could Denay's class raise exactly $31? If so, how many posters would need to be sold? If not, why not?

**28.** Nora wants to make a batch of mixed nuts for a party that she is planning. The local health food store sells nuts by weight, so Nora can measure out exactly how much she wants. The table shows the cost of each type of nut.

**Cost of Nuts**

| Nut | Amount | Price |
|---|---|---|
| Almonds | 16 oz | $4.80 |
| Cashews | 16 oz | $8.00 |
| Filberts | 16 oz | $4.32 |
| Peanuts | 16 oz | $2.00 |

**a.** Nora wants to know how much 1 ounce of each type of nut will cost. How can she figure this out? How much does 1 ounce of each type of nut cost?

**b.** Nora paid the same for the almonds as the cashews but bought different amounts of each. How can you decide which nut gave her more for her money? Which nut is that?

**c.** Nora wants to buy 10 ounces of almonds, 6 ounces of cashews, 20 ounces of filberts, and 30 ounces of peanuts for her party. What will be the total cost of her peanut purchase, not including tax?

**29.** Jack is in the grocery store comparing two sizes of his favorite toothpaste. He can buy three ounces for $1.49 or four ounces for $1.97. Without calculating unit prices, how can Jack decide whether one size is a better buy than the other?

**Mixed Review**

**30.** Which of the following are factors of 36?

| 1 | 2 | 3 | 4 | 5 | 6 | 7 | 8 |
|---|---|---|---|---|---|---|---|
| 9 | 10 | 11 | 12 | 13 | 14 | 24 | 36 |

**Fill in each ◯ with <, >, or = to make a true statement.**

**31.** $\frac{6}{5}$ ◯ 1.2

**32.** 0.37 ◯ $\frac{7}{20}$

**33.** $\frac{13}{18}$ ◯ $\frac{7}{10}$

**34.** 1.5 ◯ $\frac{25}{19}$

**35.** 0.0375 ◯ $\frac{3}{80}$

**36.** $\frac{23}{24}$ ◯ $\frac{24}{25}$

**Measurement** Fill in the blanks.

**37.** 356 cm = _____ m

**38.** 356 cm = _____ mm

**39.** 44 m = _____ mm

**40.** 5 mm = _____ m

**41.** 5 mm = _____ cm

**42.** 89,000 mm = _____ m

# Proportions

The sixth graders at Summerville Middle School are selling calendars to raise money for a class trip. The amount of money raised is *proportional* to the number of calendars sold. *Proportional* means that as one variable doubles the other doubles, as one variable triples the other triples, and so on.

| Dollars raised | 0 | 15 | 30 | 45 |
|---|---|---|---|---|
| Calendars sold | 0 | 5 | 10 | 15 |

A graph could also be used to represent the same data from the table.

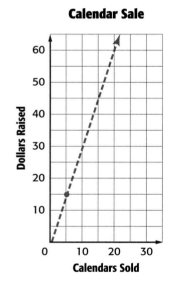

**Calendar Sale**

*Think & Discuss*

The point (5, 15) is on the graph. What does this tell you about the calendar sale?

Identify two more points on the graph. Find the ratio of dollars raised to calendars sold for all three points. How do the ratios compare?

What do the table and the graph tell you about the calendar sale? How does it relate to the ratios you found?

Now that you know the ratio of calendars sold to dollars earned, how can you calculate the amount of dollars raised if 215 calendars were sold?

# Investigation  Proportional Relationships

In Think & Discuss, you may have found that the ratio of dollars raised to calendars sold is the same for any values you choose. For example, (5, 15) and (10, 30) are both on the graph.

$$\frac{15}{5} = \frac{30}{10}$$

**Math Link**
Two ratios are equivalent ratios if they represent the same relationship.

This gives you another way to think about proportional relationships. A *proportional relationship* is a relationship in which all pairs of corresponding values have the same ratio.

We already know that when one ratio is set equal to another ratio, they are proportional.

$$\frac{1}{2} = \frac{3}{6}$$

$\frac{1}{2}$ and $\frac{3}{6}$ are proportional, just like $\frac{15}{5}$ and $\frac{30}{10}$ are proportional. There are several ways you can test for proportionality, or equality.

You can compare the cross products.

$$\frac{3}{6} = \frac{4}{8}$$

Multiply the numerator of the first fraction with the denominator of the second fraction.

$$\frac{3}{6} \diagdown \frac{4}{8} = 3 \cdot 8 = 24$$

Now multiply the numerator of the second fraction with the denominator of the first fraction.

$$\frac{3}{6} \diagup \frac{4}{8} = 4 \cdot 6 = 24$$

Using cross products, you can ask yourself if three times eight is equal to four times six. Since it does, we can state that these two fractions are proportional.

Another way to test for proportionality is to reduce each fraction to lowest terms.

$$\frac{9}{27} = \frac{18}{54}$$

Since both $\frac{9}{27}$ and $\frac{18}{54}$ reduce to $\frac{1}{3}$, we can state that these two fractions are proportional, or equal.

## ☑ *Develop & Understand: A*

**1.** Consider these triangles.

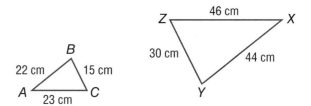

Are the side lengths of triangle *ABC* proportional to those of triangle *XYZ*? Explain how you know.

**2.** For each rectangle, find the ratio of the length to the width. Is the relationship between the lengths and widths of these rectangles proportional? Explain.

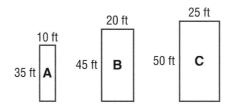

**3.** On the cards below, the top numbers are proportional to the bottom numbers. Find another card that belongs in this set. Explain how you know your card belongs.

| 1 | 1.5 | 2 | 2.5 |
|---|-----|---|-----|
| 2 | 3 | 4 | 5 |

**Math Link**

To be a pattern, the arrangement of tiles must change in a predictable way from stage to stage.

**4.** Santos created this tile pattern.

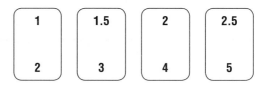

| Stage 1 | Stage 2 | Stage 3 | Stage 4 |

**a.** Describe how the pattern of blue and purple tiles changes from one stage to the next.

**b.** For the stages shown, is the number of blue tiles proportional to the number of purple tiles? Explain.

**c.** Starting with stage 1, draw the next two stages for a tile pattern in which the number of blue tiles is proportional to the number of purple tiles.

5. Jeff says, "I started with Mixture A. I created Mixture B by adding one blue container and one yellow container to Mixture A. Then I made Mixture C by adding one blue container and one yellow container to Mixture B. Since I added the same amount of blue and the same amount of yellow each time, the number of blue containers is proportional to the number of yellow containers."

Is Jeff correct? Explain.

**Mixture A**

**Mixture B**

**Mixture C**

## Share & Summarize

1. Describe at least two ways to determine whether a relationship is proportional.

2. Describe two quantities in your daily life that are proportional to each other.

3. Describe two quantities in your daily life that are not proportional to each other.

# Investigation 2 Equal Ratios

You know that in a proportional relationship, all pairs of corresponding values are in the same ratio. You may find it helpful to think about this idea as you solve the next set of exercises.

## ✓ Develop & Understand: A

1. When the sixth grade calendar sale began, Mr. Diaz bought the first six calendars. From his purchase, the class raised $18.00. By the end of the first day of the sale, $63.00 was raised. How many calendars had been sold by the end of the first day?

### Example

Luis, Kate, and Darnell each thought about Exercise 1 in a different way, but all three found the same answer.

Luis used unit rates.

I found out how much they would raise if they sold 1 calendar. I divided $63.00 by that amount.

Kate wrote and scaled a ratio.

They raised $18 for 6 calendars, so I wrote the ratio $\frac{18}{6}$. I multiplied both parts of it by the same number to get an equal ratio in the form of "$63.00 to something."

Darnell created a table.

I created a table. I know that 6 calendars were worth $18.

| Calendars | 1 | 2 | 6 | × | × |
|-----------|---|---|----|---|----|
| Cost | | 3 | 6 | 18 | × | 63 |

I completed my table to find the answer.

2. Complete each student's method and find the number of calendars sold the first day. Find the same answer with each method.

3. The rectangles below have lengths and widths that are proportional. How can you find the value of *x*? What is the value of *x*?

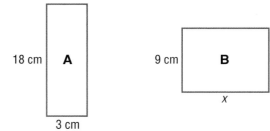

18 cm | **A**

3 cm

9 cm | **B**

*x*

4. Susan rented skis at Budget Mountain. She paid $20 for 4 hours and she used the skis for 6 hours. The rental cost is proportional to time. What will be her rental charge?

5. First-Rate Rags wants to make a large batch of the shade of dye below. The final batch must have 136 containers in all (blue and yellow). How many containers of blue dye will be needed?

6. Kyle and his parents went to Algeria during his summer break. Before they left, Kyle exchanged some money for Algerian dinar. In exchange for $12 U.S., he received 804 Algerian dinar. He returned with 201 Algerian dinar. If the exchange rate is still the same, how many dollars will Kyle receive for his dinar?

### Share & Summarize

Choose one of the exercises from 3–6. Solve it using a different method. Explain each step clearly enough that someone from another class could understand what you did.

# *Inquiry*

## Investigation 3 Solve Proportions

Proportional situations are one of the most common mathematical occurrences in life. Interior designers and landscape designers, for example, use the idea of proportions when creating a new design. In this investigation, you will be an interior designer responsible for ordering materials for a project. You will need to calculate the cost of those materials. Your client would like to spend as little money as possible on the materials.

Your client has decided to tile the floor of a rectangular room that measures 12 feet by 15 feet. The square tiles that your client likes come in three different sizes.

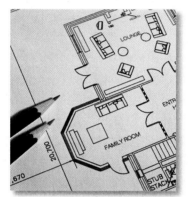

| Tile Dimensions (inches) | Cost per Tile (dollars) |
|---|---|
| 12 × 12 | $1.45 |
| 18 × 18 | $3.45 |
| 24 × 24 | $5.65 |

1. Because the tiles are different sizes, you will need to know the number of each tile size needed to cover the floor. You know that there are 12 inches in one foot. Determine the number of square feet each tile will cover. Copy and complete the table.

| Tile Dimensions | Convert Square Inches to Square Feet |
|---|---|
| 12 inches × 12 inches | 12 in. = 1 ft<br>12 in. × 12 in. = 1 ft × 1 ft<br>or 1 ft$^2$ |
| 18 inches × 18 inches | |
| 24 inches × 24 inches | |

### Try It Out

2. The floor measures 12 feet by 15 feet. How many of each size tile is needed to cover the floor? Set up and solve proportions to determine the number of each size tile needed to cover the floor.

## Try It Again

**3.** Now you know the number of each tile that you will need if you used only one size. Set up and solve proportions to determine the total cost of the project if you use only one size tile for the entire project.

**4.** Now consider a situation where the room is not rectangular. Your customer does not want you to cut tiles, so you must use whole tiles. Given the following shape of the room, what would be the most economical approach?

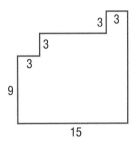

## What Have You Learned?

Discuss with a classmate what you have learned about using proportions in everyday situations. Talk about professions with which you are familiar or about a profession you hope to have when you are older. Discuss how you think proportions might be used.

- How might proportions save a life in the medical profession?

- How would a hair stylist use proportions?

# On Your Own Exercises

**Lesson 5.2**

**Practice & Apply**

1. **Ecology** One of the new energy-efficient cars will travel many miles on a gallon of fuel by using a combination of electricity and gasoline. The table shows estimates of how far the car will travel on various amounts of fuel.

| Gallons of Fuel | 0.5 | 1 | 1.5 | 2 | 2.5 | 3 |
|---|---|---|---|---|---|---|
| Miles | 30 | 60 | 90 | 120 | 150 | 180 |

Are the miles traveled proportional to the gallons of fuel? How do you know? If so, describe how they are related and write the ratio.

2. Is the number of blue dye containers in these two mixtures proportional to the total number of containers? Explain how you know.

**Mixture A**

**Mixture B**

3. The Summerville Co-op sells two types of trail mix. Here are the ingredients.

| Mountain Trail Mix | Hiker's Trail Mix |
|---|---|
| 8 oz toasted oats | 6 oz toasted oats |
| 7 oz nuts | 5 oz nuts |
| 5 oz raisins | 4 oz raisins |

   **a.** Is the amount of nuts in each mix proportional to the total ounces of mix? Why or why not?

   **b.** Is the amount of toasted oats in each mix proportional to the total ounces of mix? Why or why not?

4. Set up a proportion for the following situation. Choose any method to solve it.

   *Elena's grandmother wants to sell a gold ring that she no longer wears. The jeweler offered her $388 per ounce for her gold. This came to $97 for the ring. How much does the ring weigh?*

**5.** Mirna is following this recipe for maple oatmeal bread.

Just as she is preparing to mix the ingredients, she realizes her brother used most of the maple syrup for his breakfast. Mirna has only $\frac{1}{4}$ cup of syrup. She decides to make a smaller batch of bread. How much of each ingredient should she use?

Maple Oatmeal Bread

1 cup quick-cooking oats
3 cups bread flour
$\frac{1}{3}$ cup maple syrup
1 tablespoon cooking oil
$1\frac{1}{4}$ cups water
3 tablespoons yeast
1 teaspoon salt

**6.** Many schools have a recommended student-teacher ratio. At South High, the ratio is 17:1. Next year, South High expects enrollment to increase by 136 students. How many new teachers will need to be hired to maintain this student-teacher ratio?

**7.** A farmer wants to cut down three pine trees to use for a fence he is building. To decide which trees to cut, he wants to estimate their heights. He holds a 9-inch stick perpendicular to and touching the ground. He measures its shadow to be about 6 inches. He wants to cut down trees that are about 40 feet tall. How long are the tree shadows for which he should look?

**8.** A surveyor needs to find the distance across a lake. She makes several measurements and prepares this drawing. The ratio of *AC:CE* is equal to the ratio of *AB:DE*. What is the distance across the lake?

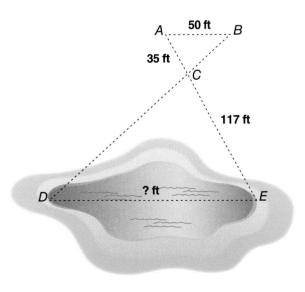

**Connect & Extend**

**9.** Captain Hornblower is out at sea and spots a lighthouse in the distance. He wants to know how far he is from the lighthouse.

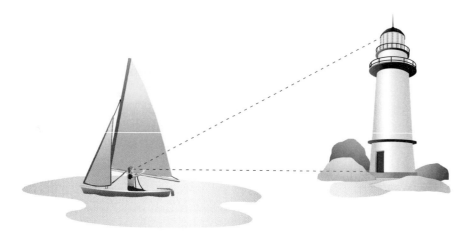

He holds up his thumb at arm's length. Then he brings his hand closer to his eye, until his thumb just covers the image of the lighthouse. His thumb is about 2.5 inches long. He measures the distance from his eye to the base of his thumb and finds that it is about 19 inches. His charts indicate that the lighthouse is Otter Point Lighthouse, which is 70 feet tall.

**a.** Make a sketch of this situation.

**b.** The ratio of the height of his thumb to the height of the lighthouse is equal to the ratio of the distance from his eye to the base of his thumb to the distance from the boat to the lighthouse. Find the distance from the boat to the lighthouse.

**10.** Ravi was born on his father's 25th birthday. On the day Ravi turned 25, his father turned 50, and they threw a big party. Then Ravi wondered, "Dad's twice as old as I am. Does that mean our ages are proportional?" Answer Ravi's question.

**11.** This recipe makes one and a half dozen peanut butter cookies. Write a new recipe that will make more cookies and has ingredients in the proper proportions.

*Peanut Butter Cookies*

$\frac{1}{2}$ cup peanut butter

1 cup flour

$\frac{1}{2}$ cup butter

$\frac{3}{4}$ cup brown sugar

$\frac{1}{4}$ cup white sugar

2 egg whites

$\frac{1}{2}$ tsp baking soda

$\frac{1}{3}$ tsp salt

Decide whether the numbers of tiles of any two colors are proportional to each other. Consider three ratios, red:purple, purple:white, and red:white, for each stage of the tile patterns. Explain your answers.

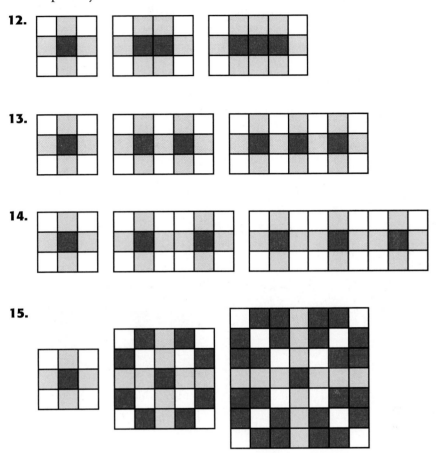

**12.**

**13.**

**14.**

**15.**

**16.** Felix made the following conjecture.

*Take any fraction. Add two to the numerator and the denominator. The new fraction will never be in the same ratio as the original.*

What do you think of his conjecture? Use examples or counter examples in your explanation.

**17. In Your Own Words** An architect has agreed to visit a sixth grade class to talk about how architects use scale drawings in their work. She is not sure all the students will know what it means when she talks about shapes being "in proportion." Write some advice telling her how to explain what this means.

**Mixed Review**

**Number Sense** Name a fraction between the given fractions.

**18.** $\frac{1}{3}$ and $\frac{1}{2}$

**19.** $\frac{1}{4}$ and $\frac{4}{15}$

**20.** $\frac{13}{16}$ and $\frac{11}{12}$

**21.** Here are the first three stages of a sequence.

| Stage 1 | Stage 2 | Stage 3 |

**a.** Describe the pattern in this sequence.

**b.** Draw the next two stages in the sequence.

**c.** Draw stage 15. Explain how you know you are correct.

**Find each product without using a calculator.**

**22.** 44 · 781

**23.** 4.4 · 0.781

**24.** 440 · 781,000

**25.** 0.44 · 7.81

**26.** 0.044 · 0.0781

**27.** 440 · 78.1

# LESSON 5.3

# Similarity and Congruence

What does it mean to say that two figures are the same?

These figures are "the same" because they are members of the same *class of objects*. They are both rectangles.

## Vocabulary

congruent

similar

Some figures have more in common than just being the same *type* of figure. One of the figures below is an enlargement of the other. They have the same *shape* but are different *sizes*. Two figures that have the same *shape* are **similar**.

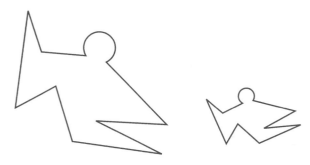

The most obvious way in which two figures can be "the same" is for them to be identical. Figures that are the same size *and* the same shape are **congruent**. The figures below are congruent.

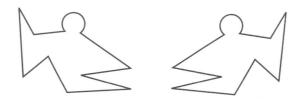

These rectangles are also congruent.

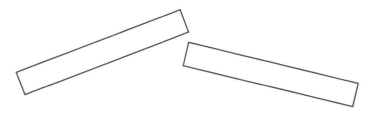

Notice that similarity and congruence do not depend on how the objects are positioned. They can be flipped and rotated from each other.

Your teacher will give you a sheet of paper with drawings of three figures. One or two other students in your class have figures that are congruent to yours. Find these students.

How did you determine which of the other students' figures were congruent to yours?

# Investigation (1) Congruent Figures and Angles

## Vocabulary

counterexample

## Materials

• ruler
• protractor

To find who had figures congruent to yours, you needed to invent a way to tell whether two figures are congruent. Now you will use your test for congruence on more figures and on angles.

### ✓ Develop & Understand: A

**Decide whether figures A and B are congruent. If they are not congruent, explain why not.**

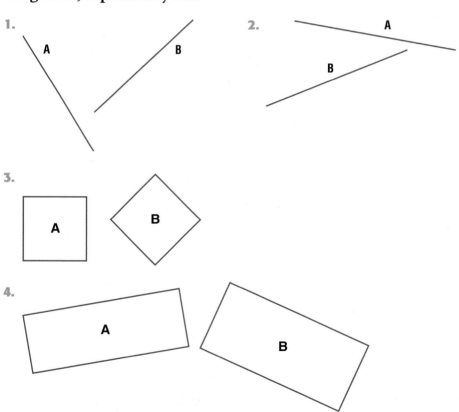

1.

2.

3.

4.

**Decide whether figures A and B are congruent. If they are not congruent, explain why not.**

5.
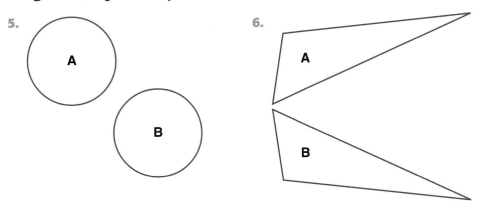

6.

You just compared several pairs of figures. One important geometrical object is the angle, which is a part of many figures. What do you think congruent *angles* look like?

> ### *Think & Discuss*
>
> The angles in each pair below are congruent.
>
>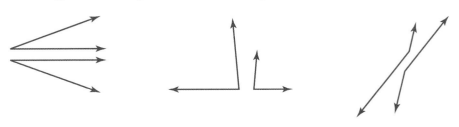
>
> What do you think it means for angles to be congruent?

## *Develop & Understand: B*

**Decide whether the angles in each pair are congruent. If they are not congruent, explain why not.**

7.

8.

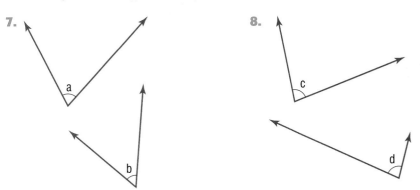

One way to test whether two figures are congruent is to try fitting one exactly on top of the other. Sometimes it is not easy to cut out or trace figures. It is helpful to have other tests for congruency.

## ✅ Develop & Understand: C

Each exercise below suggests a way to test for the congruence of two figures. Decide whether each test is good enough to be *sure* the figures are congruent. Assume you can make *exact* measurements.

If a test is not good enough, give a **counterexample**, an example for which the test would not work.

9. For two line segments, measure their lengths. The line segments are congruent if the lengths are equal.

10. For two squares, measure the length of one side of each square. The squares are congruent if the side lengths are equal.

11. For two angles, measure each angle with a protractor. They are congruent if the angles have equal measures.

12. For two rectangles, find their perimeters. The rectangles are congruent if the perimeters are equal.

> ### Share & Summarize
>
> **Decide which figures in each set are congruent. Explain how you know.**
>
> 1.
>
>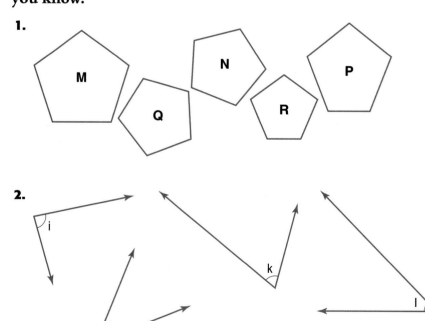
>
> 2.

# Investigation ② Similar Figures

## Materials

- metric ruler
- linkage strips and fasteners

### Math Link

Figures are similar if they have the exact same shape. They may be different sizes.

You now have several techniques for identifying *congruent* figures. How can you tell whether two figures are *similar*?

### ✅ Develop & Understand: A

Work with a partner. To begin, draw a rectangle with sides 1 cm and 3 cm long. You will need only one rectangle for the two of you.

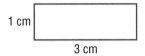

1. One partner should draw a new rectangle whose sides are 7 times the length of the original rectangle's sides. The other partner should draw a new rectangle in which each side is 7 cm longer than those of the original rectangle. Label the side lengths of both new rectangles.

2. With your partner, decide which of the new rectangles looks similar to the original rectangle.

In Exercises 1 and 2, you modified a figure in two ways to create *larger* figures. Now you will compare two ways for modifying a figure to create *smaller* figures.

### ✅ Develop & Understand: B

Work with a partner. Begin by drawing a rectangle with sides 11 cm and 12 cm long.

3. Now, one partner should draw a new rectangle whose sides are one-tenth the length of the original rectangle's sides. The other partner should draw a new rectangle in which each side is 10 cm shorter than those of the original rectangle. Label the side lengths of both new rectangles.

4. With your partner, decide which of the new rectangles looks similar to the original rectangle.

You have used two types of modifications to create rectangles larger and smaller than a given rectangle. You will now try these modifications on a triangle.

## ✓ Develop & Understand: C

Work with a partner. You each need a set of linkage strips and three fasteners. To find lengths on the linkage strips, count the gaps between holes. Each gap is one unit.

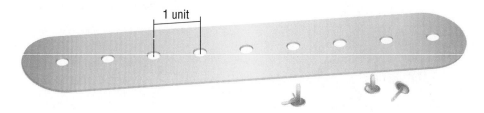

Separately, you and your partner should use your three linkage strips to construct a right triangle with legs 6 units and 8 units and hypotenuse 10 units. Trace the inside of your triangle on a sheet of paper.

5. One partner should follow the instructions in Part a, and the other should follow the instructions in Part b.

   a. Construct a triangle whose side lengths are half those of the first triangle. That is, the lengths should be 3, 4, and 5 units. Trace the inside of the triangle on your paper.

   b. Construct a triangle whose side lengths are each 2 units less than those of the first triangle. That is, the lengths should be 4, 6, and 8 units. Trace the inside of the triangle on your paper.

6. With your partner, decide which modification produces a triangle that looks similar to the original triangle.

### Share & Summarize

You have modified rectangles and triangles in two ways to create larger and smaller figures.
   - In one method, you multiplied or divided each side length by some number.
   - In the other method, you added a number to or subtracted a number from each side length.

Which method produced figures that looked similar to the original?

# Investigation ③ Ratios of Corresponding Sides

## Vocabulary

corresponding
    angles

corresponding sides

equivalent ratios

## Materials

• ruler

• protractor

• metric ruler

Now that you have had some practice thinking about different types of comparisons, turn your attention to ratios of corresponding sides.

It is possible to use different ratios to describe the same relationship.

### Example

Maya and Simon think about the ratios of the side lengths in triangles differently.

Two ratios are **equivalent ratios** if they represent the same relationship. Maya pointed out that the ratio 1:3 means that for every 1 cm of length on one segment, there are 3 cm of length on the other. Simon said the ratio 4:12 means that for every 4 cm of length on one segment, there are 12 cm of length on the other.

The length of the first segment is multiplied by 3 to get the length of the second segment. Therefore, 1:3 and 4:12 are equivalent ratios.

In the last lesson, we referred to equivalent ratios as proportional.

### ☑ *Develop & Understand: A*

1. Name at least two ratios equivalent to the ratio of the length of Segment *MN* to the length of Segment *OP*.

**Decide whether the ratios in each pair are equivalent. Explain how you know.**

2. 1:4 and 2:8

3. $\frac{2}{5}$ and $\frac{3}{9}$

4. 3:5 and 5:3

5. $\frac{1}{3}$:1 and 1:3

6. Brock and Dina were analyzing a pair of line segments. "The lengths are in the ratio 2:3," Brock said. "No," Dina replied, "the ratio is 3:2." Their teacher smiled. "You're both right. But to be clear, you need to give more information about your ratios."

   What did their teacher mean? Are 2:3 and 3:2 equivalent? How could Brock and Dina both be correct?

In Investigation 2, you created rectangles and triangles that were similar to other rectangles and triangles. For each shape, you used a part of the original figure to create the *corresponding part* of the new figure.

Corresponding parts of two similar figures are located in the same place in each figure. For example, Triangles *ABC* and *DEF* are similar. Sides *AB* and *DE* are **corresponding sides**, and ∠*B* and ∠*E* are **corresponding angles** .

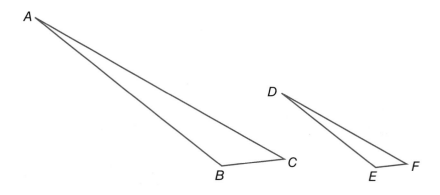

When you created similar rectangles and triangles, the ratios of the lengths of each pair of *corresponding sides* were equivalent. This is true for all similar figures. The ratios of the lengths of each pair of corresponding sides must be equivalent.

## ✓ Develop & Understand: B

**The figures in each pair are similar. Identify all pairs of corresponding sides and all pairs of corresponding angles.**

7.
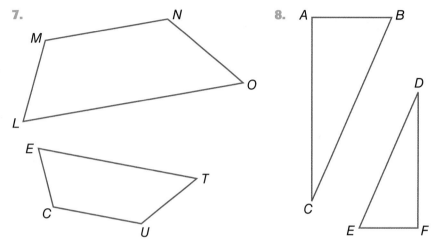

8.

9.
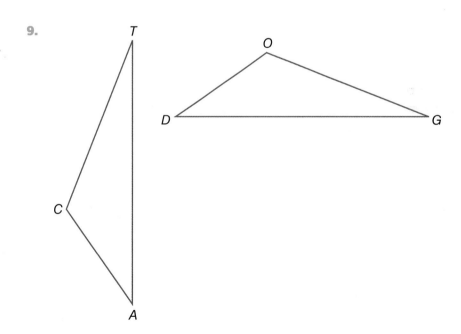

### Real-World Link

The concept of similar triangles can be used to estimate dimensions of lakes, heights of pyramids, and distances between planets.

. . . . . . . . . . . . . . . . . . . . . .

If figures are similar, each pair of corresponding sides must have the same ratio. But if ratios of corresponding sides are the same, does that mean the figures *must* be similar? You will explore this question now.

**10.** Here are two quadrilaterals.

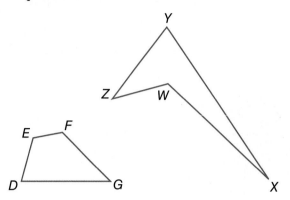

**a.** Copy and complete the table for quadrilateral *DEFG*.

| Description | Side | Length (cm) |
| --- | --- | --- |
| longest side | DG | |
| second-longest side | FG | |
| third-longest side | DE | |
| shortest side | EF | |

**b.** Now complete the table for quadrilateral *WXYZ*.

| Description | Side | Length (cm) |
| --- | --- | --- |
| longest side | XY | |
| second-longest side | WX | |
| third-longest side | YZ | |
| shortest side | ZW | |

**c.** Find the ratio of the longest side in quadrilateral *WXYZ* to the longest side in quadrilateral *DEFG*. Find the ratios of the remaining three pairs of sides in the same way.

- second longest to second longest

- third longest to third longest

- shortest to shortest

**d.** What do you notice about the ratios in Part c? Are quadrilaterals *WXYZ* and *DEFG* similar? Explain your answer.

**11.** Here is a third quadrilateral.

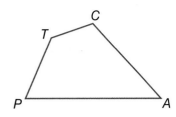

**a.** Complete the table for quadrilateral *CAPT*.

| Description | Side | Length (cm) |
|---|---|---|
| longest side | AP | |
| second-longest side | CA | |
| third-longest side | TP | |
| shortest side | TC | |

**b.** Find the ratio of the longest side in quadrilateral *CAPT* to the longest side in quadrilateral *DEFG*. Find the ratios of the remaining three sides in the same way.

**c.** What do you notice about the ratios you found in Part b? Could quadrilateral *CAPT* be similar to quadrilateral *DEFG*? Explain.

**12.** The corresponding side lengths of quadrilateral *Y* and quadrilateral *Z* and rectangle *A* are 1:2.

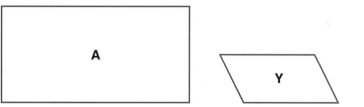

**a.** Are quadrilaterals *Y* and *Z* both similar to rectangle *A*? Explain.

**b.** How is quadrilateral *Y* different from quadrilateral *Z*?

**c.** What information, other than corresponding side lengths having the same ratio, might help you decide whether two polygons are similar?

## Share & Summarize

**1.** Describe *equivalent ratios* in your own words.

**2.** Is the fact that corresponding sides are in the same ratio enough to guarantee that two polygons are similar? Explain your answer.

# Investigation ④ Identifying Similar Polygons

## Materials

- ruler
- protractor

In Investigation 3, you found that when two figures are similar the ratio of their corresponding side lengths is always the same. Another way of saying this is that the lengths of corresponding sides share a *common ratio.*

You also discovered that *angles* are important in deciding whether two figures are similar. However, you might not have found the relationship between corresponding angles. In fact, *for two polygons to be similar, corresponding angles must be congruent.* You will not prove this fact here, but you will use it throughout the rest of this chapter.

To test whether two polygons are similar, you need to check only that corresponding side lengths share a common ratio and that corresponding angles are congruent. Two angles are congruent if they have the same measure.

## ✓ Develop & Understand: A

**Determine whether the figures in each pair are similar. If they are not similar, explain how you know.**

1.

2.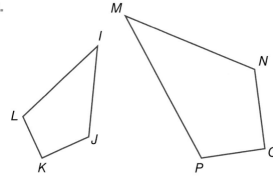

3.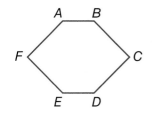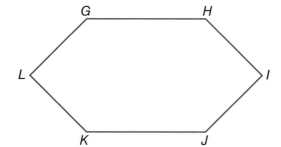

**Determine whether the figures in each pair are similar. If they are not similar, explain how you know.**

4.

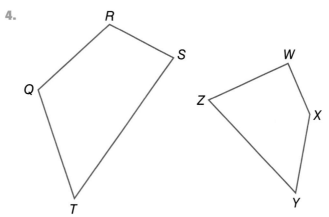

5.

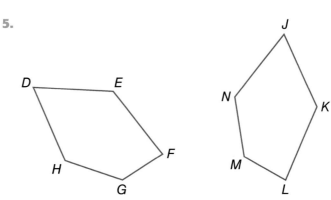

If two figures do not have line segments and angles to measure, how can you decide whether they are similar? One method is to check important corresponding segments and angles, even if they are not drawn. For example, on these two spirals, you might measure the widest and tallest spans of the figures, shown by the dashed segments, and check whether they share a common ratio.

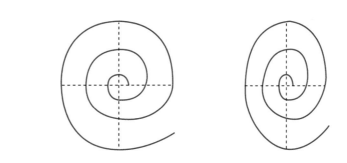

Just checking these two segments will not tell you for *sure* that the figures are similar, but it will give you an idea whether they *could* be. If the ratios are not equivalent, you will know for certain the figures are not similar.

### ✅ *Develop & Understand: B*

**Work with a partner. Try to figure out whether each pair of figures is or could be similar. Explain your decisions.**

6.

7.

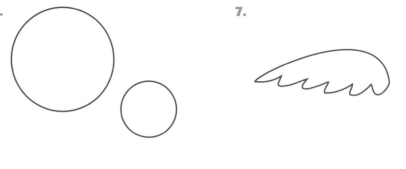

8.

### Share & Summarize

Challenge your partner by drawing two pentagons, one that is similar to this pentagon and another that is not. Exchange drawings with your partner.

Try to figure out which of your partner's pentagons is similar to the original, and explain how you decided. Verify with your partner that you each have correctly identified the similar pentagon.

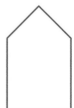

**Practice** **&** **Apply**

1. Look at the triangles below. Make no measurements.

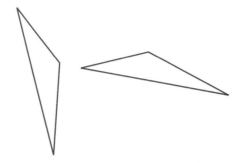

   **a.** Just by looking, guess whether the triangles are congruent.

   **b.** Check your guess by finding a way to determine whether the triangles are congruent. Are they congruent? How do you know?

2. Look at the triangles below. Make no measurements.

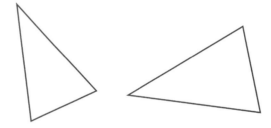

   **a.** Just by looking, guess whether the triangles are congruent.

   **b.** Check your guess by finding a way to determine whether the triangles are congruent. Are they congruent? How do you know?

3. Examine these rectangles.

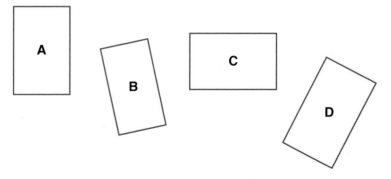

   **a.** Just by looking, guess which rectangle is congruent to rectangle A.

   **b.** Find a way to determine whether your selection is correct. Which rectangle *is* congruent to rectangle A? How do you know?

**4.** Examine the figures below.

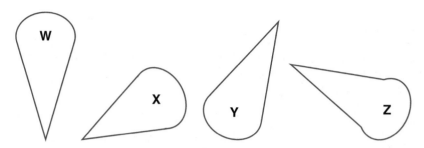

**a.** Just by looking, guess which figure is congruent to figure W.

**b.** Find a way to determine whether your selection is correct. Which figure *is* congruent to figure W? How do you know?

**5.** Rectangle R is 4.5 cm by 15 cm.

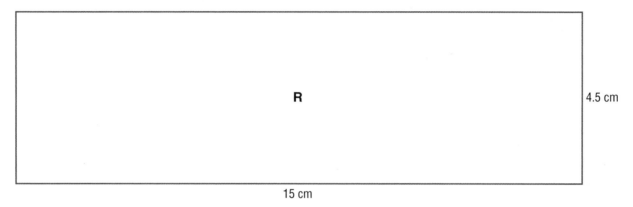

**a.** Draw and label a rectangle with sides one-third as long as those of rectangle R.

**b.** Draw and label a rectangle with sides three centimeter shorter than those of rectangle R.

**c.** Which of your rectangles is similar to rectangle R?

**6.** In Investigation 2, you explored two ways to modify rectangles and triangles. One method produces similar figures. The other does not. In this exercise, you will examine whether either of the methods will produce a similar figure when the original is a square.

**a.** Draw a square that is 6 cm on a side. This is your *original* square.

**b.** Make a new square with sides one-third as long as the sides of your original square.

**c.** Make a new square with sides three centimeters shorter than those of your original square.

**d.** Which of the methods in Parts b and c creates a square that is similar to your original? Explain.

**Decide whether the ratios in each pair are equivalent. Explain how you decided.**

**7.** 1:3 and 9:11

**8.** $\frac{1}{2}$ and $\frac{2}{3}$

**9.** 3:4 and 6:8

**10.** $a$:$b$ and $2a$:$2b$

**Name two ratios that are equivalent to each given ratio.**

**11.** 2:3

**12.** $\frac{6}{10}$

**13.** 50:50

**Real-World Link**
Of the approximately 301 million people in the United States in 2007, an estimated 31 million were born in another country. This is a ratio of 31:301, or about 1 in 10.

Exercises 14 and 15 show a pair of similar figures. Identify all pairs of corresponding sides and angles.

**14.**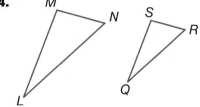

**15.**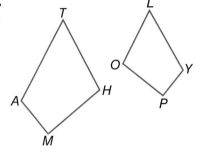

**16.** Examine these triangles.

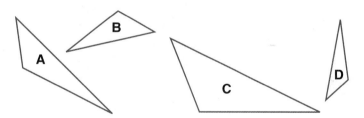

**a.** Just by looking, guess which triangle is similar to triangle A.

**b.** Make some measurements to help determine whether your selection is correct. Which triangle *is* similar to triangle A?

**17.** Examine these quadrilaterals.

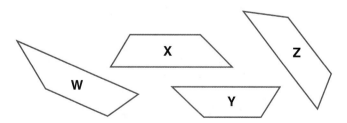

**a.** Just by looking, guess which is similar to quadrilateral Z.

**b.** Make some measurements to help determine whether your selection is correct. Which quadrilateral *is* similar to quadrilateral Z?

**Connect & Extend**   **For each pair of figures, explain what you would measure to test for congruence and what you would look for in your measurements.**

**18.** two circles

**19.** two equilateral triangles

**20.** One way to determine whether two-dimensional figures are congruent is to lay them on top of each other. This test will not work with three-dimensional figures.

   **a.** How could you determine whether two cereal boxes are congruent?

   **b.** How could you determine whether two cylindrical soup cans are congruent?

**21. In Your Own Words** If two pentagons are similar, are they congruent? Explain why or why not. If two pentagons are congruent, are they similar? Explain why or why not.

**22. Challenge** The word *bisect* means to divide into two equal parts. The steps below show how to bisect ∠*JKL* using a compass and a straightedge.

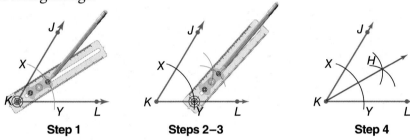

| Step 1 | Steps 2–3 | Step 4 |

**Step 1.**   Place the compass at point *K* and draw an arc that intersects both sides of the angle. Label the intersections *X* and *Y*.

**Steps 2–3.** With the compass at point *X*, draw an arc in the interior of ∠*JKL*. Using this setting, place the compass at point *Y* and draw another arc.

**Step 4.**   Label the intersection of these arcs *H*. Then draw $\overrightarrow{KH}$. $\overrightarrow{KH}$ is the *bisector* of ∠*JKL*.

   **a.** Describe what is true about ∠*JKH* and ∠*HKL*.

   **b.** Draw several angles and then bisect them using the above steps.

**23.** Maps are designed to be similar to the layout of a city's streets. This map shows a section of London.

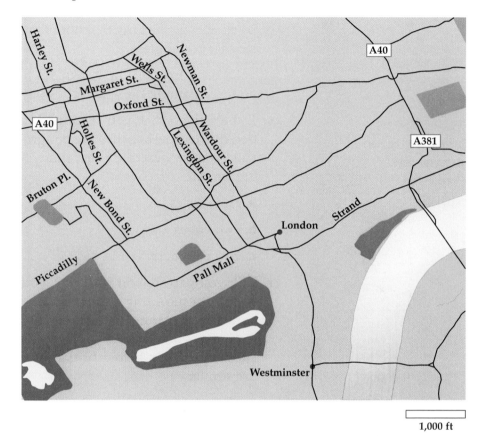

1,000 ft

**a.** The scale of the map is given at the right. How many inches on the map are the same as 1,000 feet in London? Measure to the nearest $\frac{1}{16}$ inch.

**b.** What is the distance on the map along Oxford St. between Holles St. and Newman St.?

**c.** What is the real distance (in feet) along Oxford St. between Holles St. and Newman St.?

**d.** What is the distance on the map along New Bond St. between Bruton Pl. and Piccadilly?

**e.** What is the real distance along New Bond St. between Bruton Pl. and Piccadilly?

**24.** You have two polygons that you know are similar.

**a.** What would you measure to determine whether the two similar polygons are also congruent?

**b.** What would you need to know about your measurements to be sure the polygons are congruent? Explain.

25. **Preview** Delsin proposed a conjecture. "If you are given two similar rectangles with side lengths that share a common ratio of 1 to 2, the ratio of the perimeters are also 1 to 2. That is, the perimeter of the larger rectangle is twice the perimeter of the smaller rectangle."

Is Delsin correct? If he is, explain how you know. If he is not, give a counterexample for which the conjecture is not true.

**Math Link**

Ratios can be written in several ways.

- one to two
- 1 to 2
- 1:2
- $\frac{1}{2}$

26. In Investigation 1, you examined rules for testing two figures to determine whether they are congruent. For each pair of figures in this exercise, describe a test that you could use to tell whether they are similar.

   **a.** two circles

   **b.** two cubes

   **c.** two cylinders

27. In Investigation 4 on page 332, you discovered that *similar polygons* have corresponding sides that share a common ratio and corresponding angles that are congruent. For some special polygons, you can find easier tests for similarity. For each pair of special polygons below, find a shortcut for testing whether they are similar.

   **a.** two rectangles

   **b.** two squares

**Mixed Review**

**Use the fact that $783 \cdot 25 = 19,575$ to find each product without using a calculator.**

**28.** $7.83 \cdot 25$     **29.** $78.3 \cdot 2.5$     **30.** $7,830 \cdot 250$

**Use the fact that $7,848 \div 12 = 654$ to find each quotient without using a calculator.**

**31.** $7,848 \div 0.12$     **32.** $7.848 \div 12$     **33.** $78.48 \div 1.2$

**Measurement** Convert each measurement to meters.

**34.** 32 cm     **35.** 32 mm     **36.** 32,000 cm

37. **Statistics** Monica asked her homeroom classmates which cafeteria lunch was their favorite. She recorded her findings in a table. Make a bar graph to display Monica's results.

| Lunch | Number of Students |
|---|---|
| Pizza | 10 |
| Veggie lasagna | 4 |
| Macaroni and cheese | 5 |
| Hamburger | 7 |
| Tuna casserole | 2 |
| Never buy lunch | 2 |

# Review & Self-Assessment

## Chapter Summary

Comparisons can take many forms, including differences, rates, ratios, and percentages. In this chapter, you learned to compare ratios using equivalent ratios and *unit rates*. You also used percentages as a common scale for comparisons.

In this chapter, you examined two ways in which figures can be considered the same, *congruence* and *similarity*.

You looked at characteristics of congruent and similar figures. For example, congruent figures must be exactly the same shape and size. Similar figures can be different sizes but must be the same shape.

In congruent figures, *corresponding sides* and *corresponding angles* must be congruent. In similar figures, corresponding sides must have lengths that share a common *ratio*. Corresponding angles must be congruent.

You discovered tests that allow you to decide whether two triangles are similar or congruent without finding the measurements of both the angles *and* the sides.

## Strategies and Applications

The questions in this section will help you review and apply the important ideas and strategies developed in this chapter.

### Comparing and scaling ratios and rates

**1.** The Quick Shop grocery store sells four brands of yogurt.

| Brand | Meyer's | Quick Shop | Rockyfarm | Shannon |
|---|---|---|---|---|
| Price | 2 for $1.50 | 3 for $2 | $.75 each | $.80 each |
| Size | 8 oz | 6 oz | 6 oz | 8 oz |

**a.** For each brand, find the ratio of price to ounces.

**b.** For each brand, find the unit price.

**c.** Use the unit rates to list the brands from least expensive to most expensive.

**d.** Explain how you could use the ratios in Part a instead of the unit rates to list the brands by how expensive they are.

2. Every summer, Joe makes his famous peach cooler. To make the drink, he mixes three quarts of tea with $\frac{1}{2}$ quart of peach juice.

   **a.** Find the ratio of tea to peach juice in Joe's peach cooler.

   **b.** Find the ratio of tea to peach cooler, that is, to the final drink.

   **c.** If Joe has only two quarts of tea, how much peach juice should he add?

   **d.** For a party, Joe wants to make seven quarts of peach cooler. How much tea and peach juice does he need?

   **e.** Explain how you found your answer for Part d.

3. The sandhill crane has a wingspan of approximately six feet, and is about four feet tall. If you were to draw a proportional picture of the crane with a wingspan of nine inches, what would be the height of the crane in your drawing?

## Understanding congruence and similarity

4. Consider the difference between similarity and congruence.

   **a.** Can similar figures also be congruent? Do similar figures *have* to be congruent?

   **b.** Can congruent figures also be similar? Do congruent figures *have* to be similar?

5. Explain how you can tell whether two angles are congruent.

6. Suppose you know that two triangles are similar.

   **a.** What do you know about their side lengths?

   **b.** What do you know about their angles?

## Demonstrating Skills

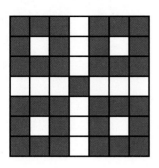

**7.** Examine the tile pattern. Write the ratio of white tiles to purple tiles.

**8.** Suppose you want to tile a large area using this pattern. Make a ratio table of possible numbers of purple and white tiles that you could use.

**Find the value of the variable in each proportion.**

**9.** $\dfrac{12}{5} = \dfrac{x}{15}$

**10.** $\dfrac{2}{y} = \dfrac{4}{7}$

**11.** $\dfrac{92}{36} = \dfrac{23}{w}$

**12.** $\dfrac{a}{4} = \dfrac{15}{60}$

## Testing figures for congruence and similarity

**13.** Consider the tests you know for congruent and similar triangles.

   **a.** Describe a congruence test involving only the sides of triangles.

   **b.** Describe a similarity test involving only the sides of triangles.

   **c.** Describe a similarity test involving only the angles of triangles.

   **d.** Compare the three tests you described.

**14.** How can you tell whether two polygons are similar? How can you tell whether two polygons are congruent?

## Demonstrating Skills

**Tell whether the figures in each pair are congruent, similar, or neither.**

**15.**

**Tell whether the figures in each pair are congruent, similar, or neither.**

**16.**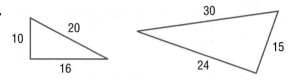

**17.**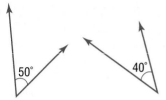

**18.**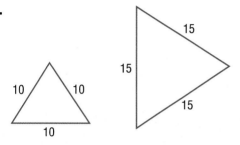

**19.**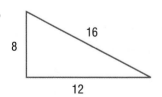

**20.**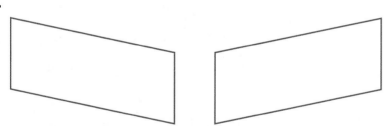

**21.**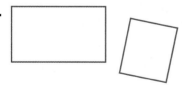

**22.**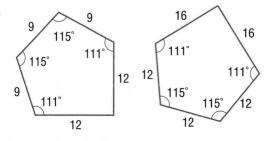

# Test-Taking Practice

### SHORT RESPONSE

**1** The table below shows the amount that different people paid for gasoline. Use proportions to find the unit price that each person paid. Who received the best buy? Round to the nearest cent.

| Name | Amount Paid | Gallons of Gas |
|---|---|---|
| Dave | $35.76 | 12 |
| Amelia | $41.44 | 14 |
| Roman | $27.45 | 9 |

**Show your work.**

**Answer** _____

### MULTIPLE CHOICE

**2** Which of the following is <u>not</u> proportional to $\frac{36}{60}$?

**A** $\frac{27}{45}$

**B** $\frac{3}{5}$

**C** $\frac{12}{25}$

**D** $\frac{9}{15}$

**3** Triangle *ABC* is similar to triangle *DEF*.

What is the missing measure *x*?

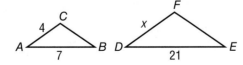

**E** 7

**F** 10

**G** 12

**H** 16

**4** Which of the following is a fraction between $\frac{1}{3}$ and $\frac{1}{4}$?

**A** $\frac{1}{2}$

**B** $\frac{3}{10}$

**C** $\frac{1}{5}$

**D** $\frac{2}{20}$

**5** Which of the following fractions is proportional to $\frac{16}{54}$?

**E** $\frac{32}{54}$

**F** $\frac{4}{9}$

**G** $\frac{18}{56}$

**H** $\frac{8}{27}$

# Percents

## Real-Life Math

**Survey Says!** Results of surveys are often reported as percents. For example, a retail association recently reported that 72%, a little less than $\frac{3}{4}$, of Americans give Mother's Day gifts. Of course, the association did not survey every American. People who conduct surveys often use a method called *sampling*. This method involves surveying a part of a population, called a *sample*, and using the results to make predictions about the entire population.

**Think About It** For survey predictions to be reliable, the sample should include all the different types of people in the population. How would you design a survey to find out the favorite lunch in your school?

### Contents in Brief

**Math Online**
Take the **Chapter Readiness Quiz** at glencoe.com.

# Dear Family,

Look in any magazine or newspaper, and you are likely to see numbers written as percents. Listen to any sporting event, and you will probably hear statistics reported using percents.

## Key Concept—Percents

The word *percent* means *for each 100*. So, a percent like 50% is the same as the fraction $\frac{50}{100}$ (or $\frac{1}{2}$), or the decimal 0.50 (or 0.5). Fractions, decimals, and percents can be used interchangeably to represent parts of a whole quantity.

Often, the word *percent* is used in connection with the *percent of* some quantity. No matter what the quantity, 100% of a quantity always means all of it, and 50% always means half of it.

The amount indicated by a certain percent changes as the size of the quantity changes. For example, 50% of 10 dogs is 5 dogs, but 50% of 100 dogs is 50 dogs.

## Chapter Vocabulary

**percent**　　　　　　　　　　**rational numbers**

## Home Activities

- Calculate the tip when you eat at a restaurant.
- Calculate the price of an item that is on sale for 25% off.
- Compare interest rates on credit cards.
- Discuss the relationship between percents, fractions, and decimals.

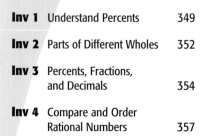

# Use Percents

You see and hear percents used all of the time.

You may not understand exactly what percents are, but you are probably familiar with them from your everyday experiences. For example, you know that 95% is a good test score while 45% is not.

### Think & Discuss

Use what you know about percents to answer these questions.

- In the last game, the Kane High School basketball team made 10% of its shots. Do you think the team played well? Explain.

- Nate is going camping this weekend. The weather report for the area to which he is traveling claims a 90% chance of rain. Do you think Nate should bring his rain gear? Explain.

- The latest Digit Heads CD normally costs $16. But this week, it is on sale for 25% off. Rosita has $15. Do you think she has enough money to buy the CD? Explain.

In this lesson, you will explore what percents are and how they can be used to make comparisons.

# Investigation 1 Understand Percents

## Vocabulary

percent

## Materials

- sheet of 100-grids
- transparent 100-grid
- supermarket receipts

Like a fraction or a decimal, a percent can be used to represent a part of a whole. The word **percent** means "out of 100." For example, 28% means 28 out of 100, or $\frac{28}{100}$, or 0.28.

### Think & Discuss

Each *100-grid* below contains 100 squares. Express the part of each grid that is shaded as a fraction, a decimal, and a percent.

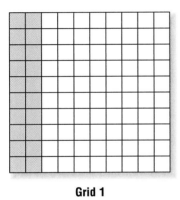

Grid 1

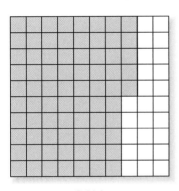

Grid 2

## Develop & Understand: A

**Shade the given percent of a 100-grid. Then express the part of the area that is shaded as a fraction and as a decimal.**

1. 10%
2. 25%
3. 1%
4. 15%
5. 50%
6. 110%

For each square in Exercises 7–12, estimate the percent of the area that is shaded.

### Real-World Link

Of the 10,000 to 15,000 cheetahs alive today, about 10% live in captivity. In the wild, a cheetah lives about 7 years. The average life span in captivity is about 70% longer.

7.

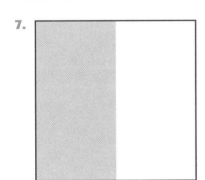

8.

**9.**

**10.**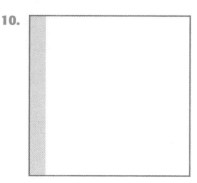

**11.**

**12.**

**13.** Describe the strategies that you used to estimate the percents of the areas that were shaded.

**14.** Now place a 100-grid over each square in Exercises 7–12. Express the exact portion that is shaded as a percent, a fraction, and a decimal.

You have seen that a percent is a way of writing a fraction with a denominator of 100. You can change a fraction to a percent by first finding an equivalent fraction with a denominator of 100. However, in many cases, it is easier to find a decimal first.

**Think & Discuss**

Write each fraction or decimal as a percent. Explain how you found your answers.

$\dfrac{13}{20}$  $\dfrac{3}{5}$  $\dfrac{73}{50}$  0.13  0.9  0.072

Write each fraction as a percent. Explain how you found your answers.

$\dfrac{5}{8}$  $\dfrac{11}{15}$  $\dfrac{87}{150}$

In Exercises 1–14, you used percents to represent part of an area. You can also use a percent to represent part of a collection or a group. Finding a percent is easy when the group is made up of 100 items. In other cases, you can apply what you know about fractions and decimals to find a percent.

## ✓ Develop & Understand: B

15. Of the 25 students in Ms. Sunseri's homeroom, 11 are in band or choir. Express the part of the class in band or choir as a fraction, a decimal, and a percent. Explain how you found your answers.

16. Of the 78 All-Star Baseball Games played between 1933 and 2007, 40 were won by the National League. Express the portion of games won by the National League as a fraction, a decimal, and a percent. Round the decimal to the nearest hundredth and the percent to the nearest whole percent. Explain how you found your answers.

17. Last winter, Reynaldo worked by shoveling driveways. He hoped to earn $200 so he could buy a new bike. At the end of the winter, he had earned $280. Express the portion of the bike's cost that Reynaldo earned as a fraction, a decimal, and a percent. Explain how you found your answers.

## ✓ Develop & Understand: C

In October 1989, *Harper's* magazine printed this fact.

*Percent of supermarket prices that end in 9 or 5: 80%*

More than 15 years have passed since this statistic was printed. In this exercise set, you will analyze some data to see whether it is still true.

18. With your group, devise a plan for testing whether the statistic is true today. Describe your plan.

19. Carry out your plan. Describe what you discovered.

20. Compare your results with those of other groups in your class. Describe how the findings of other groups are similar to or different from your findings.

21. Do you think the statistic is still true? If not, what percent do you think better describes the portion of today's supermarket prices that end in 9 or 5?

*Real-World Link*
The first price scanner was introduced at a supermarket convention in 1974. The first product ever purchased using a checkout scanner was a pack of chewing gum.

## Share & Summarize

Tell what the word *percent* means. Explain how to use a percent to represent part of a whole. Give an example to illustrate your explanation.

# Investigation 2 Parts of Different Wholes

A survey was conducted at Pioneer Middle School. Data were gathered from 160 sixth-grade students. The students were asked to respond to the following question.

*Which is your favorite sport to watch? Choose one.*

*football    soccer    basketball    baseball    ice hockey    other    none*

Below is a table of the results.

| Sport | Number of Votes | Fraction of Total Votes |
|---|---|---|
| Football | 12 | $\frac{12}{160}$ |
| Soccer | 22 | $\frac{22}{160}$ |
| Basketball | 40 | $\frac{40}{160}$ |
| Baseball | 28 | $\frac{28}{160}$ |
| Ice Hockey | 14 | $\frac{14}{160}$ |
| Other | 28 | $\frac{28}{160}$ |
| None | 16 | $\frac{16}{160}$ |

## Think & Discuss

Suppose you want to compare the popularity of a particular sport among sixth graders at Pioneer Middle School with its popularity among students in your class.

- Would comparing the numbers of votes the sport received tell you whether it was more popular at Pioneer or in your class? Explain.

- Would comparing the fraction of votes each sport received tell you whether it was more popular at Pioneer or in your class? What about comparing percents? Explain.

- Do you think comparing numbers of votes, fractions, or percents would be best for comparing popularity? Explain.

## ✅ Develop & Understand: A

1. Calculate the percent of the votes each sport received at Pioneer Middle School. Round your answers to the nearest whole percent.

2. Do the percents add to 100%? If so, why? If not, why not?

3. Write a paragraph for a newspaper article comparing the Pioneer Middle School data with the data from your class.

In Exercises 1–3, you found that percents allow you to compare parts of different groups, even if the groups are of very different sizes. The next two exercise sets will give you more practice with this idea.

## ✅ Develop & Understand: B

Mrs. Torres asked her first and second period classes a question.

*Which continent would you most like to visit?*

The results for the two classes are listed below.

| Continent | Period 1 Votes | Period 2 Votes |
|---|---|---|
| Europe | 2 | 3 |
| Antarctica | 0 | 1 |
| Asia | 3 | 8 |
| Australia | 10 | 5 |
| South America | 1 | 2 |
| Africa | 4 | 11 |

4. Ajay is in the Period 2 class. He said Europe was a more popular choice in his class than in the Period 1 class. Is he correct? Explain.

5. Luisa is in the Period 1 class. She said Australia was twice as popular in her class as in the Period 2 class. Is she correct? Explain.

6. Nolan is in the Period 2 class. He said that in his class, Asia was four times as popular as South America. Is he correct? Explain.

7. Write two true statements similar to those made by Ajay, Luisa, and Nolan comparing the data in the table.

**Real-World Link**

Rugby is the second most-played team sport in the world. Only soccer is more popular.

## ✓ Develop & Understand: C

Marathon City held a walkathon to raise money for charity. Of the 500 students at East Middle School, 200 participated in the walkathon. At West Middle School, 150 of the 325 students participated. The sponsors of the walkathon plan to give an award to the middle school with the greater participation.

8. What argument might the principal at East present to the sponsors to convince them to give the award to her school?

9. What argument might the principal at West make to convince the sponsors to give the award to his school?

10. Which school do you think deserves the award? Defend your choice.

### Share & Summarize

Suppose Ms. Wright's class has fewer students than your class. They would like to compare their results for the sports survey with the results from your class.

- Celia suggests comparing the number of votes each sport received.
- Ian thinks it would be better to compare the fraction of the votes each sport received.
- Oscar says it is best to compare percents.

Which type of comparison do you think is best? Defend your answer.

**Real-World Link**

Hockey pucks are constructed of rubber and measure 3 inches in diameter.

---

## Investigation 3 — Percents, Fractions, and Decimals

Percents, fractions, and decimals can all be used to represent parts of a whole. However, in some situations, one form may be easier or more convenient.

For example, you have seen that it is often easier to compare percents than to compare fractions. To be a good problem-solver, you need to become comfortable changing numbers from one form to another.

# ✅ *Develop & Understand: A*

In Exercises 1–3, write each given fraction or mixed number as a decimal and a percent. Write each given percent as a decimal and a fraction or mixed number in lowest terms.

1. The head of the cafeteria staff took a survey to find out what students wanted for lunch.

   a. He found that 77% of the students wanted pizza every day.

   b. He was surprised to find that $\frac{2}{5}$ of students would like to have a salad bar available, in case they did not want what was served for lunch.

   c. He found that $\frac{11}{20}$ of students favored french fries while 45% preferred mashed potatoes.

2. Mt. Everest, with an estimated elevation of 29,028 feet, is the highest mountain in Asia and in the world.

   a. Mt. Aconcagua is the highest mountain in South America. It is approximately 78% of the height of Mt. Everest.

   b. Mt. McKinley is the highest mountain in North America. It is approximately 70% of the height of Mt. Everest.

   c. Mt. Kilimanjaro is the highest peak in Africa. It is approximately $\frac{2}{3}$ of the height of Mt. Everest.

**Real-World Link**
One of the greatest challenges for mountain climbers is the "seven summits," climbing the highest mountain on each of the seven continents. Mt. Kiliminjaro in Tanzania is the highest mountain in Africa.

3. The Ob-Irtysh, with a length of approximately 3,460 miles, is the longest river in Asia and the fourth longest river in the world.

   a. The Mississippi–Missouri–Red Rock River is the longest river in North America and the third longest in the world. It is about $1\frac{2}{25}$ as long as the Ob-Irtysh.

   b. The Amazon is the longest river in South America and the second longest in the world. It is about 113% as long as the Ob-Irtysh.

You have been thinking about percents as parts of wholes. Like fractions and decimals, you can also think of percents simply as numbers.

In the next exercise set, you will label points on a number line with fractions, decimals, and percents. As you work, you will become familiar with some common percents that are often used as benchmarks.

## ✅ *Develop & Understand: B*

**Copy each number line. Fill in the blanks so that each tick mark is labeled with a percent, a fraction, and a decimal. Write all fractions in lowest terms.**

**4.**

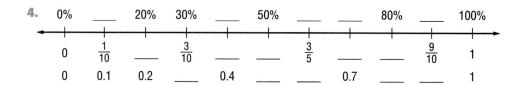

**5.**

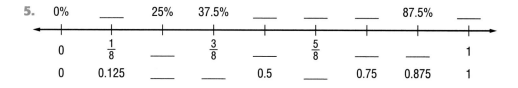

**6.**

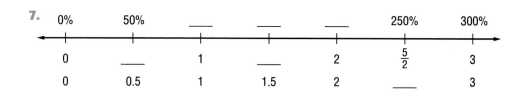

**7.**

**8.** The percent equivalent for $\frac{1}{3}$ is often written as $33\frac{1}{3}\%$. Explain why this makes sense.

**9.** What fraction is equivalent to $66\frac{2}{3}\%$?

### ✅ *Develop & Understand: C*

Sunscreens block harmful ultraviolet (UV) rays produced by the sun. Each sunscreen has a Sun Protection Factor (SPF) that tells you how many minutes you can stay in the sun before you receive one minute of burning UV rays. For example, if you apply sunscreen with SPF 15, you get one minute of UV rays for every 15 minutes you stay in the sun.

To solve Exercises 10–12, you will need to apply what you know about converting between fractions and percents.

10. A sunscreen with SPF 15 blocks $\frac{14}{15}$ of the sun's UV rays. What percent of UV rays does the sunscreen block?

11. Suppose a sunscreen blocks 75% of the sun's UV rays.

    a. What fraction of UV rays does this sunscreen block? Give your answer in lowest terms.

    b. Use your answer from Part a to calculate this sunscreen's SPF. Explain how you found your answer.

12. A label on a sunscreen with SPF 30 claims the sunscreen blocks about 97% of harmful UV rays. Assuming the SPF factor is accurate, is this claim true? Explain.

### *Share & Summarize*

1. How do you change a percent to a fraction?
2. How do you change a percent to a decimal?
3. How do you change a decimal to a percent?
4. How do you change a fraction to a percent?

## Investigation 4 Compare and Order Rational Numbers

### Vocabulary

rational numbers

In Investigation 3, you used number lines to show the fraction, decimal, and percent forms of various benchmark numbers. Number lines can be helpful when you want to use a benchmark to estimate the size of a number, compare the value of two numbers, or order several different numbers by value.

**Math Link**

All of the numbers with which you have worked to this point are called **rational numbers**. Rational numbers can be written as the ratio of two whole numbers. Fractions, percents, decimals, or repeating decimals are all rational numbers.

## Explore

- Which of these numbers are greater than one half? Which are less than one half?

$$0.593 \qquad \frac{11}{20} \qquad 50.1\% \qquad \frac{7}{15} \qquad 48\% \qquad 0.49$$

Draw a number line. Place the numbers in order on the number line.

- What is a convenient way to label the number line so that these numbers can be placed on it? What should be the units on the number line? What value will be the left most value? The right most value?

- What strategies can you use to figure out where to place the numbers?

## Develop & Understand: A

Use the number line from Explore for Exercises 1–3.

1. Which number, $0.593$, $\frac{11}{20}$, $50.1\%$, $\frac{7}{15}$, $48\%$, $0.49$, is closest to one half?

2. Which is furthest from one half?

3. Is 50% a good approximation for each of the numbers?

You can often estimate a fraction, decimal, or percent by comparing it to a *benchmark,* a number whose fraction, decimal, and percent representations you know. Benchmarks can also help you decide which of two numbers is greater. For example, because you know that $\frac{11}{20}$ is more than $\frac{1}{2}$ and 48% is less than $\frac{1}{2}$, you can tell that $\frac{11}{20}$ must be more than 48%.

## Develop & Understand: B

**For each pair of numbers, use a benchmark to decide which number is greater. Show the location of the benchmark and the approximate location of the other numbers on a number line.**

4. Which is greater, 32% or $\frac{4}{9}$?

5. Which is greater, 0.6% or $\frac{1}{98}$?

For Exercises 6–8, use benchmarks and a number line to order sets of three numbers from least to greatest.

0                    1

6. $\frac{3}{8}$, 0.21, 52%

7. 0.3, $\frac{5}{6}$, 96%

8. $\frac{3}{4}$, 0.22, 100%

9. Identify a fraction between $\frac{2}{3}$ and $\frac{5}{6}$.

10. Identify a decimal between 57% and 71%.

11. Identify a percent between $\frac{1}{3}$ and $\frac{2}{5}$.

**Mark the approximate location of each fraction on the number line. For those percents which are not whole numbers, use the closest benchmark to estimate the value of the fraction as a percent.**

0%       50%       100%       150%       200%

12. $\frac{6}{5}$ _____

13. $\frac{3}{8}$ _____

14. $\frac{5}{12}$ _____

15. $\frac{35}{25}$ _____

16. $\frac{1}{6}$ _____

17. Which of the percents in Exercises 12–16 are exact?

*Real-World Link*

Milk is approximately 87% water and 13% solids. As it comes from the cow, the solids portion of milk contains approximately 3.7% fat and 9% solids-not-fat. Milk is then processed into various forms, fat-free, 1%, 2%, and so on, before being sold.

## Explore

Play *Guess My Number* with a partner. You will need two copies of a number line marked in tenths that goes from 0 to 2.

**Step 1.** Each player secretly writes a rational number between 0 and 2. The number can be written as a fraction, a decimal, or a percent.

If it *is* a decimal, it should have no more than 3 digits to the right of the decimal point. If it *is* a percent, it should have no more than 1 digit to the right of the decimal point. If it *is* a fraction, the denominator should be either between 1 and 10, or a multiple of 10.

**Step 2.** One player asks the other a yes/no question about the secret rational number. Three types of questions are allowed.

**a.** Is your number a fraction/decimal/percent?

**b.** Is your number greater/less than _____?

**c.** Is your number exactly _____?

The second player answers the question. Using this information, the first player crosses off any part of the number line where the number cannot lie.

**Step 3.** Now it is the second player's turn to ask the first player a question about that player's secret number. The players continue taking turns until one player guesses the other's number exactly.

## ✓ Develop & Understand: C

Miguel made up a rational number puzzle.

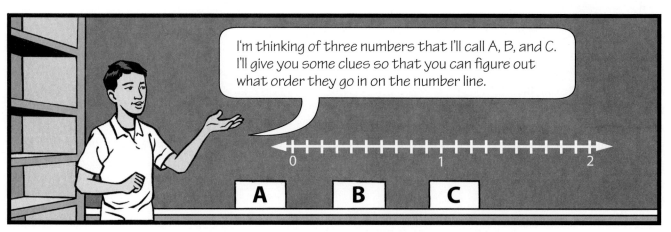

Use the number line to help you solve Miguel's puzzle.

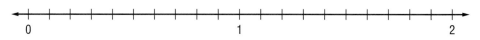

**Clue 1**   A is greater than 0.05 and less than 0.82.

B is greater than $\frac{1}{5}$ and less than $\frac{8}{5}$.

C is greater than 83% and less than 200%.

**18.** What can you tell about the relationship between A, B, and C?

**Clue 2**   A $> \frac{1}{5}$ and A $< \frac{4}{5}$.

B $> 50\%$ and B $< 145\%$

C $> 1.50$ and C $< 1.85$

**19.** Given clue 2, now what can you tell about the order of A, B, and C?

**Clue 3**   A is greater than 0.25 and less than 0.4.

B is greater than $\frac{9}{10}$ and less than $\frac{7}{5}$.

C is greater than 159% and less than 180%.

**20.** Given clue 3, now what can you tell about the order of A, B, and C?

**21.** What values could A be?

**22.** What values could B be?

**23.** What values could C be?

**24.** Could A, B, or C be $\frac{3}{10}$?

**25.** Give one value that each of the secret numbers could have. Mark each of your choices on the number line. Check students' number lines.

## Share & Summarize

**1.** Which quantity is greatest: 45%, $\frac{14}{25}$, or 0.532?

**2.** What strategies can you use to compare fractions, percents, and decimals?

**Practice & Apply**

**Shade the given percent of a 100-grid. Then express the shaded portion as a fraction and a decimal.**

**1.** 37%          **2.** 72%          **3.** 4%          **4.** 125%

**Estimate the percent of each square that is shaded. Describe how you made your estimate.**

**5.**

**6.**

**7.**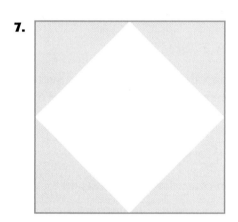

**8.**

**Write each fraction or decimal as a percent. Round to the nearest tenth of a percent.**

**9.** $\frac{4}{5}$          **10.** 0.32          **11.** 0.036          **12.** $\frac{3}{71}$

**13.** 1          **14.** $\frac{19}{20}$          **15.** 2.7          **16.** 0.004

**17.** Seven of the 20 people at Sondra's birthday party are in her math class. Express the portion of the guests who are in her math class as a fraction, a decimal, and a percent. Explain how you found your answers.

**Real-World Link**

In 2005, the United States recycled about 30% of its garbage, more than triple the percent recycled 30 years ago.

**18. Social Studies** Of the 232 million tons of garbage generated in the United States every year, about 87 million tons are paper products. Express the portion of U.S. garbage that is paper as a fraction, a decimal, and a percent. Explain how you found your answers.

**19.** Last season, Brian set a goal of hitting 9 home runs. When the season was over, he had hit 12 home runs. Express the portion of his goal Brian reached as a fraction, a decimal, and a percent. Round to the nearest whole percent. Explain how you found your answers.

**20.** A sports magazine asked 3,600 of its subscribers this question.

*Which of these sports do you think is most dangerous? Choose one.*

☐ *football*      ☐ *ice hockey*      ☐ *skiing*
☐ *sky diving*    ☐ *rock climbing*   ☐ *other*

Ms. Johnson's math students decided to conduct the same survey in their class. The table shows the results of the magazine's survey and Ms. Johnson's class survey.

| Sport | Fraction of Votes in Magazine Survey | Fraction of Votes in Ms. Johnson's Class |
|---|---|---|
| Football | $\frac{524}{3,600}$ | $\frac{3}{25}$ |
| Ice Hockey | $\frac{320}{3,600}$ | $\frac{0}{25}$ |
| Skiing | $\frac{870}{3,600}$ | $\frac{8}{25}$ |
| Skydiving | $\frac{607}{3,600}$ | $\frac{11}{25}$ |
| Rock Climbing | $\frac{959}{3,600}$ | $\frac{2}{25}$ |
| Other | $\frac{320}{3,600}$ | $\frac{1}{25}$ |

**Real-World Link**

The ancient Chinese and Leonardo da Vinci are both credited with conceiving the idea of a parachute. The first recorded parachute jump was made in France in 1797 by Andre Jacques Garnerin who jumped from a hot air balloon.

**a.** Find the percent of votes each sport received in the magazine survey. Round to the nearest percent.

**b.** Find the percent of votes each sport received in Ms. Johnson's class. Round to the nearest percent.

**c.** Write a short newspaper article comparing the results of the magazine survey with the results of the survey conducted in Ms. Johnson's class.

**21.** Mr. Gordon asked his first and fifth period classes to vote for their favorite type of movie. The results are below.

| Movie Type | Period 1 Votes | Period 5 Votes |
|---|---|---|
| Action | 3 | 6 |
| Suspense | 0 | 3 |
| Drama | 8 | 9 |
| Comedy | 11 | 7 |
| Animation | 2 | 2 |
| Other | 1 | 6 |

**a.** Chloe is in Period 5. She said drama was a more popular choice in her class than in the Period 1 class. Is she correct? Explain.

**b.** Meliah is in Period 5. She said comedy was more than twice as popular as suspense in her class. Is she correct? Explain.

**c.** Ricky is in Period 1. He said animation was just as popular in Period 5 as it was in his class. Is he correct? Explain.

**d.** Write two true statements comparing the data in the table.

**22. Nutrition** Harvest Granola has 6.5 grams of fat per 38-gram serving. Crunch Granola has 8 grams of fat per 50-gram serving.

**a.** The makers of Harvest Granola claim that their product has less fat than Crunch Granola. How can they defend their claim?

**b.** The makers of Crunch claim that their granola has less fat than Harvest. How can they defend their claim?

For Exercises 23–26, write each given percent as a decimal and a fraction in lowest terms. Write each given fraction as a decimal and a percent.

**23. Social Studies** About 7% of Americans are under age 5, and about 13% are over age 65.

**24.** About $\frac{2}{3}$ of U.S. households with televisions subscribe to a cable-television service.

**25.** In 1820, almost 72% of U.S. workers were farmers. By 1994, only $\frac{1}{40}$ of U.S. workers were employed in farming.

**26.** The number of students in band is about 115% of the number in orchestra and about $\frac{5}{4}$ of the number in choir.

**Real-World Link**

Cable-television signals are received from antennas and satellites by cable companies. The signals are then sent out to customers along coaxial and fiber-optic cables.

**27.** Copy and complete the table so the numbers in each row are equivalent.

| Fraction | Decimal | Percent |
|----------|---------|---------|
| $\frac{1}{2}$ | 0.5 | 50% |
|  |  | 7.8% |
|  | 5.2 |  |
| $\frac{7}{16}$ |  |  |
|  | 0.37 |  |

Use information from Develop and Understand: C on pages 360 and 361 to solve Exercises 28 and 29.

**28. Science** What percent of the sun's UV rays are blocked with SPF 25 sunscreen?

**29.** A sunscreen blocks 95% of the sun's UV rays. What is the SPF of the sunscreen?

**Use a benchmark to order the numbers from least to greatest in each set.**

**30.** 18%, $\frac{2}{18}$, 0.08      **31.** 31%, 0.355, $\frac{3}{6}$      **32.** 84%, 0.845, $\frac{6}{7}$,

**Use the closest benchmark to estimate the value of the fraction as a percent. Mark the approximate location of each fraction on the number line. Put a star next to any answers that are exact.**

**33.** $\frac{9}{10}$ _____      **34.** $\frac{5}{19}$ _____

**35.** $\frac{7}{6}$ _____      **36.** $\frac{2}{9}$ _____

*Connect & Extend*   **37.** Every January, Framingham Middle School holds its annual Winter Event. An article in the school paper reported that 45% of seventh graders voted that this year's event should be an ice-skating party. The president of the seventh grade class said that $\frac{9}{20}$ of seventh graders voted for ice skating. Could both reports be correct? Explain.

**38.** Imagine that you are in charge of planning a town park. The park will be shaped like a square. The community council has given you these guidelines.

- At least 12% of the park must be a picnic area.
- Between 15% and 30% of the park should be a play area with a sandbox and playground equipment.
- A goldfish pond should occupy no more than 10% of the park.

On a 100-grid, sketch a plan for your park. You may include any features you want as long as the park satisfies the council's guidelines. Label the features of your park, including the picnic area, play area, and goldfish pond. Tell what percent of the park each feature will occupy.

**39. Economics** Suni needs a new winter coat. The Winter Warehouse advertises that everything in the store is 75% of the retail price. Coats Galore advertises that of all its coats are on sale for $\frac{7}{10}$ of the retail price. If the stores carry the same brands at the same prices, where will Suni find better prices? Explain.

**40.** At Valley Middle School, the sixth grade has 160 students, the seventh grade has 320 students, and the eighth grade has 240 students. The student congress is traditionally made up of eight representatives from each grade.

**a.** For each grade, find the percent of students in the congress. Give your answers to the nearest tenth of a percent.

**b.** Tom is in the seventh grade class. He believes that since his class has more students, it should have more representatives. He suggests that each grade be represented by the same *percent* of its students.

Devise a plan for setting up the student congress this way. Tell how many representatives each grade should elect and the percent of each grade that is represented.

**c.** Which plan do you think is more fair, the original plan or the plan you devised in Part b? Defend your answer.

**41. In Your Own Words** Give an example to illustrate how percents can be used to compare data for two groups of different sizes.

**42.** Six students agreed to sell tickets for a raffle. After two weeks, they met to report how sales were going. Some of them had sold more tickets than they had promised. Others had sold fewer. Each of them reported the percent of the target number of tickets they had sold. Some of them expressed the percent as a fraction or a decimal.

Students: Cora, Juanita, Jase, John, Chandra, Stuart
Percent reported: $\frac{5}{16}$, 86%, 0.13, $\frac{12}{9}$, 126%, 0.49

**a.** Use the clues to match the students to the percent they sold.

**Clues:** A. Cora sold more than 100% of the target number.

B. Juanita sold a greater percent of the target number than Jase did.

C. Jase sold a greater percent of the target number than John did.

D. Chandra's percent was the closest to $\frac{1}{8}$.

E. Stuart's percent was greatest.

**b.** On a line labeled in increments of 10%, place each student's reported percent. You may need to approximate.

**c.** Which numbers were you able to place exactly?

**d.** If you had a line labeled in increments of 1%, which numbers would you be able to place exactly?

*Mixed Review*

**Statistics** Find the mean, median, and mode of each set of test scores. Then tell which measure you think best represents the data.

**43.** 85, 99, 73, 64, 99, 80, 69, 72, 70

**44.** 0, 90, 93, 6, 85, 97, 84

**45.** 52, 94, 73, 81, 65, 88

**Write a rule that fits all the input/output pairs in each table.**

**46.**

| Input | 1 | 2 | 4 | 6 | 10 | 11 |
|---|---|---|---|---|---|---|
| Output | 4 | 7 | 13 | 19 | 31 | 34 |

**47.**

| Input | 2 | 4 | 6 | 3 | 7 | 5 |
|---|---|---|---|---|---|---|
| Output | 3 | 11 | 19 | 7 | 23 | 15 |

**Describe the pattern in each sequence. Use the pattern to find the next three terms.**

**48.** 12, 6, 3, $\frac{3}{2}$, $\frac{3}{4}$, $\frac{3}{8}$, ...    **49.** a, c, e, g, i, ...    **50.** 1, 2, 4, 7, 11, 16, ...

# Percent of a Quantity

You often read and hear statements that mention the "percent of" a particular quantity.

**73% of Voters Favor New Park**

**70% of U.S. Schools Have Internet Access**

13% of Americans Are Age 65 or Over

No matter what the quantity, 100% is all of it, and 50% is half of it. The specific amount a given percent represents depends on the quantity. For example, 50% of 10 tree frogs is 5 tree frogs, while 50% of 100 tree frogs is 50 tree frogs.

## Think & Discuss

Use what you know about fractions and percents to answer each of these questions.

- What is 100% of 500?
- What is 50% of 500?
- What is 25% of 500?
- What is 1% of 500?
- What is 100% of 40?
- What is 50% of 40?
- What is 25% of 40?
- What is 1% of 40?

# Investigation ① Model Percents

## Materials
- sheet of 100-grids

Imagine that this 100-grid represents a value of 200 and that this value is divided evenly among the 100 small squares.

| 2 | 2 | 2 | 2 | 2 | 2 | 2 | 2 | 2 | 2 |
|---|---|---|---|---|---|---|---|---|---|
| 2 | 2 | 2 | 2 | 2 | 2 | 2 | 2 | 2 | 2 |
| 2 | 2 | 2 | 2 | 2 | 2 | 2 | 2 | 2 | 2 |
| 2 | 2 | 2 | 2 | 2 | 2 | 2 | 2 | 2 | 2 |
| 2 | 2 | 2 | 2 | 2 | 2 | 2 | 2 | 2 | 2 |
| 2 | 2 | 2 | 2 | 2 | 2 | 2 | 2 | 2 | 2 |
| 2 | 2 | 2 | 2 | 2 | 2 | 2 | 2 | 2 | 2 |
| 2 | 2 | 2 | 2 | 2 | 2 | 2 | 2 | 2 | 2 |
| 2 | 2 | 2 | 2 | 2 | 2 | 2 | 2 | 2 | 2 |
| 2 | 2 | 2 | 2 | 2 | 2 | 2 | 2 | 2 | 2 |

**Value: 200**

### Think & Discuss

What is the value of each small square in the grid? What percent of 200 does each small square represent?

What percent of 200 do 10 small squares represent? What is the value of 10 small squares?

How could you use the grid to find 20% of 200?

## ✓ Develop & Understand: A

1. Refer to the grid above.

   a. What is 15% of 200?

   b. What is 50% of 200?

2. Imagine that a 100-grid represents 300 and that this value is divided evenly among the small squares. So, each small square is worth 3. Use a new grid for each part of this exercise. Label each grid "Value: 300."

   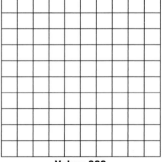

   **Value: 300**

   a. Shade 25% of a grid. What is 25% of 300?

   b. Shade 10% of a grid. What is 10% of 300?

   c. Shade 17% of a grid. What is 17% of 300?

   d. Shade 75% of a grid. What is 75% of 300?

   e. Shade 120% of a grid. What is 120% of 300?

3. Imagine that a 100-grid has value 50 and that this value is divided evenly among the small squares. Use a new grid for each part of this exercise. Label each grid "Value: 50."

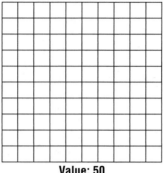

**Value: 50**

a. Shade 1% of a grid. What is 1% of 50?

b. Shade 10% of a grid. What is 10% of 50?

c. Shade 50% of a grid. What is 50% of 50?

4. Imagine that a 100-grid has value 24 and that this value is divided evenly among the small squares. Use a new grid for each part of this exercise. Label each grid "Value: 24."

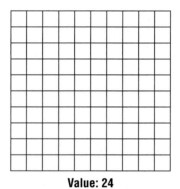

**Value: 24**

a. Shade 1% of a grid. What is 1% of 24?

b. Shade 10% of a grid. What is 10% of 24?

c. Shade 80% of a grid. What is 80% of 24?

5. Look back at your work in Exercises 1–4. Describe a shortcut for finding a percent of a number without using a grid. Give an example to show how it works.

## ✅ Develop & Understand: B

**Find each result without using a grid.**

6. 3% of 400

7. 15% of 600

8. 44% of 50

9. 12% of 250

**Real-World Link**

Silver and white are the most popular car colors in the United States.

· · · · · · · · · · · · · · · · · · · · · · ·

10. In a recent year, about 8,000,000 new cars were sold in the United States. About 16% of these cars were luxury cars.

    **a.** About how many new luxury cars were sold?

    **b.** About 10% of the new luxury cars sold were green. How many green luxury cars were sold?

11. You know that 10% of 360 is 36.

    **a.** Use this fact to find 20% of 360 and 40% of 360. Explain the calculations that you did.

    **b.** Use the fact that 10% of 360 is 36 to find 5% of 360 and 15% of 360. Explain your calculations.

12. The Palmer family went out for dinner. The total cost for the meal was $40. Calculate a 15% tip in your head. Explain what you did.

## Share & Summarize

Describe two methods for finding 15% of 400.

# Investigation 2  Calculate Percents with a Shortcut

## Materials

- slips of paper numbered 1–20
- paper bag

In the last investigation, you used a grid to help find a given percent of a number. You probably discovered a shortcut or two for finding a percent of a number without using a grid. The example shows how Conor and Rosita thought about finding 23% of 800.

### Example

In Exercises 1–11, you will practice Rosita's method. Estimate each answer by using benchmark fractions and percents. This will check whether your answer is reasonable.

For example, 23% of 800 is a little less than 25%, or $\frac{1}{4}$, of 800. So, the result should be a little less than 200. Conor and Rosita's answer of 184 is reasonable.

### ✅ Develop & Understand: A

**Estimate each result using benchmarks. Then find the exact value using Rosita's shortcut.**

1. 30% of 120
2. 45% of 400
3. 9% of 600
4. 72% of 1,100

**Find each result using any method you like.**

5. 3% of 45
6. 15% of 64
7. 44% of 125
8. 2% of 15.4
9. 125% of 40
10. 12.5% of 80

11. In 2006, there were 124,521,886 households in the United States.

   a. About 64% of U.S. households had at least one computer. How many U.S. households had a computer?

   b. About 57% of U.S. households had access to the Internet. How many U.S. households had Internet access?

When stores have sales, they often advertise that items are a certain "percent off." In the rest of this investigation, you will practice calculating the sale price when a percent discount is taken.

## ✅ Develop & Understand: B

A department store is having its annual storewide sale.

12. All athletic shoes are on sale for 20% off the original price. Caroline bought a pair of cross-trainers with an original price of $80. What was the sale price for this pair of shoes? Explain how you found your answer.

13. All fall jackets are on sale for 35% off the original price. Miguel bought a jacket originally priced at $60. How much did he pay? Explain how you found your answer.

14. A CD player, originally priced at $120, is on sale for 25% off. What is the sale price?

Caroline and Miguel have different ways of calculating the sale price for the items that they bought.

As you work on Exercises 15–17, try both of their methods to see which you prefer.

**Real-World Link**

The number of Internet users grew from 3 million in 1994 to over 200 million in 2000, a 6,667% increase!

## ✅ *Develop & Understand: C*

**15.** At K.C. Nickel's back-to-school sale, everything is 40% off the price marked on the tag. Find the sale price of each item.

a.

b.

**Real-World Link**

The average U.S. resident receives 20 greeting cards each year, a third of which are birthday cards.

**16.** At Celebrate! card store, birthday cards are on sale for 25% off. What is the sale price for a card originally marked $1.60?

**17.** Zears and GameHut both usually charge $75 for Quasar-Z, a hand-held electronic game. This week, Quasar-Z is on sale for $60 at GameHut and is 25% off at Zears. At which store is the game less expensive? Explain.

## ✅ *Develop & Understand: D*

The If the Shoe Fits shoe store is having a "Draw a Discount" sale. For the sale, the numbers 1 through 20 are put in a box. Each customer draws two numbers and adds them. The result is the percent the customer will save on his or her purchase.

**18.** To test how the sale works, pretend to be five customers. The price of each customer's purchase is given in the table. Place slips of paper numbered 1 through 20 in a bag. For each customer, draw two slips of paper, record the numbers, and return the two slips to the bag.

| Customer | Original Price | First Number | Second Number | Percent Off | Sale Price |
|---|---|---|---|---|---|
| 1 | $37.00 | | | | |
| 2 | 20.50 | | | | |
| 3 | 12.98 | | | | |
| 4 | 45.79 | | | | |
| 5 | 79.99 | | | | |

**19.** Complete the table by finding the percent off and the sale price for each customer's purchase.

**20.** What is the total amount the five customers paid for their purchases?

**21.** Now figure out the price each customer would have paid if, instead of the "Draw a Discount" sale, the store had offered 20% off all purchases.

| Customer | Original Price | Percent Off | Sale Price |
|----------|----------------|-------------|------------|
| 1 | $37.00 | 20% | |
| 2 | 20.50 | 20% | |
| 3 | 12.98 | 20% | |
| 4 | 45.79 | 20% | |
| 5 | 79.99 | 20% | |

***Real-World Link***

People in Canada and the United States spend approximately $35 billion per year on shoes, averaging five pairs per person.

**22.** What is the total amount the five customers would have paid during a 20% off sale? How does this compare to the total for the "Draw a Discount" sale?

**23.** If you were the store manager, which type of sale would you hold? Explain.

## Share & Summarize

**1.** Describe a method for calculating a given percent of a number. Demonstrate your method by finding 67% of 320.

**2.** Write an exercise about a "percent off" sale. Explain how to solve it.

**Practice & Apply**

1. Imagine that a 100-grid has value 150 and that this value is divided evenly among the small squares.

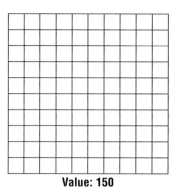

Value: 150

a. What is the value of 25 small squares?

b. What is the value of 1% of the grid?

c. What is the value of $\frac{1}{10}$ of the grid?

d. What is 40% of 150?

e. What is 17% of 150?

f. What is 150% of 150?

**Find each result without using a grid.**

2. 22% of 700                           3. 90% of 120

4. 30% of 15                            5. 65% of 210

6. In a recent year, Americans spent about 313 billion dollars on food prepared away from home. Of this total, almost 48% was spent on fast food.

a. About how much money did Americans spend on fast food? Round your answer to the nearest billion dollars.

b. Of the total dollars spent on fast food, about 64% was spent on takeout food. About how many fast-food dollars were spent on takeout food?

7. **Economics** Mrs. Diaz took her mother out for dinner. The total for the items they ordered was $20.

a. Mentally calculate the 5% sales tax on the order.

b. Mrs. Diaz wants to leave a 20% tip on the food cost plus the sales tax. Mentally calculate how much the tip should be.

**Estimate each result using benchmarks. Then find the exact value.**

**8.** 75% of 80

**9.** 60% of 90

**10.** 65% of 60

**11.** 57% of 80

**Find each result using any method you like.**

**12.** 19% of 43

**13.** 45% of 234

**14.** 67% of 250

**15.** 112% of 70

**16.** 0.55% of 100

**17.** 72% of 3.7

**18. Nutrition** If a 64-ounce carton of fruit juice contains 10% real fruit juice, how many ounces of fruit juice does the carton contain?

**19.** A hockey arena has a seating capacity of 30,275. Of these seats, about 31% are taken by season ticket holders. About how many seats are taken by season ticket holders?

**20.** Of the 2,000 students at Franklin High School, 28% are freshmen. How many Franklin students are freshmen?

**21. Economics** At Sparks electronic store, all CD players are reduced to 66% of the original price.

**a.** What is the sale price for a CD player that originally cost $90?

**b.** How much money would you save on a $90 CD player? What "percent off" is this?

**22.** The Fountain of Youth health products store is going out of business. To help clear out the remaining merchandise, the store is having a "Save Your Age" sale. Each customer saves the percent equal to his or her age on each purchase.

**a.** Andrew bought a case of all-natural soda originally priced at $18. Andrew is 12 years old. How much did he pay for the case of soda?

**b.** Andrew's father bought some soap and shampoo originally priced at $24. He is 36 years old. How much did he pay for the items?

**c.** Andrew's grandmother is 63 years old. She bought a juicer originally priced at $57. How much did she pay?

23. **In Your Own Words** Describe the difference between finding a "percent of" a given price and the "percent off" a given price. Give examples to show how to do each calculation.

**Connect & Extend**

24. You can use 100-grids to model real-world situations involving percents. In Parts a and b, show how you could use a 100-grid to model and solve the exercise.

    **a.** Students in the sixth grade have raised $880 of the $2,000 they need for a class trip. What percent of the $2,000 do they still need to raise?

    **b. Challenge** This year, tickets to the dance cost $15. This is 125% of last year's cost. How much did tickets cost last year?

25. This exercise will give you practice thinking about percents greater than 100%.

    **a.** Could there be a 125% chance of rain tomorrow? Explain.

    **b.** Could a fundraiser bring in 130% of its goal? Explain.

    **c.** Could a drink be 110% fruit juice?

    **d.** Could a candidate get 115% of the votes in an election?

    **e.** Could prices in a store increase by 120%?

26. **Economics** You have learned two ways to compute the sale price when a percent discount is taken. You can use similar methods to solve exercises involving a percent increase.

    **a.** Last year, Irene bought an antique radio for $28. Since then, the radio's value has increased by 25%. By how many dollars did the value increase? What is the new value of the radio?

    **b.** In Part a, you computed the value of the radio in two steps. You calculated the number of dollars the value increased, then you added the increase to the original value.

    How could you calculate the value in one step? Explain why your method works. Show that it gives the same answer that you found in Part a.

**27.** Which is greater, 300% of 8 or 250% of 10?

**28.** Tinley's department store is having a 25% off sale. Victor has a $15 Tinley's gift certificate that he wants to use toward a chess set with an original price of $32. He is unsure which method the sales clerk will use to calculate the amount owed.

- Method 1: Subtract $15 from the price and take 25% off the resulting price.

- Method 2: Take 25% off the original price and then subtract $15.

**a.** Do you think both methods will give the same result? If not, predict which method will give a lower price.

**b.** For each method, calculate the amount Victor would have to pay. Show your work.

**c.** Which method do you think stores actually use? Why?

*Mixed Review*

**Find the next three terms or stages in each sequence.**

**29.** 99, 98, 96, 93, 89, 84, …

**30.** 729, 243, 81, 27, 9, …

**31.** 1, 1, 2, 6, 24, 120, 720, …

**32.** ❀, ✿, ❀, ✿, ✿, ❀, ✿, ✿, ✿, ❀, ✿, ✿, …

**33.** Of the 80 acres on Ms. Cole's farm, 28 acres are devoted to growing corn. What percent of the farm's area is devoted to corn?

**34.** Of the animals on Ms. Cole's farm, 12.5% are goats. If Ms. Cole has 7 goats, what is the total number of animals on her farm?

**35.** At the farmer's market, Ms. Cole sold 60% of the 42 pounds of tomatoes she picked last week. How many pounds of tomatoes did she sell at the market?

**Measurement** Fill in the blanks.

**36.** 429 cm = _____ m

**37.** 862 cm = _____ mm

**38.** 16 m = _____ mm

**39.** 1 mm = _____ m

**40.** 7 mm = _____ cm

**41.** 47,000 mm = _____ m

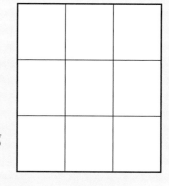

# LESSON 6.3

# Percents and Wholes

Think about these mathematical sentences.

$$44\% \text{ of } 125 = 55 \qquad 15\% \text{ of } 200 = 30$$

Both sentences are in this form.

a percent **of** the whole **=** the part

In the last lesson, you were given a percent and the whole. You found the part. This is like filling in the blank in sentences like these.

$$44\% \text{ of } 125 = \underline{\hspace{1cm}} \qquad 15\% \text{ of } 200 = \underline{\hspace{1cm}}$$

In this lesson, you will fill in the blanks in sentences like these.

$$\underline{\hspace{1cm}}\% \text{ of } 125 = 55 \qquad 15\% \text{ of } \underline{\hspace{1cm}} = 30$$

As you work, you will find it helpful to use benchmark fractions, decimals, and percents. The *Percent Bingo* game below will help refresh your memory about equivalent fractions, decimals, and percents.

### Explore

Choose nine of the numbers listed below. Write one in each square of a grid like the one at the right.

| $\frac{1}{8}$ | $\frac{1}{4}$ | $\frac{1}{2}$ | $\frac{1}{3}$ | $\frac{1}{5}$ | $\frac{1}{6}$ |
|---|---|---|---|---|---|
| $\frac{2}{3}$ | $\frac{2}{5}$ | $\frac{3}{4}$ | $1$ | $\frac{1}{10}$ | $\frac{5}{8}$ |

| 0.125 | 0.25 | 0.5 | $0.\overline{3}$ | 0.2 | $0.1\overline{6}$ |
|---|---|---|---|---|---|
| $0.\overline{6}$ | 0.4 | 0.75 | 0.1 | 0.625 | |

When your teacher calls out a percent, look for an equivalent decimal or fraction on your grid. If you find one, circle it. If your grid contains both a fraction and a decimal equal to the given percent, circle both.

If you circle three numbers in a row, call out "Bingo!" Rows can be horizontal, vertical, or diagonal. The first student who gets bingo wins.

# Investigation 1 Find the Percent

## Materials

- page from a telephone book

The question "What percent of 75 is 20?" gives you the part, 20, and the whole, 75, and asks you to find the percent. Answering this question is like filling in the blank in the sentence below.

$$\_\_\_\_\_\% \text{ of } 75 = 20$$

You already solved several exercises like this in Lesson 6.1.

### Example

In a sports survey, 14 out of 160 students said they like watching ice hockey best. What percent of the students is this? In other words, what percent of 160 is 14?

In this case, the part is 14 and the whole is 160. To find the percent, write "14 out of 160" as a fraction. Then change the fraction to a percent.

$$14 \text{ out of } 160 = \frac{14}{160} = 0.0875 = 8.75\%$$

Estimating with benchmarks can help you make sure your answer is reasonable. The fraction $\frac{16}{160}$ is equivalent to $\frac{1}{10}$. Since $\frac{14}{160}$ is a little less than $\frac{16}{160}$, or $\frac{1}{10}$, the percent should be a little less than 10%. Therefore, 8.75% is reasonable.

# ✅ *Develop & Understand: A*

1. Your teacher will give you a page from a telephone book. Quickly scan the last four digits of the phone numbers on your page. Which digit do you think appears most often?

2. Starting with a phone number near the top of the page, analyze 30 phone numbers in a row. Use a table like the one below to keep a tally of the last four digits. For example, if one of the numbers you choose ends 2329, make two tally marks next to the 2, one next to the 3, and one next to the 9.

| Digit | Tally | Number of Tallies | Fraction of Tallies | Estimated Percent | Exact Percent |
|-------|-------|-------------------|---------------------|-------------------|---------------|
| 0 | | | | | |
| 1 | | | | | |
| 2 | | | | | |
| 3 | | | | | |
| 4 | | | | | |
| 5 | | | | | |
| 6 | | | | | |
| 7 | | | | | |
| 8 | | | | | |
| 9 | | | | | |

3. Count the number of tally marks for each digit. Record the results in your table. You should have a total of 120 tally marks.

4. Find the fraction of the 120 tally marks each digit received. Record the results.

5. Use benchmarks to estimate the percent of the 120 tally marks each digit received. Record your results.

6. Calculate the exact percent of the 120 tally marks each digit received. Record your results.

7. Choose one of the digits. Explain how you found the estimated percent and the exact percent for that digit.

8. Which digit occurred most often? For what percent of the 120 digits does this digit account?

9. Which digit occurred least often? For what percent of the 120 digits does this digit account?

Exercises 10–13 will give you more practice finding percents when you know the part and the whole.

### ☑ Develop & Understand: B

For Exercises 10–13, first write a fraction and then calculate the percent. Round your answers to the nearest whole percent.

10. What percent of numerals from 1 through 40 are formed at least partially with curved lines? Assume the digits are written like the digits below.

0   1   2   3   4   5   6   7   8   9

11. Consider the whole numbers from 1 through 80.

a. What percent of these numbers have two digits?

b. What percent are multiples of 9?

c. What percent are even and prime?

d. What percent are greater than 9?

e. What percent are factors of 36?

12. Now think about the whole numbers from 1 through 26.

a. What percent contain only even digits?

b. What percent contain only odd digits?

c. What percent contain one even digit and one odd digit?

13. Consider the whole numbers from 1 through 52.

a. What percent are common multiples of 2 and 3?

b. What percent are common factors of 24 and 42?

*Real-World Link*

Early phone operators knew the names of customers in their area, and users just gave them the name of the person they wanted to call.

## Math Link

The area of a rectangle is its length times its width.

## ☑️ *Develop & Understand: C*

**Fill in the blanks. Round each answer to the nearest whole percent.**

14. _____ % of 65 = 45

15. _____ % of 23 = 17

16. _____ % of 9 = 4.5

17. _____ % of 93 = 75

18. _____ % of 45 = 60

19. _____ % of 250 = 500

20. What percent of the perimeter of the large rectangle is the perimeter of the small rectangle? Explain how you found your answer.

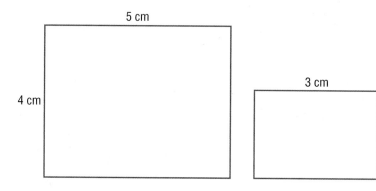

5 cm

4 cm

3 cm

2.4 cm

## *Share & Summarize*

Describe a method for finding a percent when you are given the part and the whole. Demonstrate your method by finding what percent 25 is of 60.

# Investigation ② Find the Whole

The question "50% of what number is 8?" gives you the percent, 50%, and the part, 8, and asks you to find the whole. Answering this question is like filling in the blank in the sentence below.

$$50\% \text{ of } \underline{\hspace{1cm}} = 8$$

To find the whole in this exercise, try using what you know about fraction equivalents for percents.

## Think & Discuss

Find each missing whole. Explain how you found your answer.

50% of _____ = 8          20% of _____ = 10

**Math Link**

$33\frac{1}{3}\% = \frac{1}{3}$

## ✅ Develop & Understand: A

In Exercises 1–6, find the missing whole.

1. 50% of _____ = 23
2. 100% of _____ = 65
3. 25% of _____ = 3
4. 10% of _____ = 7
5. 25% of _____ = 20
6. 150% of _____ = 9
7. $33\frac{1}{3}\%$ of what number is 30?
8. 75% of what number is 30?

9. It's a Party! caterers rents equipment for parties. The cost to rent an item for one week is 25% of the price the caterers paid to purchase the item. The rental costs are listed in the table. Find the purchase price of each item. Explain how you found your answers.

| Item | Rental Cost |
| --- | --- |
| Punchbowl | $10.00 |
| Cup and saucer | 0.75 |
| Table | 12.00 |
| Folding chair | 4.00 |
| Sound system | 250.00 |

Familiar percents whose fraction equivalents are easy to work with appeared in Exercises 1–8. For more complicated exercises, you can use what you know about the relationship between multiplication and division.

## Example

Find the missing whole.

$$55\% \text{ of } \underline{\hspace{1cm}} = 20$$

First, make an estimate. Since 55% is close to 50%, or $\frac{1}{2}$, the missing number must be close to 40.

To find the exact answer, rewrite the sentence as a multiplication equation.

$$0.55 \cdot \underline{\hspace{1cm}} = 20$$

You know that this multiplication equation is equivalent to the division equation below.

$$20 \div 0.55 = \underline{\hspace{1cm}}$$

Now you can just divide to find the answer.

$$20 \div 0.55 = 36.\overline{36}$$

So, the missing whole is $36.\overline{36}$, or about 36.

## ✓ Develop & Understand: B

In Exercises 10–12, first estimate the missing whole. Then calculate the exact value.

10. $40\% \text{ of } \underline{\hspace{1cm}} = 70$

11. $12\% \text{ of } \underline{\hspace{1cm}} = 3$

12. $124\% \text{ of } \underline{\hspace{1cm}} = 93$

13. 90% of what number is 99.9?

14. 4% of what number is 30?

## ✅ *Develop & Understand: C*

Tammy's Food Emporium is having a sale.

**15.** Spices are on sale for 35% of the original price.

    **a.** A jar of cinnamon costs $3 on sale. What was the original price?

    **b.** Salt is on sale for $0.20 per pound. What was the original price?

    **c.** Pepper costs $4 per package. What was the original price?

**16.** Fruit is on sale for 45% of the original price.

    **a.** Apples are on sale for $4.00 per bag. What was the original price?

    **b.** Dried apricots cost $2.50 per bag. What was the original price?

**17.** Snacks are to be marked down to 75% of the original price. Tammy might have marked some incorrectly. For each item, tell whether the sale price is correct. If not, give the correct price.

    **a.**

    **b.**

    **c.**

    **d.**

### Share & Summarize

Describe a method for finding the whole when you are given the part and the percent. Demonstrate your method by answering this question, "54 is 6% of what number?"

# Inquiry

## Investigation 3 Play *Percent Ball*

### Materials

- trash can or bucket
- 6 sheets of paper
- score sheets

In this investigation, you will play a game. You will explore how your score for each turn affects your cumulative score for the game.

Play this game with a partner. Crumple each sheet of paper into a ball. On your turn, take the following steps.

- Try to toss six paper balls into a trash can from about 5 feet away.
- Record your *turn score* as a number out of 6 and as a percent. Round to the nearest whole percent.
- Record your *cumulative score* for the game so far. For example, if you made 4 shots on the first turn and 2 on the second, you have made 6 shots out of 12. So, record $\frac{6}{12}$ and 50%.

**Player 1's Score Sheet**

| | Turn Score | | Cumulative Score | |
|---|---|---|---|---|
| **Turn 1** | $\frac{4}{6}$ | 67% | $\frac{4}{6}$ | 67% |
| **Turn 2** | $\frac{2}{6}$ | 33% | $\frac{6}{12}$ | 50% |

After 10 turns, the highest cumulative score wins.

### Try It Out

1. Look at the turn score percents for you and your partner. What pattern do you see? (Hint: Are there certain scores that occur over and over?) Explain why the pattern makes sense.

2. Look at the cumulative score percents. Do you see the same pattern? Explain why or why not.

3. Look at the turns for which your cumulative score increased. Can you see a pattern that could help you predict whether your score for a particular turn will make your cumulative score rise? If so, describe it.

### Try It Again

Play again. After each turn, predict whether your cumulative score will go up, down, or stay the same. Play until your prediction method works every time.

4. How can you predict, based on your turn score, whether your cumulative score will go up, go down, or stay the same?

## Take It Further

Play a new version of the game in which the object is to make your cumulative score alternate between going up and going down. As you play the game, record a D next to a cumulative score if it went down from the previous turn and a U if it went up. Here is how to determine the winner.

- A player earns one point each time his or her score changes from D to U or from U to D.
- Subtract one point for every turn score of 0 out of 6.
- The player with the greatest point total wins.

For this scoresheet, the cumulative score changes between U and D three times, but the player got 0 out of 6 once. So, the final score is $3 - 1 = 2$.

| | Turn Score | | Cumulative Score | |
|---|---|---|---|---|
| **Turn 1** | $\frac{2}{6}$ | 33% | $\frac{2}{6}$ | 33% |
| **Turn 2** | $\frac{4}{6}$ | 67% | $\frac{6}{12}$ | 50%  U |
| **Turn 3** | $\frac{0}{6}$ | 0% | $\frac{6}{18}$ | 33%  D |
| **Turn 4** | $\frac{5}{6}$ | 83% | $\frac{11}{24}$ | 46%  U |
| **Turn 5** | $\frac{2}{6}$ | 33% | $\frac{13}{30}$ | 43%  D |
| **Turn 6** | $\frac{1}{6}$ | 17% | $\frac{14}{36}$ | 39%  D |

## What Did You Learn?

**5.** Suppose you are playing the game with the original rules. In your first six turns, you have made 27 shots. What is the fewest number of shots you must make on your seventh turn to make your cumulative score rise?

**6.** In which situation below does making 1 out of 6 cause a greater change in your cumulative score? Why?

- After two turns, you have a cumulative score of 50%. On your third turn, you make 1 out of 6 shots.
- After nine turns, you have a cumulative score of 50%. On your tenth turn, you make 1 out of 6 shots.

# On Your Own Exercises

## Lesson 6.3

**Practice & Apply**

1. Consider the whole numbers from 1 to 64. In Parts a–c, round your answer to the nearest tenth of a percent.

   **a.** What percent of the numbers are greater than 56?

   **b.** What percent have an even tens digit?

   **c.** What percent are multiples of 6?

2. Ivan has three dogs, two cats, three parakeets, and eight fish. In Parts a–c, round your answer to the nearest tenth of a percent.

   **a.** What percent of his pets have four legs? What percent have no legs?

   **b.** What percent of Ivan's pets have beaks? What percent have fins?

   **c.** What percent of the total number of pet legs belong to birds? What percent belong to cats?

**Fill in the blanks. Round each answer to the nearest whole percent.**

3. ____% of 15 = 3

4. ____% of 120 = 17

5. ____% of 41 = 20

6. ____% of 132 = 80

7. ____% of 45 = 60

8. ____% of 16 = 2.4

9. Of the 27 girls on the varsity soccer team, 18 are seniors. What percent of the players are seniors?

10. Last year, 235 seniors out of 346 in the graduating class went on to college. What percent went to college?

**Find each missing whole.**

11. 50% of ____ = 342    12. 100% of ____ = 9    13. 25% of ____ = 5

**Estimate each missing whole. Then find the value to the nearest hundredth.**

14. 12% of ___ = 7    15. 28% of ___ = 20    16. 98% of ___ = 85

17. **Economics** Jodi and her friends went out to lunch. They left a $5 tip, which was 20% of the bill. How much was the bill?

18. **Life Science** Scientists have named about 920,000 insect species. This is about 85% of all known animal species. How many known animal species are there?

19. About 2,600 bird species live in the rain forest. This is about $33\frac{1}{3}\%$ of the world's bird species. About how many bird species are there?

20. **In Your Own Words** Write an exercise that requires finding the whole when you know the part and the percent of the whole that part is. Then explain how to solve it.

 **21.** An aquarium with an original price of $95 is on sale for $80.

    **a.** What *percent of* the original price is the sale price? Explain how you found your answer.

    **b.** What *percent off* the original price is the sale price? Explain how you found your answer.

22. **Preview** The students in Mr. Turner's class were asked how many siblings they had. The results are shown in this plot. An X over a number indicates one student with that number of siblings. For example, the three X's over the 4 indicate that three students have four siblings.

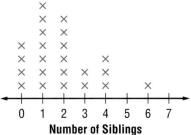

    **a.** How many students are in the class? Explain how you know.

    **b.** What percent of the students are only children?

    **c.** What percent of the students have more than five siblings?

    **d.** What percent of the students have fewer than three siblings?

    **e.** What percent of the students are from a three-child family?

**Real-World Link**

American Lance Armstrong won the 2,287-mile Tour de France between the years 1999–2005. Just three years before his first victory, Armstrong was diagnosed with cancer and given a 50% chance of survival.

In Exercises 23–25, solve if possible. If it is not possible, tell what additional information you would need to solve it.

**23.** The Tour de France bicycle race has been won by a French cyclist 36 times. What percent of the races have been won by a French cyclist?

**24.** Of the new trucks and vans sold in the United States in a recent year, 22.5% were white and 11.5% were black. How many of the trucks and vans were black or white?

**25. Nutrition** A serving of asparagus contains 2 grams of protein. What percent of the asparagus' weight is protein?

**26.** Renee is a lifeguard at the local pool. She thinks there may be more than 40 swimmers, the maximum number allowed. She starts to count, but Emilio says, "Just count the swimmers with red bathing suits. I've already figured out that 20% of the people in the pool have red bathing suits." Renee counts 8 people wearing red bathing suits. How many people are in the pool?

**27. Economics** The label on a bottle of shampoo states, "20% More Than Our Regular Size!" The bottle contains 18 ounces of shampoo. How many ounces are in a regular-sized bottle? Explain how you found your answer.

**28. Geometry** Isabella's family has a plot in the community garden that measures 9 feet by 12 feet.

    **a.** A section measuring 6 feet by 2 feet is devoted to tomatoes. What percent of the garden's area is planted in tomatoes?

    **b.** The green bean section has 75% of the area of the tomato section. What is the area of the green bean section?

    **c.** The green bean section is 90% of the area of the squash section. What is the area of the squash section?

**Mixed Review**

**Number Sense** Fill in each ○ with >, <, or =.

**29.** $33\frac{1}{3}$ ○ $\frac{1}{3}$

**30.** $\frac{7}{8}$ ○ 85%

**31.** 0.398 ○ $\frac{2}{5}$

**32.** −5 ○ −1

**33.** $0.\overline{5}$ ○ $\frac{5}{9}$

**34.** $\frac{31}{40}$ ○ 75%

**35.** $\frac{347}{899}$ ○ $\frac{347}{900}$

**36.** $\frac{6}{7}$ ○ $\frac{7}{8}$

**37.** 80% ○ $\frac{45}{60}$

**38.** 0.01 ○ 0.1%

# Review & Self-Assessment

## Chapter Summary

In this chapter, you learned that, like a fraction or a decimal, a percent can be used to represent a part of a whole. You used the fact that percent means "out of 100" to convert fractions and decimals to percents and to convert percents to fractions and decimals.

You saw that percents are useful for comparing parts of different groups, even when the groups are of very different sizes. Then you learned how to find a given percent of a quantity and to compute a sale price when a percent discount is taken. Finally, you solved situations that involved finding the percent when you know the part and the whole and finding the whole when you know the part and the percent that part represents.

## Strategies and Applications

The questions in this section will help you review and apply the important ideas and strategies developed in this chapter.

### Converting among fractions, decimals, and percents

**1.** Explain how to convert a decimal to a percent and how to convert a fraction to a percent. Give examples to illustrate your methods.

**2.** Explain how to convert a percent to a fraction and to a decimal. Give examples to illustrate your methods.

### Using a percent to represent part of a whole

**3.** Estimate the percent of the square that is shaded. Explain how you made your estimate.

**4.** Of the 20 students in Dulce's ballet class, 17 took part in the spring recital. What percent of the students participated in the recital? Explain how you found your answer.

**5.** The school band held a carnival to raise $750 for new uniforms. When the carnival was over, the band had raised $825. What percent of its goal was reached? Explain how you found your answer.

### Using percents to compare groups of different sizes

**6.** This summer, the 96 fifth graders and 72 sixth graders at Camp Maple Leaf were asked this question.

*Which is your favorite camp activity? Choose one.*

☐ *swimming*      ☐ *hiking*        ☐ *arts and crafts*
☐ *volleyball*    ☐ *canoeing*      ☐ *other*

The results are given in the table.

| Activity | Fifth-Grade Votes | Sixth-Grade Votes |
|---|---|---|
| Swimming | 34 | 24 |
| Hiking | 5 | 10 |
| Arts and crafts | 18 | 6 |
| Volleyball | 14 | 12 |
| Canoeing | 21 | 16 |
| Other | 4 | 4 |

**a.** Coty said arts and crafts is three times as popular among fifth graders as among sixth graders. Is he correct? Explain.

**b.** Maya said volleyball is more popular among sixth graders than among fifth graders. Is she correct? Explain.

**c.** Dante said that, among sixth graders, swimming is four times as popular as arts and crafts. Is he correct? Explain.

**d.** Kylie said the choice "other" was equally popular among the two grades. Is she correct? Explain.

### Calculating a percent of a whole

**7.** Describe a method for computing a given percent of a quantity. Demonstrate your method by finding 72% of 450 and 125% of 18.

**8.** A CD originally priced at $15.99 is on sale for 20% off. Calculate the sale price. Explain the method that you used.

### Finding the whole from the part and the percent

**9.** Last year, Eliza's hourly wage was 75% of her hourly wage this year. She made $12 per hour last year. How much does she make this year? Explain how you found your answer.

**10.** Write an equation that requires finding the whole when you know the percent and the part. Explain how to solve your equation.

## Demonstrating Skills

**Convert each fraction or decimal to a percent.**

**11.** 0.56

**12.** $\frac{7}{8}$

**13.** 0.3

**14.** $\frac{90}{125}$

**15.** 7.25

**16.** $\frac{67}{20}$

**17.** $\frac{2}{3}$

**18.** 0.008

Convert each percent to a decimal and a fraction or mixed number in lowest terms.

**19.** $33\frac{1}{3}\%$          **20.** 99%          **21.** 25%

**22.** 7.6%          **23.** 0.4%          **24.** 325%

**Fill in the blanks.**

**25.** _____% of 25 = 10          **26.** 34% of 650 = _____

**27.** 10% of _____ = 5.3          **28.** _____% of 54 = 81

**29.** What is 83% of 320?          **30.** What percent of 65 is 26?

**31.** A tricycle originally priced at $76 is on sale for 20% off. What is the sale price?

**For each pair of numbers, use a benchmark to decide which number is larger.**

**32.** 33% or $\frac{1}{3}$          **33.** 45% or $\frac{4}{9}$          **34.** 0.5% or $\frac{1}{4}\%$

**35.** 25% or $\frac{1}{4}$          **36.** 0.21 or $\frac{3}{10}$          **37.** .01 or $\frac{1}{9}$

**Use a number line to compare the sets of three numbers from least to greatest.**

**38.** $\frac{2}{5}$, 0.5, 45%          **39.** $\frac{1}{9}$, $\frac{1}{3}$, 22%

**40.** Identify a decimal between 23% and 34%.

**41.** Identify a percent between $\frac{1}{2}$ and $\frac{8}{11}$.

## Test-Taking Practice

**SHORT RESPONSE**

**1** Derek is buying a shirt that originally cost $25. If the shirt is on sale at 30% off, how much will Derek save?

*Show your work.*

*Answer* _____

**MULTIPLE CHOICE**

**2** Amber made $\frac{6}{8}$ of her free-throws in the last basketball game. What percent of free-throws did Amber make?

   **A** 25%

   **B** 68%

   **C** 75%

   **D** 86%

**3** There are 16 sixth graders who want to stay inside for recess. This is 32% of the sixth graders. How many sixth graders are in this school?

   **F** 20

   **G** 45

   **H** 48

   **J** 50

# Area, Volume, and Capacity

## Real-Life Math

**Olympic Proportions** The Olympic Games feature sports that are played on rectangular courts, fields, mats, and pools. The table below lists some dimensions, perimeter, and area.

| Sport | Length (meters) | Width (meters) | Perimeter (meters) | Area (square meters) |
|---|---|---|---|---|
| Football (soccer) field | 100 | 70 | 340 | 7,000 |
| Field hockey field | 91.4 | 55 | 292.8 | 5,027 |
| Swimming pool | 50 | 25 | 150 | 1,250 |
| Handball court | 40 | 20 | 120 | 800 |
| Water polo pool | 30 | 20 | 100 | 600 |
| Volleyball court | 18 | 9 | 54 | 162 |
| Badminton court (singles) | 13.4 | 5.18 | 37.16 | 69.41 |
| Gymnastics floor | 12 | 12 | 48 | 144 |

**Think About It** How do the dimensions of your classroom compare to the dimensions of the gymnastics floor?

**Math Online**
Take the **Chapter Readiness Quiz** at glencoe.com.

# Dear Family,

The next chapter is about calculating area, surface area, volume, and capacity.

### Key Concept—Area

Area is measured in square units. For example, a square centimeter is the area inside a square with sides one centimeter long.

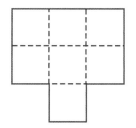

1 cm
1 cm
1 square centimeter

The area of a shape is the number of square units that fit inside it.

Area = 7 square centimeters

Finding the area of a shape by counting squares can be tedious. Fortunately, there are shortcuts for some shapes. Some formulas that students will learn in this chapter are listed to the right.

| Area of a Rectangle | $A = l \cdot w$ |
|---|---|
| Area of a Parallelogram | $A = b \cdot h$ |
| Area of a Triangle | $A = \frac{1}{2} \cdot b \cdot h$ |
| Area of a Circle | $A = \pi \cdot r^2$ |

### Chapter Vocabulary

| | | |
|---|---|---|
| arc | circle sector | rectangular prism |
| area | parallelogram | surface area |
| capacity | perfect square | trapezoid |
| central angle | prism | volume |

### Home Activities

• Ask your student for examples of area in his or her daily life.
• Help your student figure out if it is more cost effective to order a circular pizza with a 10-inch diameter for $8, or to order a rectangular pizza that is 16 inches by 10 inches for $14.

# Squares

You know that the perimeter of a two-dimensional shape is the distance around the shape. The **area** of a two-dimensional shape is the amount of space inside the shape.

## Vocabulary

area

## Materials

- copies of the two shapes
- scissors

### *Real-World Link*

Shape 1 is made with tangram pieces. A *tangram* is a Chinese puzzle consisting of a square cut into five triangles, a square, and a parallelogram that can be put together to form various shapes.

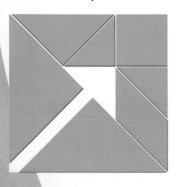

**Explore** . . . . . . . . . . . . . . . . . . . . . . . . . . . . . . . . . . . . . . . .

Consider these shapes.

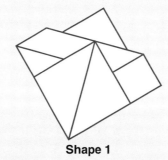   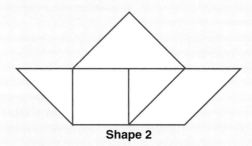

**Shape 1**                                    **Shape 2**

Which shape do you think is larger? That is, which shape do you think has the greater area?

Cut out shape 1 along the lines. Rearrange the pieces to make a square. Do the same for shape 2.

Of the two squares you made, which has the greater area? How can you tell?

Do the original shapes have the same areas as the squares? Why or why not?

When determining which shape has the greater area, is it easier to compare the original shapes or the squares? Why?

Squares are the basic unit used for measuring areas. In this lesson, you will look closely at areas of squares and at a special operation associated with the areas of squares.

# Investigation  Count Square Units

## Materials

- 1-inch tiles
- 1-inch dot paper
- page from a newspaper
- metric ruler

Area is measured in *square units*, such as square inches and square centimeters. A *square inch* is the area inside a square with sides 1 inch long. A *square centimeter* is the area inside a square with sides 1 centimeter long.

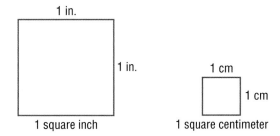

The area of a shape is the number of square units that fit inside it.

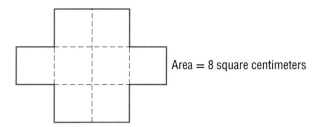

## ✓ Develop & Understand: A

1. Use your tiles to create two rectangles with perimeters of 12 inches but different areas. Sketch your rectangles. Label them with their areas.

2. Now create two rectangles with areas of 12 square inches but different perimeters. Sketch your rectangles. Label them with their perimeters.

3. Now use your tiles to create this shape.

   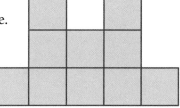

   a. Find the perimeter and area of the shape. Do not forget to give the units.

   b. Move one tile to create a shape with a smaller perimeter. Sketch the new shape. Give its perimeter. How does the new shape's area compare to the original area?

   c. Reconstruct the original shape. Move one tile to create a shape with a greater perimeter. Sketch the new shape. Give its perimeter. How does the new area compare to the original area?

4. Use your tiles to create two new shapes so that the shape with the smaller area has the greater perimeter. Sketch your shapes. Label them with their perimeters and areas.

### ✅ *Develop & Understand: B*

The shapes in Exercises 5–8 are drawn on dot grids. Find the area of each figure. Consider the horizontal or vertical distance between two dots to be 1 unit.

5.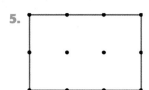

6.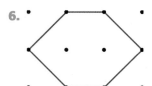

7.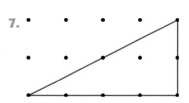

8.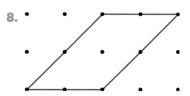

In Exercises 9–12, draw the shape by connecting dots on a sheet of 1-inch dot paper.

9. a square with area 4 square inches

10. a rectangle with area 2 square inches

11. a shape with an area of at least 15 square inches and a perimeter of no more than 25 inches

12. **Challenge** a square with an area of 2 square inches

**Find the area of each shape.**

13.  7 in.

7 in.

14. $\frac{1}{2}$ mi

$\frac{1}{2}$ mi

15.  50 cm

70 cm

16.

2 in.

$\frac{1}{4}$ in.

17. If you know the length and width of a rectangle, how can you find the rectangle's area without counting squares?

Finding the area of a shape by counting squares is not always easy or convenient. Fortunately, there are shortcuts for some shapes.

To find the area of a rectangle, just multiply the length by the width.

| Area of a Rectangle |
| --- |
| $A = L \cdot W$ |

In this formula, $A$ represents the area of a rectangle, and $L$ and $W$ represent the length and width.

### Think & Discuss

On dot or grid paper, draw a rectangle with side lengths 5 units and $7\frac{1}{2}$ units.

Use the formula above to find the area of your rectangle. Check that your answer is correct by counting the squares.

## ✓ Develop & Understand: C

**18.** On your newspaper page, draw rectangles around the major items, such as photographs and art, advertisements, articles, and headlines.

   **a.** Measure the sides of each rectangle to the nearest tenth of a centimeter.

   **b.** Calculate the area of each rectangle.

   **c.** Calculate the area of the entire page.

**19.** What percent of your newspaper page is used for the following items?

   **a.** photographs and art

   **b.** advertisements

   **c.** articles

   **d.** headlines

**Real-World Link**

The first successful daily newspaper in the United States was the *Pennsylvania Packet & General Advertiser,* which was first printed on September 21, 1784.

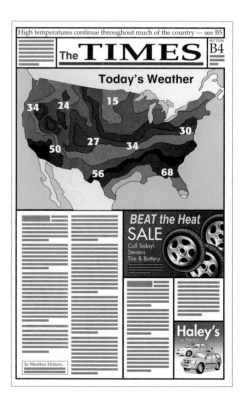

## Share & Summarize

1. Give an example of something you would measure in inches. Then give an example of something you would measure in square inches.

2. If one shape has a greater area than another, must it also have a greater perimeter? Explain or illustrate your answer.

3. Describe two ways to find the area of a rectangle.

# Investigation 2 Perfect Squares

## Vocabulary

perfect square

## Materials

- 1-inch dot paper
- 1-inch tiles

Recall that an exponent tells you how many times a number is multiplied by itself. You can write the product of a number times itself using the exponent 2.

$$5 \cdot 5 = 5^2$$

Multiplying a number by itself is called *squaring* the number. The expression $5^2$ can be read "5 squared."

### Think & Discuss

Evaluate $5^2$. Then, on a sheet of dot paper, draw a square with an area equal to that many square units.

How long is each side of the square?

Why do you think $5^2$ is read "5 squared"?

You can use the $\boxed{x^2}$ key on your calculator to square a number. To calculate $5^2$, press these keys.

$$[5] \ \boxed{x^2} \ \boxed{\text{ENTER}}$$

The exponent 2 is often used to abbreviate square units of measurement. For example, *square inch* can be abbreviated $in^2$ and *square centimeter* can be abbreviated $cm^2$.

### ✅ *Develop & Understand: A*

**Fill in the blank to find the area of each square. The first one is done for you.**

2 in.

2 in.     Area = ___2___ $^2$ in$^2$ = ___4___ in$^2$

**1.**     13 ft

13 ft     Area = _____ $^2$ ft$^2$ = _____ ft$^2$

**2.** 1.25 cm

1.25 cm     Area = _____ $^2$ cm$^2$ = _____ cm$^2$

**3.** $\frac{7}{4}$ in.

$\frac{7}{4}$ in.     Area = _____ $^2$ in$^2$ = _____ in$^2$

**4.** Write a formula for finding the area *A* of a square if you know the side length *s*. Use an exponent in your formula.

**Find the area of a square with the given side length.**

**5.** 1 in.          **6.** $\frac{1}{3}$ in.          **7.** 19 cm

**Find the side length of a square with the given area.**

**8.** 144 ft$^2$          **9.** 10,000 in$^2$          **10.** 53.29 cm$^2$

In Exercises 11–15, you will look at the squares that can be made with square tiles.

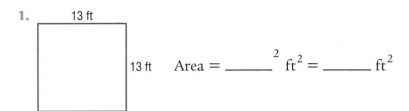

# ✅ Develop & Understand: B

11. Find every square that can be made from 100 tiles or fewer. Give the side length and area of each square.

12. Is it possible to make a square with 20 tiles? If so, explain how. If not, explain why not.

13. Is it possible to make a square with 625 tiles? If so, explain how. If not, explain why not.

14. **Challenge** Hakan tried to make a square with area 8 $in^2$ using tiles. After several tries, he said, "I don't think I can make this square using my tiles. But I know I can make it on dot paper."

    On dot paper, draw a square with area 8 $in^2$.

15. How can you tell whether a given number of tiles can be made into a square without actually making the square?

A number is a **perfect square** if it is equal to a whole number multiplied by itself. In other words, a perfect square is the result of *squaring* a whole number.

| Whole Number Squared | $1^2$ | $2^2$ | $3^2$ | $4^2$ | $5^2$ |
|---|---|---|---|---|---|
| Perfect Square | 1 | 4 | 9 | 16 | 25 |

In Exercises 11–15, the perfect squares were the numbers of tiles that could be formed into squares.

# ✅ Develop & Understand: C

16. Find three perfect squares greater than 1,000.

17. Is 50 a perfect square? Why or why not?

**Tell whether each number is a perfect square. Explain how you know.**

18. 3,249
19. 9,196.81
20. 12,225
21. 184,041

22. Find two perfect squares whose sum is also a perfect square.

23. Find two perfect squares whose sum is not a perfect square.

## Share & Summarize

1. How is the idea of squaring a number related to the area of a square?

2. Can *any* number be squared? Why or why not?

3. Can *any* number be a perfect square? Why or why not?

**1.** On dot paper or grid paper, draw a rectangle with an area of 20 square units, whole-number side lengths, and the greatest possible perimeter. What is the perimeter of your rectangle?

**2.** On dot paper or grid paper, draw a rectangle with an area of 20 square units, whole-number side lengths, and the least possible perimeter. What is the perimeter of your rectangle?

**These shapes are drawn on centimeter dot grids. Find the area of each shape.**

**3.**     **4.**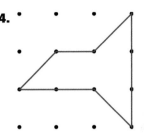

**5.**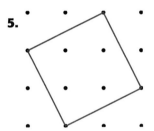

**6.** Find the area of a rectangle with length 7.5 feet and width 5.7 feet.

**7.** Find the length of a rectangle with width 11 centimeters and the given area.

    **a.** 165 square centimeters    **b.** 60.5 square centimeters

**8.** Find the length of a rectangle with area 484 square inches and the given width.

    **a.** 10 inches           **b.** 22 inches

**9.** A square garden has an area 289 square feet. How long is each side of the garden?

**10.** If one rectangle has a greater perimeter than another, must it also have a greater area? Explain your answer.

**Square each number.**

**11.** 14

**12.** 21.5

**13.** $\frac{9}{10}$

**14.** 0.3

**15.** List five perfect squares between 100 and 500.

**Tell whether each number is a perfect square. Explain how you know.**

**16.** 40

**17.** 81

**18.** 125

**19.** 256

**20.** If a square has area 30.25 square feet, how long is each side?

**Real-World Link**

The length of a soccer field can vary from 100 yards to 130 yards. The width can vary from 50 yards to 100 yards. So, the least possible area is 100 · 50, or 5,000, square yards. The greatest possible area is 130 · 100, or 13,000, square yards.

**21.** Ms. Dixon built this tile patio around a square fountain. The tiles measure 1 foot on each side. The patio is constructed of white, light green, dark green, and blue tiles.

**a.** What is the total perimeter of the patio? Add the inner and outer perimeters.

**b.** What is the area of the patio?

**c.** Express the portion of the patio that each color makes up as a fraction and as a percent.

**22.** Daniel wants to build a fenced-in play area for his rabbit. He has 30 feet of fencing. Give the dimensions and area of the largest rectangular play area that he can fence.

**23.** Each of these rectangles has whole-number side lengths and an area of 25 square units.

Below is the only rectangle with whole-number side lengths and an area of 5 square units.

   **a.** How many different rectangles are there with whole-number side lengths and an area of 36 square units? Give the dimensions of each rectangle.

   **b.** Consider every whole-number area from 2 square units to 30 square units. For which of these areas is there only one rectangle with whole-number side lengths?

   **c.** What do the areas you found in Part b have in common?

   **d.** For which area from 2 square units to 30 square units can you make the greatest number of rectangles with whole-number side lengths? Give the dimensions of each rectangle you can make with this area.

**24.** Katia squared a number. The result was the same as the number with which she started. What number might she have squared? Give all of the possibilities.

**25.** Rashid squared a number. The result was 10 times the number with which he started. What was his starting number?

**26.** Meera squared a number. The result was less than the number with which she started. Give two possible starting numbers for Meera.

**27.** In this exercise, you will explore what happens to the area of a square when you double its side lengths.

   **a.** Draw and label four squares of different sizes. Calculate the areas.

   **b.** For each square you drew, draw a square with sides twice as long. Calculate the areas of the four new squares.

   **c.** When you doubled the side lengths of your squares, did the areas double as well? If not, how did the areas change? Why do you think this happened?

   **d.** If you double the side lengths of a rectangle that is not a square, do you think the same pattern would hold? Why or why not?

   **e.** If you triple the side lengths of a square, what do you think will happen to the area? Test your hypothesis on two or three squares.

**28. In Your Own Words** Explain how squaring whole numbers is different for numbers greater than 1 than for numbers less than 1.

*Mixed Review*    **Find each product or quotient.**

**29.** $\frac{3}{4} \cdot \frac{4}{3}$                    **30.** $\frac{3}{4} \div \frac{4}{3}$

**31.** $\frac{12}{21} \cdot \frac{7}{16}$                   **32.** $\frac{27}{32} \cdot \frac{24}{45}$

**33.** $2\frac{2}{5} \cdot \frac{1}{3}$                    **34.** $3\frac{5}{8} \div \frac{1}{4}$

**35.** $1\frac{3}{8} \cdot 4\frac{1}{2}$                   **36.** $4\frac{4}{7} \div 1\frac{1}{2}$

**37.** $5 \div \frac{1}{9}$

**Evaluate each expression.**

**38.** $0.6 \cdot 0.6$

**39.** $0.3 \cdot 0.3$

**40.** $0.02 \cdot 0.02$

**41.** The 180 sixth-grade girls at Wright Middle School were asked to name their favorite activity in gym class. The results are shown in the table.

**Favorite Gym Activity**

| Activity | Percent of Girls |
|----------|------------------|
| Softball | 15 |
| Track | 8 |
| Gymnastics | 4 |
| Basketball | 28 |
| Soccer | 23 |
| Volleyball | 22 |

**a.** Which activity is most popular? About how many girls chose that activity?

**b.** Which activity is least popular? About how many girls chose that activity?

**c.** What is the difference in the *percent* of girls who chose volleyball and the percent who chose track? What is the difference in the *number* of girls who chose these sports?

In Exercises 4–8, you will explore how the base and height of a parallelogram are related to its area.

## ✅ Develop & Understand: B

4. Complete Parts a–c for each parallelogram in Exercise 1.

a. Choose a side of the parallelogram as the base. Draw a segment perpendicular to the base that extends to the side opposite the base. The segment should be completely inside the parallelogram. For parallelogram A, you might draw the segment shown using the pink dashed line.

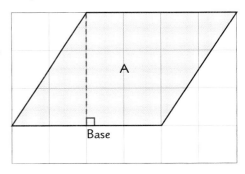

b. Find the lengths of the base and the height. The height is the length of the segment you drew in Part a. Record these measurements in a table like the one below.

| | Parallelogram | | Rectangle | |
|---|---|---|---|---|
| | **Base** | **Height** | **Length** | **Width** |
| **A** | | | | |
| **B** | | | | |
| **C** | | | | |

c. Divide the parallelogram into two pieces by cutting along the segment you drew in Part a. Then reassemble the pieces to form a rectangle. Record the length and width of the rectangle in your table.

5. How do the base and height of each parallelogram compare with the length and width of the rectangle formed from the parallelogram?

6. How does the area of each parallelogram compare with the area of the rectangle formed from the parallelogram?

7. How can you find the area of a parallelogram if you know the length of a base and the corresponding height? Use what you have discovered to explain why your method works.

8. Find the area of this parallelogram without forming it into a rectangle. Explain each step of your work.

You can find the area of a parallelogram by multiplying the length of the base by the height. This can be stated using a formula.

| Area of a Parallelogram |
| :---: |
| $A = b \cdot h$ |
| In this formula, $A$ represents the area, $b$ represents the base, and $h$ represents the height. |

## ✅ Develop & Understand: C

**Find the area of each parallelogram to the nearest hundredth of a square unit.**

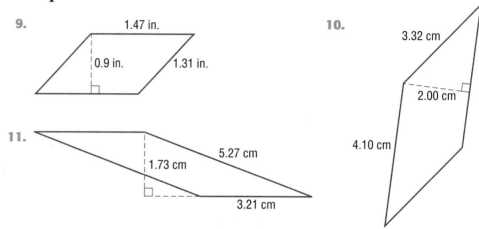

9. 1.47 in.    0.9 in.    1.31 in.

10. 3.32 cm    2.00 cm    4.10 cm

11. 5.27 cm    1.73 cm    3.21 cm

12. The area of the parallelogram below is 12.93 cm$^2$. Find the value of $b$ to the nearest hundredth.

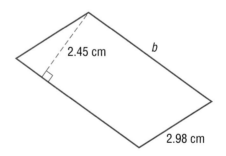

2.45 cm    $b$    2.98 cm

## Share & Summarize

How is finding the area of a parallelogram similar to finding the area of a rectangle? How is it different?

# Investigation  Areas of Triangles

## Vocabulary

**base of a triangle**

**height of a triangle**

## Materials

- copies of the triangle
- 3 copies of the triangle dot paper
- copies of the triangles
- scissors
- tape
- protractor
- ruler

You have looked at areas of rectangles and parallelograms. Now you will turn your attention to triangles.

### ✅ Develop & Understand: A

Find the area of each triangle. Explain the method that you used.

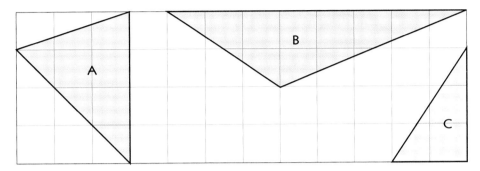

The **base of a triangle** can be any of its sides. The **height of a triangle** is the distance from the base to the vertex opposite the base. The height is always measured along a segment perpendicular to the base or the line containing the base.

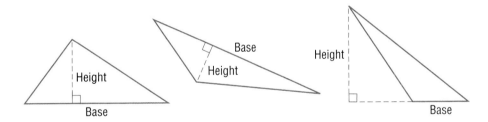

You may have used a variety of methods for finding the areas of the triangles. You will now see how you can find the area of a triangle by relating it to a parallelogram.

**Cut out two copies of each triangle like those shown above.**

1. Complete Parts a–d for each triangle.

   a. Make as many different parallelograms as you can by putting together the two copies of the triangle. Do not tape them together. Make a sketch of each parallelogram.

   b. How does the area of the triangle compare to the area of each parallelogram?

**c.** Tape the two copies of the triangle together to form one of the parallelograms you sketched in Part a. Choose one side of the parallelogram as the base. Draw a segment perpendicular to the base extending to the opposite side.

**d.** Do the base and height of the parallelogram correspond to a base and height of the triangle?

**2.** Think about what you learned in Exercise 1 about the relationship between triangles and parallelograms. How can you find the area of a triangle if you know the length of a base and the corresponding height?

Find the area of each triangle to the nearest hundredth.

**3.**

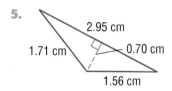

1.34 cm
0.90 cm
1.35 cm
2.00 cm

**4.**

1.37 cm  1.42 cm
0.87 cm
0.90 cm

**5.**

2.95 cm
1.71 cm  0.70 cm
1.56 cm

In Exercises 1–5, you probably discovered that the area of a triangle is half the length of the base times the height. You can state this using a formula.

| Area of a Triangle |
|---|
| $$A = \frac{1}{2} \cdot b \cdot h$$ |
| In this formula, $A$ represents the area, $b$ represents the base, and $h$ represents the height. |

**Real-World Link**

Triangles are rigid shapes. If you build a triangle out of a strong material, it will not collapse or change shape when you press on its sides or vertices. Because of this property, triangles are used frequently as supports for buildings, bridges, and other structures.

## Develop & Understand: B

Three students found the area of this triangle. Seth used the 4.5-centimeter side as the base, Margo used the 4.2-centimeter side, and Yori used the 3.8-centimeter side.

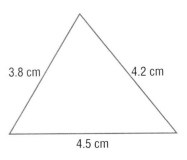

6. Assuming the students did the calculations correctly, do you think they found the same area or different areas? Explain.

7. Complete Parts a–c to find the area using the 4.5-cm side as the base.

   a. Draw a segment perpendicular to the base from the vertex opposite the base. Use your protractor to make sure the base and the segment form a right angle.

   b. Measure the height to the nearest tenth of a centimeter.

   c. Use the base and height measurements to calculate the area of the triangle.

8. Repeat Parts a–c of Exercise 7 using the 4.2-cm side as the base.

9. Repeat Parts a–c of Exercise 7 using the 3.8-cm side as the base.

10. Compare your results for Exercises 7, 8, and 9. Did the area you calculated depend on the base you used? Explain.

## Develop & Understand: C

△ABD and △ABE were created by shearing △ABC. *Shearing* a triangle means "sliding" one of its vertices along a line parallel to the opposite side. In this case, △ABC was sheared by sliding vertex C to vertex D and then to vertex E.

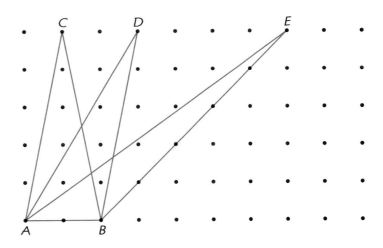

11. How are $\triangle ABC$, $\triangle ABD$, and $\triangle ABE$ alike? How are they different?

12. Draw two more triangles by shearing $\triangle ABC$.

13. Does shearing $\triangle ABC$ change its area? Explain.

14. Does shearing $\triangle ABC$ change its perimeter? Explain.

**Share & Summarize**

Describe how finding the area of a triangle is related to finding the area of a parallelogram.

# Investigation 3 Areas of Trapezoids

## Vocabulary

trapezoid

bases of a trapezoid

height of a trapezoid

In this investigation, you will explore the area of a special quadrilateral, a **trapezoid**.

These figures are trapezoids.

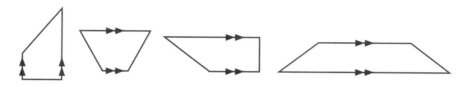

These figures are not trapezoids.

What do you notice about the figures that are trapezoids and the figures that are not trapezoids? How are they different?

Can you state the definition of a trapezoid by examining the figures above?

You may have noticed that a trapezoid has one pair of parallel sides, called the **bases of a trapezoid**. The **height of a trapezoid** is the length of a perpendicular segment between the bases.

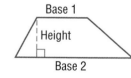

**Think & Discuss**

Can the bases of a trapezoid be different lengths? The same length?

Can the non-base side of a trapezoid be the same length? Draw an example or explain why not.

How many heights can you draw in a trapezoid?

## Math Link

Mathematicians sometimes use a different definition for a trapezoid. "A trapezoid is a quadrilateral that has *at least* one pair of parallel sides." Using this definition, what would be the difference between a trapezoid and a parallelogram?

1. Use the trapezoids below for Parts a–c.

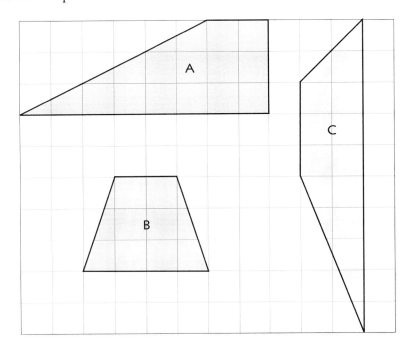

   a. Trace each trapezoid twice on a sheet of paper. Cut out the trapezoids.

   b. Label the top of each trapezoid "base 1" and the bottom "base 2".

   c. Place the two matching trapezoids together to form a parallelogram.

2. In Investigation 1, you learned the formula for the area of a parallelogram. Find the area of this parallelogram in terms of base 1, base 2, and its height, $h$.

3. If the area of the parallelogram is the same as the area of two of the original trapezoids, what would be the formula for the area of the trapezoid?

In Exercise 1, you discovered that you can find the area of a trapezoid by adding the lengths of the two bases, multiplying by the height, and dividing by 2. This can be stated using a formula.

## Math Link

The lengths of the two bases in a trapezoid, base 1 and base 2, can also be written using the subscripts, $b_1$ and $b_2$. You may see the formula for the area of a trapezoid written as $A = \frac{1}{2}h(b_1 + b_2)$.

| Area of a Trapezoid |
| --- |

$$A = \frac{1}{2}h(\text{base 1} + \text{base 2})$$

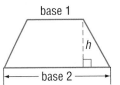

In this formula, $A$ represents the area, $h$ represents the height of the trapezoid, and base 1 and base 2 represent the two bases.

### ✅ Develop & Understand: B

**Find the area of each trapezoid.**

4.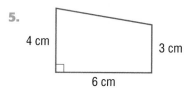
13 inches
10 inches
25 inches

5.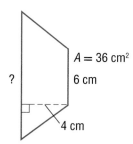
4 cm
3 cm
6 cm

### ✅ Develop & Understand: C

Quincy found another way to develop the formula for the area of a trapezoid. Here are his instructions.

**Step 1.** Find the midpoints of the two non-base sides. Connect them with a line segment.

**Step 2.** Cut along this line segment.

**Step 3.** Rotate the upper half of the trapezoid clockwise so that the two congruent half-sides meet.

**Step 4.** The new quadrilateral is a parallelogram.

6. Perform Quincy's instructions on a copy of one of the trapezoids you made in Exercise 1. Before you cut along the line, label the bases as base 1 and base 2.

7. How does the height of this parallelogram compare to the height of the original trapezoid?

8. How does the base of this parallelogram compare to the two bases of the original trapezoid?

9. What formula did Quincy use for the area of this parallelogram, if base 1 and base 2 are the bases of the original trapezoid and $h$ is the height of the original trapezoid?

**Find each missing length or area for the trapezoids below.**

10.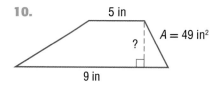
5 in
? | A = 49 in²
9 in

11. **Challenge**
A = 36 cm²
? | 6 cm
4 cm

12.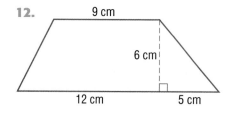
9 cm
6 cm
12 cm | 5 cm

**13.**

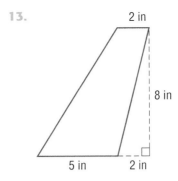

**14.** A trapezoid and a parallelogram each have the same area and the same height. Can the base of the parallelogram be equal to one of the bases of the trapezoid? Why or why not?

## Share & Summarize

How is finding the area of a trapezoid similar to finding the area of a parallelogram? How is it different?

# Investigation 4 Areas of Circles

Finding the area of a figure with curved sides often requires counting grid squares or using another estimation method. However, there is a surprisingly simple formula for calculating the area of a circle.

## ✅ Develop & Understand: A

These circles are drawn on 1-centimeter grid paper.

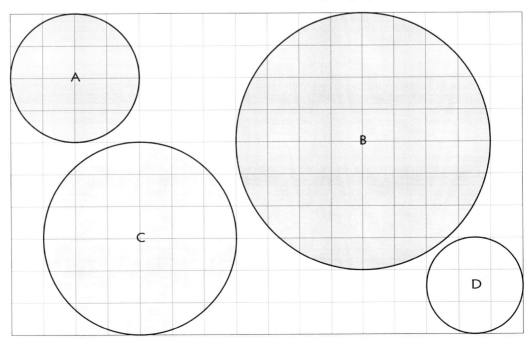

1. Copy the table. Find the radius of each circle. Record your results in the "Radius" column.

| Circle | Radius, $r$ (cm) | Estimated Area, $A$ (cm²) | $A \div r$ | $A \div r^2$ |
|--------|------------------|---------------------------|------------|--------------|
| A      |                  |                           |            |              |
| B      |                  |                           |            |              |
| C      |                  |                           |            |              |
| D      |                  |                           |            |              |

2. Estimate the area of each circle by counting grid squares. Record your estimates in your table.

3. For each circle, divide the area by the radius. Record the results.

4. For each circle, divide the area by the radius squared. Record the results.

5. Look at the last two columns of the table. Do the values in either column show an obvious pattern? If so, does it remind you of other patterns you have seen?

6. Cesar estimated that the area of a circle with a radius of 10 centimeters is about 40 cm².

   a. Explain why Cesar's estimate is not reasonable.

   b. What is a reasonable estimate for the area of a circle with a radius of 10 centimeters?

   c. What is a reasonable estimate for the radius of a circle with an area of 40 cm²?

**Math Link**

$\pi$ is a decimal number with digits that never end or repeat. It can be approximated as 3.14.

Previously, you learned about the number $\pi$ and how it is related to the circumference of a circle. You found that if $C$ is the circumference of a circle and $d$ is the diameter, the following is true.

$$\pi = C \div d$$

The number $\pi$ is also related to the area of a circle. If $A$ is the area of any circle and $r$ is the radius, the following is true.

$$\pi = A \div r^2$$

You can use this fact to develop the formula for the area of a circle.

| **Area of a Circle** |
|:---:|
| $A = \pi \cdot r^2$ |
| In this formula, $A$ is the area and $r$ is the radius. |

# ✅ Develop & Understand: B

For Exercises 7–11, express your answer in terms of π and as a decimal rounded to the nearest hundredths place.

**7.** What is the area of a circle with a radius of 15 inches?

**8.** What is the area of a circle with a radius of 10.15 centimeters?

**9.** Which has the greater area, a circle with a radius of 7.2 centimeters or a circle with a diameter of 12.75 centimeters? Explain your answer.

**Math Link**

Remember to use the π key on your calculator to approximate π. If your calculator does not have a π key, use 3.14 to approximate π.

**10.** A pizza parlor makes pizzas in two shapes. The circular pizza has a diameter of 10 inches. The rectangular pizza measures 16 inches by 10 inches. A circular cheese pizza costs $8, and a rectangular cheese pizza costs $14. Which shape gives you more pizza for your money? Explain how you found your answer.

**11.** This is a diagram of the inner lane of the track at Walker Middle School. The lane is made of two straight segments and two semicircles, or half circles. The area inside the track is covered with grass. What is the area of the grass inside the track? Explain how you found your answer.

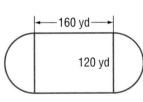

## Share & Summarize

**1.** Give the formula for the area of a circle. Tell what the letters in the formula represent.

**2.** How can you calculate the area of a circle if you know only its diameter?

**3.** How can you calculate the area of a circle if you know only its circumference?

**Real-World Link**

There are more than 60,000 pizzerias in the United States, accounting for about 15% of all restaurants.

# Investigation 5  Areas of Circle Sectors

## Vocabulary

arc

central angle

sector

## Materials

- protractor

- scissors

### Math Link

An *arc* of a circle is a segment of the circle's circumference.

In the previous investigation, you calculated the area of a circle. If someone asked you to find the area of a wedge of a circle with a radius of 2 inches, the shaded region drawn to the right, what strategies would you use?

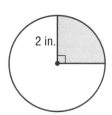

In answering the question above, you probably first thought about what fraction of the wedge was the circle's area. In mathematics, we refer to a wedge as the **sector** of a circle, or the area enclosed by two radii and the **arc**, which is the part of the circle that connects them.

You may have thought that since the sector was $\frac{1}{4}$ of the area of the whole circle, the area of the sector would be $\frac{1}{4}$ of $\pi r^2$, or $\frac{1}{4} \cdot \pi \cdot 2^2 = \pi \text{ in}^2$.

Now consider how to find the area of any sector.

Clearly, the area of a sector of a circle depends upon the measure of its **central angle**, which is the angle formed by the two radii.

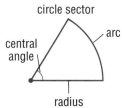

## ✓ Develop & Understand: A

1. The central angle of a sector of a circle is 45°. What fraction of the circle's area is the sector's area?

2. The central angle of a sector of a circle is 120°. What fraction of the circle's area is the sector's area?

3. The central angle of a sector of a circle is $n$°. What fraction of the circle's area is the sector's area?

Once you have determined what fraction of the circle's area the sector is, you can determine a general formula for the area of a sector of a circle.

---

**Area of a Circle Sector**

$$A = \frac{m}{360} \cdot \pi r^2$$

In this formula, $A$ represents the area of the sector, $m$ represents the central angle measured in degrees, and $r$ represents the radius.

---

## ✅ Develop & Understand: B

4. Find the area of this sector. Express your answer in terms of π and as a decimal rounded to the nearest hundredth of a centimeter.

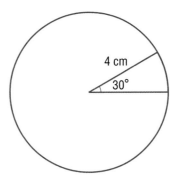

5. A circle has a radius of $3\frac{1}{2}$ feet. The central angle of a sector of this circle is 15°. Find the area of the sector. Express your answer in terms of π and as a decimal rounded to the nearest tenth.

6. The sector of a circle has an area of $6\pi$ cm$^2$ and the radius of the circle is 6 cm. Find the central angle.

7. The area of a sector of a circle is $\frac{9\pi}{5}$ in$^2$. Its central angle is 72 degrees. Find the radius of the circle.

## ✅ Develop & Understand: C

8. Find the area of the shaded region. You are given the length of the radius of the circle. Express your answer in terms of π and as a decimal rounded to the nearest hundredth.

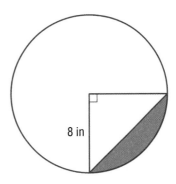

9. Find the area of the shaded region and each white region formed by the square and the circle. Express your answer as a decimal rounded to the nearest hundredth.

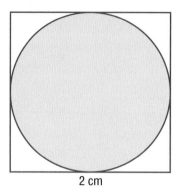

2 cm

10. Find the area of the shaded region and each white region formed by the square with side 4 centimeters.

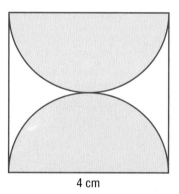

4 cm

11. Suppose a pizza slice is a circle sector with a 24° angle and a radius of 15 inches. If a can of pizza sauce will cover 94 square inches of pizza, how many cans of sauce are needed to cover this slice?

## Share & Summarize

How is the area of a sector of a circle related to the central angle of the sector?

How is the area of a sector related to the area of a circle?

**Practice & Apply**

1. Choose an object in your home with a nonrectangular surface that will fit on a piece of grid paper. A can of soup, your shoe, and an iron are some ideas. Trace the surface onto the grid paper. Estimate its area.

2. These parallelograms are drawn on a centimeter grid.

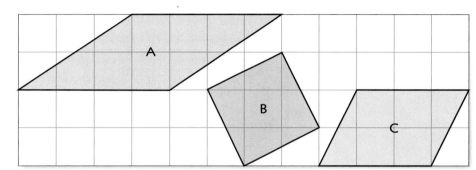

   **a.** Find the area of each parallelogram.

   **b.** Sketch a rectangle that has the same area as parallelogram A.

   **c.** Sketch a rectangle that has the same area as parallelogram C.

3. Find the area of this parallelogram.

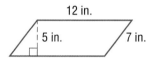

4. Can you use the area formula for a parallelogram, $A = b \cdot h$, to find the area of a rectangle? If so, where are the base and height on the rectangle? If not, why not?

5. A parallelogram has an area of 42.6 cm$^2$. The height of the parallelogram is 8 cm. What is the length of the base?

6. These triangles are drawn on a centimeter grid.

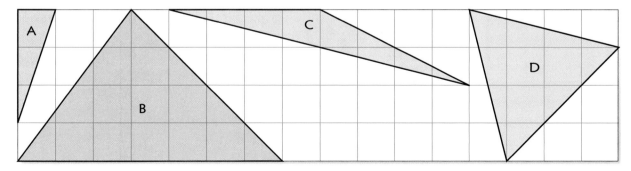

   **a.** Find the area of each triangle.

   **b.** For each triangle, sketch a parallelogram with twice the area of the triangle.

**Find the area of each triangle. Round your answers to the nearest hundredth.**

7.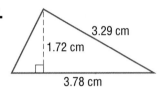

8.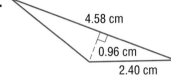

9. Consider this triangle.

    a. Which of the given measurements would you use to find the area of the triangle? Why?

    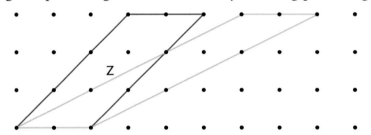

    b. What is the triangle's area? Round your answer to the nearest hundredth.

10. The green parallelogram was created by shearing parallelogram Z.

    a. Create two more parallelograms by shearing parallelogram Z. In each case, "slide" the top of the parallelogram.

    b. Does shearing a parallelogram change its area? Explain.

    c. Now shear parallelogram Z to create a parallelogram with the smallest possible perimeter. What does this new parallelogram look like?

    d. Tessa sheared parallelogram Z to create this figure. She says she has drawn the sheared parallelogram with the greatest possible perimeter. Do you agree with her? Explain.

    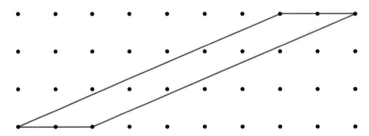

In Exercises 11–13, express your answer in terms of π and as a decimal rounded to the nearest hundredths place.

**11.** Calculate the area of a circle with radius 8.5 inches.

**12.** Calculate the area of a circle with diameter 15 feet.

**13.** Calculate the area of a circle with radius 9 feet.

**14.** A dog is tied to a 15-foot leash in the center of a yard.

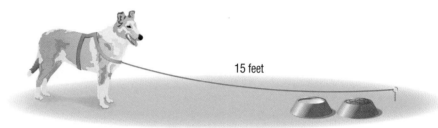

15 feet

**a.** What is the shape of the area in which the dog can play?

**b.** To the nearest square foot, what is the area of the space in which the dog can play?

**c.** Suppose that, instead of being tied in the center of the yard, the dog is tied to the corner of the house. To the nearest square foot, what is the area of the space in which the dog can play? The sides of the house are longer than the leash.

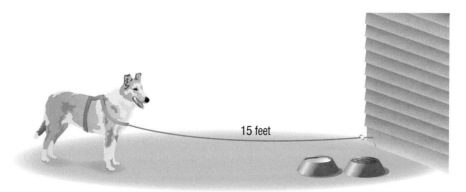

15 feet

**Find the area of each trapezoid.**

**15.**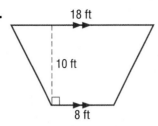

**16.**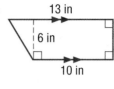

**Find the missing lengths and areas for each trapezoid.**

**17.**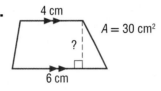

**18.**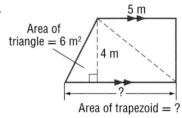

**19.**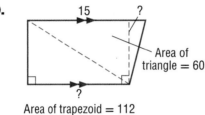

**20.**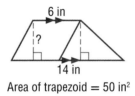

In Exercises 21–24, calculate the area of each circle sector. Use the units given and round to the nearest hundredth of a square unit. Also, express your answer in terms of π.

**21.** A sector with a central angle of 240° and a radius of 2 cm.

**22.** A sector with a central angle of 12° and a radius of 7 in.

**23.** A sector with a central angle of 36° and a radius of 12 in.

**24.** A sector with a central angle of 40° and a radius of 5 cm.

**Connect & Extend**

**25.** This parallelogram has an area of 20.03 cm². Find the values of $a$, $b$, $c$, and $d$ to the nearest hundredth of a centimeter.

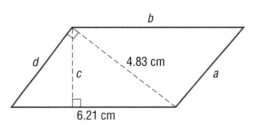

**26.** In this exercise, you will draw parallelograms.

    **a.** Draw three different parallelograms with base length 15 cm and height 7 cm.

    **b.** Which of your parallelograms has the least perimeter? Which has the greatest perimeter?

    **c.** Could you draw a parallelogram with the same base and height and an even smaller perimeter? If so, draw it. If not, explain why not.

**27.** A deck of cards has been pushed as shown. Notice that the sides of the deck are shaped like parallelograms.

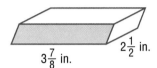

The deck contains 52 cards. Each card is $\frac{1}{48}$ of an inch thick, $3\frac{7}{8}$ inches long, and $2\frac{1}{2}$ inches wide. Find the area of the shaded parallelogram.

**28.** Below is a floor plan for a museum divided into four parallelograms and a rectangle. Find the area of the floor to the nearest hundredth of a square meter.

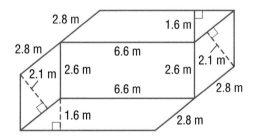

**29.** The area of this triangle is 782 square centimeters. Find $a$ and $b$ to the nearest tenth of a centimeter.

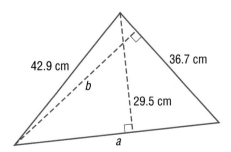

**30.** In an *equilateral triangle,* all three sides are the same length. Suppose the area of an equilateral triangle is 27.7 cm$^2$ and the height is 6.9 cm. How long are each of the triangle's sides?

**31.** Any regular polygon can be divided into identical triangles. This hexagon is divided into six identical triangles.

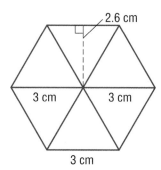

a. Find the area of each triangle and the area of the hexagon to the nearest tenth of a square centimeter. Explain how you found the areas.

b. This formula can be used to find the area of a regular polygon.

$$A = \frac{1}{2} \cdot \text{polygon perimeter} \cdot \text{height of one triangle}$$

Show that this formula gives you the correct area for the hexagon above.

c. Why do you think the formula works?

d. A stop sign is in the shape of a regular octagon. This sketch of an octagon has been divided into eight identical triangles.

Use the formula from Part b to find the area of a stop sign.

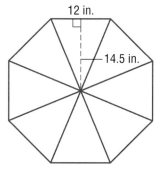

e. Brett found the area of the stop sign by surrounding it with a square.

How long are the sides of the square? How long are the perpendicular sides of the small triangles in the corners of the square?

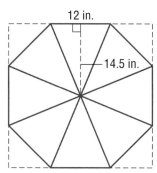

f. Explain how Brett might have calculated the area. Show that this method gives the same area that you found in Part d.

**Real-World Link**

The world's highest fountain, located in Fountain Hills, Arizona, is capable of sending 8-ton streams of water 560 feet into the air. This is 10 feet higher than the Washington Monument.

· · · · · · · · · · · · · · · · ·

**32.** The Smallville town council plans to build a circular fountain surrounded by a square concrete walkway. The fountain has a diameter of 4 yards. The walkway has an outer perimeter of 28 yards.

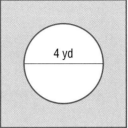

Find the area of the walkway to the nearest tenth of a square yard.

**33. Preview** The *surface area* of a three-dimensional figure is the sum of the areas of its faces. For example, this cube consists of six faces, each with area 9 in². So, its total surface area is 9 · 6, or 54 in².

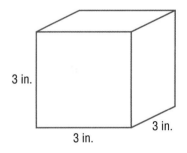

**a.** Find the surface area of this rectangular box.

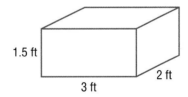

**b.** To find the surface area of a cylinder, you can imagine it as three separate pieces, the circular top and bottom and the rectangle wrapped around them. Find the surface area of this cylinder. (Hint: You need to figure out what the length of the rectangle is. To do this, think about how this length is related to the circles.)

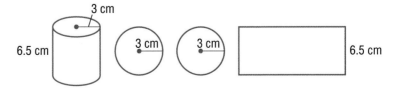

**34. In Your Own Words** Explain the similarities and differences among the area formulas you learned in this lesson.

**35.** This parallelogram has been divided
into a trapezoid and a triangle.
Find the area of the trapezoid.

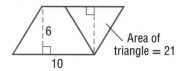
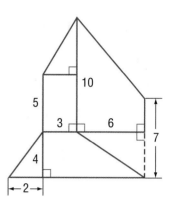

**36.** Find the total area of this polygon.
You will need to add up the areas of the
smaller polygons that make up the big
one. Show which smaller polygons you
used and what each area is.

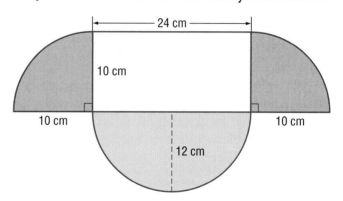

**For each mosaic, calculate the area covered by each color.**

**37.**

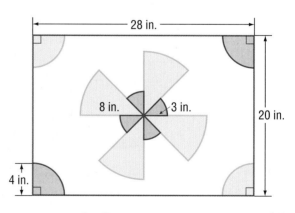

    **a.** Green         **b.** Grey         **c.** White

**38.**

    **a.** Green         **b.** Grey         **c.** White

Calculate the area of each circle sector. For Exercises 39–41, express your answers in terms of π and as a decimal rounded to the nearest hundredths place. Use the units given in each picture. Remember, the pictures are not drawn to scale.

**39.**

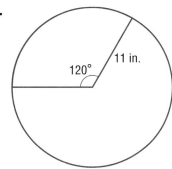

**40.**

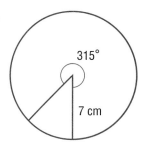

**41.**

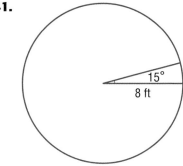

*Mixed Review*

**Evaluate each expression.**

**42.** $0.25 \times 0.4$    **43.** $0.25 \times 4$    **44.** $0.25 \times 0.1$

**45.** $0.04 \times 0.04$    **46.** $0.02 \times 4$    **47.** $0.2 \times 0.7$

**48. Ecology** The table lists the number of endangered species for five groups of animals.

**a.** Extend this table showing what the percentage each category is of the total number of species listed. Round to the nearest tenth.

**b.** About what percentage of the endangered species in the five groups are mammals or birds?

| Group | Number of Endangered Species |
|---|---|
| Mammals | 316 |
| Birds | 253 |
| Fishes | 82 |
| Reptiles | 78 |
| Snails | 22 |

**c.** Write three statements comparing the number of endangered fish species to the number of endangered snail species.

# Surface Area and Volume

In the previous lesson you discovered that area, the space inside a two-dimensional figure, is measured in *square units.*

## Vocabulary

surface area

volume

## Materials

- cubes
- graph paper or dot paper

### *Math Link*

The volume of a one-block structure is 1 cubic unit.

You can say that the area of the figure shown here is 8 square units. Or, if you know the size of the squares, you can use it to state the area exactly. For example, these squares have a side length of 1 centimeter, so the area of the figure is 8 cm$^2$.

The **surface area** of a three-dimensional object is the area of the region covering the object's surface. If you could open up the object and flatten it so you could see all sides at once, the area of the flat figure would be the surface area. Do not forget to count the bottom surface. Surface area is measured in square units.

**Volume**, the space inside a three-dimensional object, is measured in *cubic units.* If you build a structure with blocks that are each 1 cubic unit, then the volume of a block structure is equal to the number of blocks in the structure. For example, a structure made from eight blocks has a volume of 8 cubic units. If the blocks have an edge length of 1 cm, the structure's volume is 8 cm$^3$.

In this lesson, when an exercise refers to blocks, the blocks each have an edge length of 1 unit, faces of area 1 square unit, and a volume of 1 cubic unit.

### Think & Discuss

What is the surface area of a single block in square units?

If the edge lengths of a block are 2 cm, what is the block's surface area?

What is the volume of the structure at the right in cubic units?

What is the surface area of the structure above in square units? Count only the squares on the *outside* of the structure.

# Investigation ⓵ Volume

## Vocabulary

prism

rectangular prism

## Materials

• cubes

• graph paper or dot paper

In the exercises below, you will find the surface area and volume of block structures.

## ✅ Develop & Understand: A

1. Find the volume and the surface area of each three-block structure.

   a. 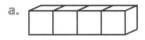   b.

2. Find the volume and the surface area of each four-block structure.

   a. 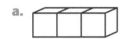   b.

   c.    d.

3. Do the structures in Exercise 2 all have the same volume? Explain your answer.

4. Which of the structures in Exercise 2 have the greatest surface area? Which has the least surface area?

5. Build two block structures with at least six blocks each that have the same volume but different surface areas.

   a. For each structure, draw a view of its top.

   b. Record the volume and the surface area of each structure.

If a block structure has a constant height, that is, has the same number of blocks in every column, that structure is a **prism**. If the top view of such a structure is a rectangle, the structure is a **rectangular prism**.

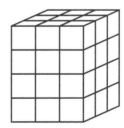

## Math Link

The *surface area* of a rectangular prism is the sum of the areas of its six faces. The *volume* of a three-dimensional figure is the amount of space inside it. Surface area is measured in square units, and volume is measured in cubic units.

## ✅ Develop & Understand: B

By using its dimensions, you can describe a rectangular prism exactly. For example, a prism with edge lengths 3 units, 2 units, and 4 units is a 3 × 2 × 4 prism.

6. Make all the rectangular prisms you can that contain 8 blocks.

   a. Record the dimensions, volume, and surface area of each prism you make.

**b.** Do the rectangular prisms all have the same volume?

**c.** Which of the rectangular prisms has the greatest surface area? Give its dimensions.

**d.** Which of the prisms has the least surface area?

7. Make all the rectangular prisms you can that have a volume of 12 cubic units. Repeat Parts a, c, and d of Exercise 6 for this prism.

8. Now find all the rectangular prisms that have a volume of 20 cubic units. Try to do it without using your blocks. Repeat Parts a, c, and d of Exercise 6 for this prism.

## ✅ Develop & Understand: C

**Math Link**

The length, width, and height are often referred to as *dimensions*.

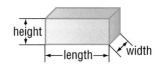

9. Here is a view of the top of a prism.

**a.** Build a prism 1 unit high with this top view. What is its volume?

**b.** Build a prism 2 units high with this top view. What is its volume?

**c.** What would be the volume of a prism 10 units high with this top view?

**d.** Write an expression for the volume of a prism with this top view and height $h$.

**e.** What is the area of this top view?

10. Here is another top view.

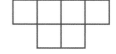

**a.** Build a prism 1 unit high with this top view. What is its volume?

**b.** Build a prism 3 units high with this top view. What is its volume?

**c.** Suppose you built a prism 25 units high with this top view. What would be its volume?

**d.** Write an expression for the prism's volume, using $h$ for height.

**e.** What is the area of this top view?

11. Here is a third top view.

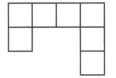

a. Build a prism 1 unit high with this top view. What is its volume?

b. Build a prism 5 units high with this top view. What is its volume?

c. Suppose you cut your blocks in half to build a structure half a unit high with this top view. What would be its volume?

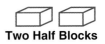
**Two Half Blocks**

d. Write an expression for the prism's volume, using *h* for height.

e. What is the area of this top view?

## Share & Summarize

1. If someone gave you some blocks, how could you use all of them to build a rectangular prism with the greatest surface area? How could you use all of them to build a rectangular prism with the least surface area?

2. Suppose the view of the top of a prism contains 8 squares. What is the volume of a prism that is:

   a. 1 unit high?        b. 10 units high?

   c. $\frac{1}{2}$ unit high?        d. *h* units high?

3. Write a general rule for finding the volume of a prism made from blocks.

# Investigation 2 Volume of a Rectangular Prism

**Vocabulary**

base of a prism

The identical top and bottom faces of a prism are called the **bases of a prism**. The other sides of a prism are always rectangles, but as you saw in Investigation 1, the bases can be any shape. A rectangular prism is a special kind of prism whose bases are rectangles.

At the end of Investigation 1, you developed a rule for finding the volume of a rectangular prism. The volume is the area of the base of the prism multiplied by the height of the prism. This can be stated using a formula.

| Volume of a Rectangular Prism |
| --- |
| $V = h(A)$ |
| In this formula, *V* represents the volume, *h* represents the height of the prism, and *A* represents the area of the base. There is another formula that is also used to find the volume of a rectangular prism: $V = h(l \times w)$. |

## Think & Discuss

Why do these two different formulas both give the same volume?

Could either of these formulas be used to calculate the volume of a prism whose bases are not rectangular?

## ✓ Develop & Understand: A

In Exercises 1–3, calculate the volume of each rectangular prism. Include the appropriate measurement units in your answer.

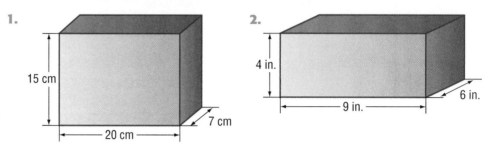

1.
2.

3. A rectangular prism with height = 2 mm, length = 13 mm, and width = 8 mm

4. A rectangular prism's volume is 96 ft³. The area of its base is 16 ft. What is the height of the prism?

5. A rectangular prism's volume is 320 cm³. Its height is 5 cm.

   a. What is the area of the base of the prism?

   b. What could be the length and width of the base? Give two possible length-width pairs.

6. Give the dimensions of two other prisms whose volume is 320 cm³.

7. Consider these four prisms.

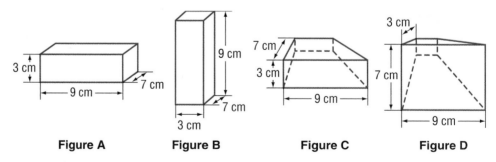

Figure A          Figure B          Figure C          Figure D

   a. Are A and B different prisms? Explain your reasoning.

   b. Are C and D different prisms? Explain your reasoning.

   c. Explain why your answers to Parts a and b were different or the same.

### ✅ *Develop & Understand: B*

8. What is the volume of this rectangular prism?

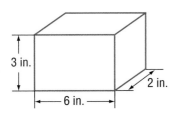

3 in.

2 in.

6 in.

9. Imagine taking two prisms like the one in Exercise 8 and stacking one on top of the other to form a new, larger prism.

   a. What would be the height, length, and width of the new prism?

   b. What would be the prism's volume?

10. Imagine stacking together four of the prisms like the one in Exercise 8 to make a larger rectangular prism.

    a. What would be the height, length, and width of the new prism?

    b. What would be the volume of the new prism?

    c. Give the dimensions of another rectangular prism you could create of the same four prisms.

    d. Is the volume of this prism different from the volume of the prism created in Part a? Explain why or why not.

11. How many of the prisms from Exercise 9 would you need to combine to create a rectangular prism whose volume is 360 in$^3$?

12. What could be the length, width, and height of the prism created in Exercise 11?

13. Consider these two rectangular prisms.

|         | Prism A | Prism B |
|---------|---------|---------|
| **Height** | 2 mm | 5 mm |
| **Length** | 2 mm | 8 mm |
| **Width**  | 2 mm | 4 mm |

   a. What is the volume of prism A? Of prism B?

   b. How many copies of prism A would you need to combine to equal the volume of prism B?

   c. Can you create prism B by stacking together copies of prism A? Explain how to do it, or why it is impossible.

   d. Prism C has these dimensions: height = 1 mm, length = 2 mm, width = 2 mm. Describe how you would create prism B by stacking together copies of prism C. How many copies of prism C would you need?

## ✅ Develop & Understand: C

Carla, Grant, and Tyra boarded an airplane in Boston bound for San Francisco. Grant found out that each passenger was allowed to bring one piece of carry-on luggage. The rule was that the height, length, and width of the bag must add up to 45 inches or less.

**Carla:**
20 in. × 14 in. × 8 in.

**Grant:**
25 in. × 10 in. × 7 in.

**Tyra:**
14 in. × 14 in. × 14 in.

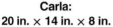

14. Whose bag holds the most? The least? Explain your answer.

15. Which bags meet the size restriction?

When they got to the airport, they saw this sign.

16. Which bags will the airline allow onto the plane as carry-on luggage, following the second size restriction?

17. Carla said, "It would be simpler if the rule gave a maximum for a bag's volume, instead of dimensions."

   a. What is the volume of the largest allowable bag?

   b. Which of the three friends could bring their bags onto the plane if the airline used this volume rule?

   c. Why might the airline not want to use this rule? Explain your reasoning.

**Maximum Carry-On Baggage Size**

Only one item per person.

Carry-on item must fit within the space below. If it does not fit, please check in your carry-on item at the front desk.

> ### Math Link
> The dimensions of a rectangular prism are often written as 15 ft · 23 ft · 4 ft. This notation can be read "15 feet by 23 feet by 4 feet." This notation can also be used for rectangles, for example, 3 cm · 2 cm. (A "two by four" is a piece of wood whose base is a rectangle 2 inches wide and 4 inches long and whose length varies as needed.)

## Share & Summarize

1. If you double the height of a rectangular prism, what happens to the volume?

2. If you double the area of the base of a rectangular prism, what happens to the volume?

3. What happens to the volume of a rectangular prism if you double two of its dimensions, for example, length and width, or length and height?

# Investigation ③ Polygons to Polyhedra

## Materials

• paper polygons

• tape

In this Inquiry investigation, you will continue to explore three-dimensional shapes.

A *polyhedron* is a closed, three-dimensional figure made of polygons. The shapes below are polyhedra. You have probably seen some of these shapes.

**Pentagonal Prism**    **Square Pyramid**    **Rectangular Prism**    **Cube**

The polygons that make up a polyhedron are called *faces*. The segments where the faces meet are called *edges*. The corners are called *vertices*.

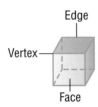

Edge

Vertex

Face

### Math Link

A *regular polygon* has sides that are all the same length and angles that are all the same size.

In a *regular polyhedron*, the faces are identical regular polygons. The same number of faces meet at each vertex. The cube shown above is a regular polyhedron. It has identical square faces and three faces meet at each vertex. None of the other shapes above is a regular polyhedron. Can you see why?

There is an infinite number of regular polygons. You can always make one with more sides. However, there is a very small number of regular polyhedra. In this Inquiry investigation, you will find them all.

## Construct the Polyhedra

**1.** Start with the equilateral triangles. Follow these steps.

**Step 1.** Tape three triangles together around a vertex as shown.

Vertex

*Go on*

**Step 2.** Bring the two outside triangles together. Tape them in place to create a three-dimensional shape.

Vertex

**Step 3.** Notice that, at one of the vertices, three triangles meet. At the other vertices, only two triangles meet. At a vertex with only two triangles add another triangle, so the vertex now has three triangles around it.

Now see if you can create a closed shape with three triangles at each vertex. If not, continue to add triangles until you can create a closed shape.

2. Repeat the process that you used in Question 1. This time, start with four triangles around a vertex. Add triangles until the figure closes and there are four triangles around each vertex.

Vertex

Vertex

3. Repeat the process again, starting with five triangles around a vertex.

4. Repeat the process once again, starting with six triangles around a vertex. What happens?

5. Now start with squares. Create a polyhedron with three squares around each vertex. What polyhedron did you make?

6. Try to create a polyhedron with four squares around each vertex. What happens?

7. Now start with pentagons. Try to create a polyhedron with three regular pentagons around each vertex. Can you do it?

8. Try to create a polyhedron with four regular pentagons around each vertex. What happens?

9. Now start with hexagons. Try to create a polyhedron with three regular hexagons around each vertex. What happens?

10. What happens when you try to create a polyhedron from regular heptagons?

You have just created all of the regular polyhedra.

**Tetrahedron**  **Octahedron**  **Icosahedron**  **Cube**  **Dodecahedron**

An interesting pattern relates the number of faces, edges, and vertices of all polyhedra. Looking at the regular polyhedra that you created can help you find that pattern.

## Find a Pattern

**11.** On each of your polyhedra, find the number of faces, the number of vertices, and the number of edges. Record your results in a table.

| Polyhedron | Faces | Vertices | Edges |
|---|---|---|---|
| Tetrahedron | | | |
| Octahedron | | | |
| Icosahedron | | | |
| Cube | | | |
| Dodecahedron | | | |

**12.** Can you find a way to relate the number of faces and vertices to the number of edges?

## What Did You Learn?

**13.** Use what you learned while building the polyhedra to explain why there are only five regular polyhedra.

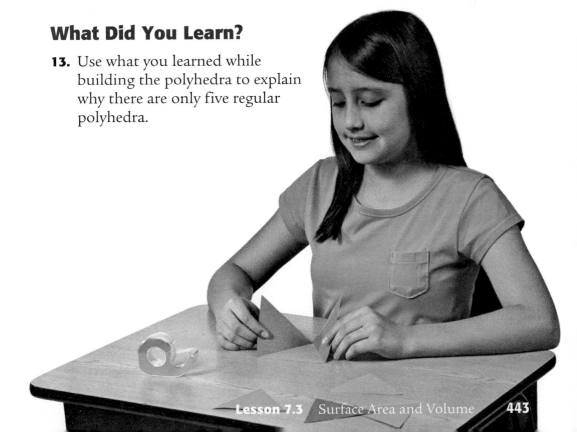

**Math Link**

Regular polyhedra are also called *Platonic solids* for the Greek philosopher Plato, who believed they were the building blocks of nature. He thought fire was made from tetrahedra, earth from cubes, air from octahedra, water from icosahedra, and planets and stars from dodecahedra.

**Practice & Apply**  Determine the volume of each block structure.

**1.**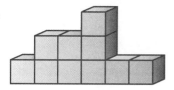

**2.**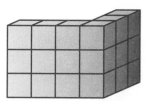

**3.**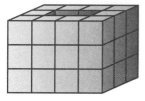

**4.**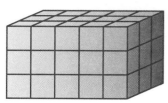

For Exercises 5–8, calculate the volume of each rectangular prism from its dimensions. Include appropriate measurement units in your answers.

**5.** Height = 5 cm   Length = 9 cm   Width = 4 cm

**6.** Height = 6 ft    Length = 14 ft   Width = 20 ft

**7.** Height = 2 m    Length = 3 m    Width = 1.5 m

**8.** Height = $\frac{3}{4}$ in.   Length = 5 in.   Width = $\frac{1}{3}$ in.

**9.** A prism has a volume of 300 in³ and a base of 25 in². What is the height of the prism?

**10.** A prism has a volume of 140 cm³, a height of 4 cm, and a length of 5 cm. What is the width?

**11.** Volume = 324 cm³, Height = 6 cm

   **a.** What is the area of the base?

   **b.** What could be the length and width?

   **c.** Give another length-width pair that would give the prism the same volume.

**12.** This swimming pool is 4 feet deep at the shallow end and 10 feet deep at the deep end. The shallow end is 75 feet long.

**Math Link**
*Volume* is measured in cubic units. *Surface area* is measured in square units.

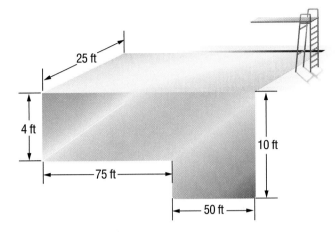

**a.** What is the total volume of the pool?

**b.** Another pool has the same length and width as this one but is the same depth everywhere. If the volume is the same as the volume of the first pool, what is the depth?

**13.** Consider all the rectangular prisms that can be made with 27 blocks.

**a.** Give the dimensions of each prism.

**b.** Which of your 27-block prisms has the greatest surface area? Which has the least surface area?

**14.** **In Your Own Words** Give two examples of structures with the same volume and different surface areas. Suppose you are making one of these structures for art class, and you want to paint them. Which one will need less paint? Explain.

**Lesson 7.3** Surface Area and Volume    **445**

*Connect* **&** *Extend*   **15.**

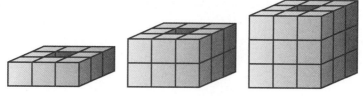

    **a.** What is the volume of each figure?

    **b.** What will the next figure in this sequence look like?

    **c.** What will be the volume of that figure?

    **d.** Write a rule to calculate the volume of the *n*th figure in the sequence.

    **e.** Explain how you know your rule is correct. Use diagrams if necessary.

**16.**

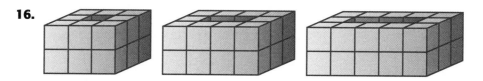

    **a.** What is the volume of each figure?

    **b.** What will the next figure in this sequence look like?

    **c.** What will be the volume of that figure?

    **d.** Write a rule to calculate the volume of the *n*th figure in the sequence.

    **e.** Explain how you know your rule is correct. Use diagrams if necessary.

**17.** Each pair of pictures shows a prism seen from the side and from the top.

**Prism A**

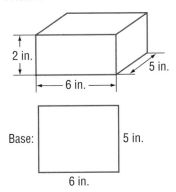

Base:

**Prism B**

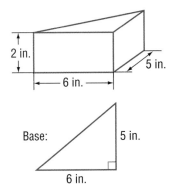

Base:

**Prism C**

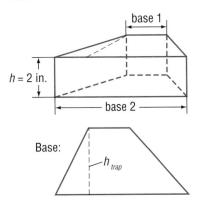

Base:

**Cylinder D**

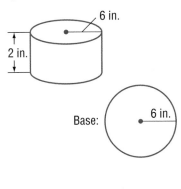

Base:

**a.** What is the volume of prism A?

**b.** What is the volume of prism B?

**c.** The base of prism C is a trapezoid. The height of the trapezoid has been labeled $h_{trap}$, to distinguish it from $h$, the height of the prism. ($h = 2$.) Find values for base 1 $b_1$, base 2 $b_2$, and $h_{trap}$ that will make the volume of prism C equal to the volume of prism A.

**d.** Cylinder D is a cylinder. Like a prism, it has two identical bases, which are parallel to each other. You could imagine making a cylinder by stacking up circles on top of each other. Does the formula for the volume of a rectangular prism also work to find the volume of a cylinder? Explain your reasoning.

**e.** Which volume is larger: a prism with $h = 4$ cm, $l = 15$ cm, and $w = 5$ cm, or a cylinder with $h = 4$ cm and $r = 5$ cm? Explain your reasoning.

**18.** The FreshStuff grocery store buys cereal boxes with dimensions of 15 in. × 12 in. × 2 in.

   **a.** Cereal boxes arrive at the store in big containers called flats. The volume of a flat is 51,840 in.$^3$. What is the maximum number of cereal boxes that could fit into a flat, judging by volume?

   **b.** The dimensions of a flat are 2.5 ft × 4 ft × 3 ft. Can the number of boxes you calculated in Part a actually be arranged so that they will fit into the flat? Explain your reasoning.

   **c.** Cereal is delivered in a truck whose shipping compartment is 18 ft × 5 ft × 6 ft. What is the maximum number of flats that will actually fit in the truck? Assume that all the flats have to be stacked facing in the same direction. Explain how you know.

   **d.** What is the volume of the truck that will be empty if it is packed as in Part c?

**Mixed Review**

**19.** Stewart answered 14 out of 18 questions correctly on his math test. What percent did he answer correctly? Round your answer to the nearest whole percent.

**20. Statistics** The pictograph shows the numbers of new dogs of seven breeds that were registered with the American Kennel Club during 2006.

**Dogs Registered During 2006**

🦴 = 10,000 dogs

Number Registered

Source: www.akc.org/reg/dogreg_stats_206.cfm

   **a.** About how many Labrador Retrievers were registered with the AKC during 2006?

   **b.** In 2001, about 35,000 Chihuahuas were registered with the AKC. About how many fewer Chihuahuas were registered in 2006?

   **c.** The number of German Shepherds registered is about how many times the number of Chinese Shar-Peis registered?

**LESSON**

# 7.4

# Capacity

## Vocabulary

capacity

**Real-World Link**

The world consumes about 500 million gallons, or 1.9 billion liters, of milk a day.

In the previous lesson, you studied the concept of volume. **Capacity**, the amount of liquid a container can hold, is closely related to volume. A brick has volume, but no capacity to hold liquid.

The metric system of measurement can be used to measure capacity. Since it was created to be consistent with our decimal number system, the rule for converting from one unit to another is always the same. Ten of a unit equals one of the next larger unit, whether you are measuring length, mass, or capacity.

| Prefix | Meaning |
|--------|---------|
| milli- | Base unit ÷ 1000 |
| centi- | Base unit ÷ 100 |
| deci- | Base unit ÷ 10 |
| no prefix | Base unit • 1 |
| deca- | Base unit • 10 |
| hecto- | Base unit • 100 |
| kilo- | Base unit • 1000 |

The metric system always uses the same pattern for naming the units. A *prefix* added to the name of the base unit tells you the size of the unit.

In this activity, you will investigate the capacities of some everyday containers to develop a sense of various metric measurement units.

**Explore**

Pick a container. Fill it with water. Use a funnel to transfer the water from the container into a 1-liter bottle. Estimate how many times you would have to do this in order to completely fill the 1-liter bottle. Record your estimate in a table.

| Container | |
|-----------|---|
| **How many in 1 liter? (estimate)** | |
| **Capacity in milliliters (measure)** | |
| **Capacity in liters** | |
| **How many in 1 liter? (calculate)** | |
| **Closest metric unit** | |

How many milliliters do you think your container will hold?

Fill the container again. This time use the funnel to transfer the water into a graduated cylinder. The graduated cylinder is marked in milliliters. How many milliliters of water did you pour into the cylinder? Record your answer in the table.

Express the container's capacity in liters, using decimals as needed.

How many times would you need to empty your container into a 1-liter bottle to fill the bottle all the way? Is the capacity of your container closer to 1 milliliter, 1 centiliter, 1 deciliter, or 1 liter?

Name something whose capacity it would make most sense to measure in milliliters.

Name something whose capacity it would make most sense to measure in liters.

# Investigation 1 Metric Units for Capacity

## Materials

- 1-liter bottle (for example, a soda bottle)
- one or more different-sized drinking glass
- one or more different-sized eating spoons
- small containers such as an empty aspirin bottle or a dental floss box
- graduated cylinder marked in milliliters
- funnel
- 2 nets to create cubes
- scissors
- tape
- half-gallon milk/orange juice carton (square bottom, straight sides)
- string

The *base unit* for measuring capacity is the liter. A liter has the capacity of about a quart. You are probably familiar with 1-liter or 2-liter drink bottles. When items in the store are labeled with metric units, capacity is usually expressed either in liters or milliliters (thousandths of a liter).

## ✅ Develop & Understand: A

1. Complete this ratio table showing the relationship between milliliters and liters.

| Liters | 1 | | 45 | 1.5 | | 0.5 | |
|---|---|---|---|---|---|---|---|
| Milliliters | | 3,000 | | | 7,453 | | 386 |

2. Write a ratio to express the relationship between milliliters and liters. Include units so that it is clear what the numbers represent.

## ✅ Develop & Understand: B

You are the manager of a small factory that manufactures kitchen cleaning products. The active ingredient in your Eco-Fresh Cleaning Liquid is citric acid. One batch of Eco-Fresh requires 2,750 milliliters of citric acid.

3. How many liters of citric acid are required for one batch of Eco-Fresh? For five batches? Write your answer as a decimal.

4. Your supplier sells citric acid in 1-liter jugs. You do not want to waste any ingredients.

   a. How many batches of Eco-Fresh do you need to make in order to use up all the citric acid you buy?

   b. How many jugs of citric acid will you need to do this?

### Real-World Link

A *prefix* is a piece of a word that has its own meaning and can be attached to the front of other words. For example, the prefix "re-" means "again." "Revisit" means "visit again," "reapply" means "apply again."

A *suffix* is just like a prefix, except that it goes at the end of a word instead of the beginning.

One batch of Spring Rinse dish soap contains 3,850 milliliters of coconut oil and 650 milliliters of citric acid. These are the only two ingredients. Use this information to answer Exercises 5–9.

5. How many liters of each ingredient does a batch contain?

6. How many liters of Spring Rinse does one batch make?

7. How many batches of Spring Rinse can you make with 23.1 liters of coconut oil?

8. If you use 11550 milliliters of coconut oil, how many liters of citric acid do you need?

9. The new and improved formula for Spring Rinse calls for 983 milliliters of coconut oil for each liter.

   a. How many milliliters of citric acid are in one liter of Spring Rinse made according to this formula?

   b. How many liters is that?

## ✅ Develop & Understand: C

How much water does a rectangular baking pan like this one hold?

You already know how to calculate the volume of a rectangular prism. The volume and capacity of a container are related as follows.

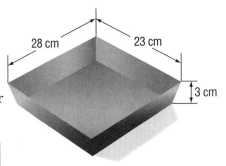

28 cm    23 cm    3 cm

> A 1-centimeter cube has a capacity of 1 milliliter.

**10.** Each of these squares is 1 cm on each side. Cut out and assemble a net like this one to make a cube 1 cm on each side. What is the capacity of this cube?

**11.** How long should the sides of a cube be so that it will hold 1 liter of water?

**12.** When you have answered Exercise 11, get a second net from your teacher. Check the length of the sides of the squares and compare them to your answer for Exercise 11. Assemble the new cube.

**13.** Describe how you could arrange 1,000 copies of the smaller cube inside the larger one.

**14.** Examine a milk carton with a square bottom.

    **a.** Measure and record the length in centimeters of the sides of the square.

    **b.** Calculate the area of the bottom of the carton.

    **c.** If the carton were the same height as your 1-liter cube, what would be the carton's volume?

    **d.** Approximately how many centimeters high would the carton have to be for its capacity to be 1 liter? Round your answer to the nearest centimeter.

    **e.** Cut the carton down to the height you calculated in Part c.

**15.** You now have three different containers with capacities of 1 liter. Compare the carton, cube, and bottle. Do they look as though they all hold the same amount of liquid? Explain how they can all have the same capacity even though they are different shapes.

**16.** How long should the sides of a cube be, so that it can hold 1 kiloliter of water?

    **a.** How many of your 1-liter cubes would fit inside such a cube?

    **b.** How many of your 1-milliliter cubes would fit inside the 1-kiloliter cube?

**17.** Simon bakes with a rectangular baking pan that is 28 cm long, 23 cm, and 3 cm deep.

    **a.** What is the volume of the pan in cubic centimeters?

    **b.** What is the capacity of the pan in milliliters?

    **c.** What is the capacity of the pan in liters?

**18.** Simon's kitchen sink is 55 cm long, 40 cm wide, and 18 cm deep.

    **a.** What is the volume of the sink in cubic centimeters? In milliliters? In liters?

    **b.** Which metric unit would you choose to describe the capacity of the sink? Explain your reasoning.

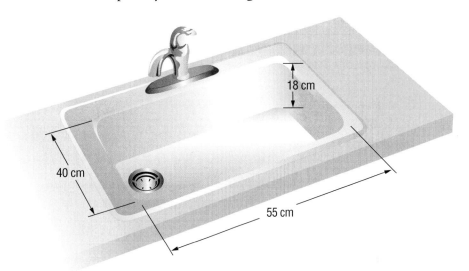

**19.** Simon's bathtub is 48 cm wide, 140 cm long, and 30 cm deep.

    **a.** What is the volume of the bathtub in cubic centimeters?

    **b.** Simon filled his sink and then transferred all the water into the bathtub. About how many times would he need to do this to fill the bathtub?

    **c.** Use your answer from Part b to estimate how many liters of water the bathtub holds.

    **d.** Use your answer from Part a to calculate the capacity of the bathtub in liters exactly.

    **e.** About how many bathtubs would be needed to hold 1 kiloliter of water?

## Share & Summarize

**1.** How many milliliters are in a liter? How many centiliters are in a liter? How many milliliters are in a centiliter?

**2.** What metric units would be reasonable to use for measuring the capacity of a drinking glass? A soup pot? A bathtub? A swimming pool?

# Investigation 2   Customary Units for Capacity

## Materials

- juice or milk cartons or bottles (gallon, half-gallon, quart, pint, cup)
- drinking glasses and water bottles of assorted sizes
- small and medium-sized saucepans, or other appropriately-sized containers
- measuring cups
- funnel
- water

In Investigation 1, you measured capacity using metric units. In this investigation, you will learn about U.S. customary units for capacity.

The customary units for capacity are *cups*, *pints*, *quarts*, and *gallons*. You have probably seen drinks sold in containers of all these sizes, as well as half-gallons.

### ✅ Develop & Understand: A

1. Find the capacity of each carton on the label.

   a. Arrange the five cartons in order of size.

   b. Estimate how many of the 1-cup cartons it would take to fill the pint carton.

2. Estimate:

   a. How many pints it would take to fill a quart container?

   b. How many quarts it would take to fill a half-gallon container?

   c. How many half-gallons it would take to fill a gallon container?

3. Use the cartons as benchmarks to help you to estimate the capacities of other containers.

   a. Which container holds about a cup?

   b. Which holds less than a cup?

   c. Which holds about a pint?

   d. Which holds about a quart?

   e. Which holds more than a gallon?

4. Fill a 1-cup measuring cup with water. Use a funnel to transfer the water into one of the containers. Repeat as many times as you need to fill the container. How much water does the container hold? Was your estimate of its capacity close to the real value?

5. Now check your estimates for Exercise 2a–c by the same method.

### Math Link

Fluid ounces are another unit for measuring capacity that you may have seen on labels. Fluid ounces also belong to the U.S. customary system. There are 8 fluid ounces in 1 cup.

6. Complete the ratio table showing the conversion ratios among customary units of capacity.

| Cups | | 16 | | | 4 | 3 | 2 | 1 |
|---|---|---|---|---|---|---|---|---|
| **Pints** | 16 | | 6 | 5 | 4 | 2 | $1\frac{1}{2}$ | 1 |
| **Quarts** | | | 3 | $2\frac{1}{2}$ | 2 | | | |
| **Gallons** | 2 | 1 | | | $\frac{1}{2}$ | $\frac{1}{4}$ | | |

7. Express each relationship as a ratio.

   a. Cups : Pints

   b. Pints : Quarts

   c. Quarts : Gallons

8. How many cups are in one gallon?

9. What fraction of a quart is a cup?

10. Martina has a soup recipe that measures all the ingredients in cups. If she substitutes pints for cups for each ingredient, how will the final volume of punch compare to the volume the original recipe made?

### Real-World Link

Drinking glasses come in many sizes and shapes, but two common sizes hold 1 cup or $1\frac{1}{2}$ cups. When you buy an individual serving of milk or juice at the store, it often comes in $1\frac{1}{2}$ cup, $1\frac{3}{4}$ cup, or 2-cup portions. Take a look at the drinks in your cafeteria or at the store. How much is in an individual serving?

## ✅ *Develop & Understand: C*

Christopher and Ana mixed juice for a party. They mixed frozen juice concentrate with water. Each can of concentrate is $1\frac{1}{2}$ cups.

**11.** The instructions on a can of concentrate state that one can makes $\frac{1}{2}$ gallon of juice. How many cups of water do Christopher and Ana need to add to one can of concentrate?

**12.** How many cups of concentrate should they use if they want to make 3 gallons of juice?

**13.** How many gallons of juice can they make with 9 cups of concentrate?

**14.** How many cups of concentrate should they use to make 1 quart of juice?

**15.** Ana's drinking glasses each held 1 cup of juice. She knew that there would be 20 guests at the party. She wanted to estimate how much juice to make.

**a.** If each guest drinks 2 glasses of juice, how many gallons of juice do Ana and Christopher need to make?

**b.** How many cups of concentrate do they need?

**c.** How many cups of concentrate do they need if each guest drinks 3 glasses of juice?

**d.** Ana decided to make 4 gallons of juice. At the end of the party, 2 quarts of juice were left. How much juice did each guest drink?

### *Math Link*

For measuring capacities smaller than a cup, the U.S. customary system uses either fluid ounces (16 fluid ounces = 1 cup) or teaspoons (8 teaspoons = 1 cup). A tablespoon equals 2 teaspoons.

You can also use fractions of a cup (for example, kitchen measuring cups usually come in 1-cup, half-cup, third-cup, and quarter-cup sizes).

### *Share & Summarize*

**1.** How many cups are in a pint? In a quart? In a gallon?

**2.** What fraction of a gallon is 3 quarts?

**3.** What are some similarities and differences between the metric system for measuring capacity and the customary system?

**Practice & Apply**

1. Complete each ratio table. Write your answers as decimals.

| Milliliters | 3,000 | | | | 125 | | 33,000 |
|---|---|---|---|---|---|---|---|
| Centiliters | | 187 | | | | | |
| Liters | | 1.87 | | 746 | | 0.009 | |
| Kiloliters | | | 2.09 | | | | 0.033 |

**Which metric unit is most reasonable to measure the capacity of the following containers?**

2. a coffee mug

3. a refrigerator

4. a small spoon

5. a cooking pot

6. The label on a bottle of Rainbow Juice advertises that the drink contains 12% fruit juice.

   a. How many milliliters of fruit juice are in a 2-liter bottle of Rainbow Juice?

   b. How many milliliters of fruit juice are in a $\frac{1}{2}$ liter bottle?

   c. What is the ratio of fruit juice to other ingredients in Rainbow Juice? Write your answer as a ratio with the smallest whole numbers possible.

   d. How many liters of Rainbow Juice can be made with 12 milliliters of fruit juice?

   e. What is the proportion of fruit juice in the final mixture? Write your answer as a ratio with the smallest whole numbers possible.

   f. How many liters of Rainbow Juice can be made with 600 milliliters of fruit juice?

7. Complete each ratio table. Write your answers as fractions or mixed numbers.

| Cups | | | | 5 | $\frac{1}{2}$ | | |
|---|---|---|---|---|---|---|---|
| Pints | | | 24 | | | | |
| Quarts | | 7 | | | | $3\frac{3}{4}$ | |
| Gallons | 5 | | | | | | $1\frac{5}{8}$ |

**Which customary unit is most reasonable to measure the capacity of the following containers?**

**8.** a bathtub

**9.** a drinking glass

**10.** a swimming pool

**11.** a cereal bowl

**12.** Doug and Arleta repainted Doug's bedroom. First, they mixed different colors of paint to make the shade of green Doug wanted. The sample mixture contained 1 cup of white paint, $\frac{1}{2}$ cup of blue paint, and $\frac{3}{4}$ cup of yellow paint. They used the same recipe to make larger quantities of paint for their project.

   **a.** If they use 3 cups of white paint, how much yellow paint do they need?

   **b.** If they use 1 quart of blue paint, how much white paint do they need?

   **c.** If they use 1 gallon of white paint, how much blue paint do they need?

   **d.** If they use 3 gallons of yellow paint, how much white paint do they need?

   **e.** If they use 2 gallons of blue paint, how much paint will they have when it is all mixed together?

**13.** Doug and Arleta had 5 pints of blue paint, 1 gallon of yellow paint, and 3 quarts of white paint.

   **a.** How much of the paint mixture from Exercise 12 could they make?

   **b.** How many quarts of each color was left over after they made the mixture?

**14.** Teacups often hold $\frac{3}{4}$ cup of tea. The capacity of a teapot is often described in terms of how many teacups it can fill. For example, a 5-cup teapot holds enough tea to fill 5 teacups.

   **a.** What is the capacity, in U.S. customary units, of a teapot that can fill 5 teacups?

   **b.** How many teacups can you fill from a teapot that holds 3 pints of tea?

   **c.** How many quarts of tea do you need to fill 16 teacups?

**d.** Rachel likes to drink her tea out of a standard coffee mug, which holds $1\frac{1}{2}$ cups. How many times can she fill her mug from a teapot that can fill 5 teacups?

**e.** Rachel wants to buy a teapot that holds enough tea for 5 mugs. So far, she has found a teapot that holds 3 pints, one that holds 2 quarts, one that holds 7 cups, and one that holds enough for 10 teacups. Are any of these pots big enough for Rachel? If so, which?

**15.** Adrian paid $2.70 for 3 pints of milk. Mariella bought 2 gallons of milk and paid a total of $9.80. Jerome bought a quart of milk for $1.60. Who got the best deal? Explain your reasoning.

**Connect & Extend**

**16.** Remember that a cube whose sides are 1 cm long has a volume of 1 cubic centimeter and a capacity of 1 milliliter. A cube with 10 cm sides has a volume of 1,000 cubic centimeters and a capacity of 1 liter.

**a.** Complete the table below.

| Side length of cube (centimeters) | 1 | 2 | 3 | 4 | 5 | 6 | 7 | 8 | 9 | 10 |
|---|---|---|---|---|---|---|---|---|---|---|
| Capacity of cube (milliliters) | 1 | 8 | | | | | | | | 1,000 (1 liter) |

**b.** Approximately how long are the sides of a cube whose capacity is $\frac{1}{2}$ liter?

**c.** Approximately how long are the sides of a cube whose capacity is $\frac{1}{4}$ liter?

**d.** What fraction of a liter is the capacity of a cube with 5 cm sides?

**e.** Compare the capacity of a cube with 2 cm sides to the capacity of a cube with 4 cm sides. How many times larger is the capacity of the larger cube?

**f.** Now compare the cube with 3 cm sides to the cube with 6 cm sides. How many times larger is the capacity of the larger cube?

**g.** What happens to the volume of a cube when you double the length of the sides?

**h.** Why doesn't the capacity of a cube double when you double the length of the sides?

**17.** Use the table from Exercise 16 to help you solve this exercise.

    **a.** Which of the cubes in the table has a capacity closest to 1 centiliter (10 milliliters)?

    **b.** Use your calculator to find a closer approximation of the length of the sides of a cube whose capacity is 1 centiliter. Approximate the side length to the closest tenth of a centimeter.

    **c.** Now approximate the side length to the closest hundredth of a centimeter.

    **d.** Which of the cubes in the table has capacity closest to 1 deciliter (100 milliliters)?

    **e.** Use your calculator to approximate the side length of a cube whose capacity is 1 deciliter. Approximate the side length to the nearest hundredth of a centimeter.

**18.** Jason has three mixing bowls. Use the following clues to figure out the capacity of each bowl. Explain your reasoning.

    **Clue 1:** The red bowl holds 1 quart more than the yellow bowl.

    **Clue 2:** The silver bowl holds the same number of quarts as the red bowl holds pints.

    **Clue 3:** When Jason poured milk from a gallon jug into the silver bowl, he was able to fill the bowl and had a quart of milk left.

**19.** The small jug's capacity is $1\frac{1}{2}$ pints. The volume of the big jug is twice the volume of the small jug. What is the capacity of the big jug in cups? Explain your reasoning.

**20. In Your Own Words** You are helping Carisa plan for a party at school. You have purchased several different containers to hold the punch. Carisa has called one evening to discuss how much punch the containers will hold. Write your side of the conversation that explains to Carisa what size the containers are and approximately how much punch each will hold.

*Mixed Review*

**21.** Here are the first four stages of a sequence.

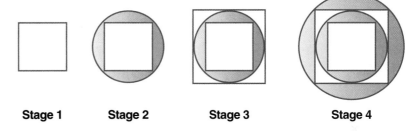

Stage 1     Stage 2     Stage 3     Stage 4

**a.** Describe the pattern in this sequence.

**b.** Draw the next stage in the sequence.

**c.** Draw stage 15. Explain how you know you are correct.

**22.** The number of students in the Science club is 87.5% of the number on the Math team. If there are 21 students in the Science club, how many are on the Math team?

**23.** Of the 21 students in the Science club, 8 are also on the Math team. What percent of the Science club is this?

# Review & Self-Assessment

## Vocabulary

- arc
- area
- capacity
- central angle
- circle sector
- parallelogram
- perfect square
- prism
- rectangular prism
- surface area
- trapezoid
- volume

## Chapter Summary

You learned that the area of a shape is the number of square units that fit inside it. You estimated areas of shapes by counting squares, and you learned formulas for calculating areas of rectangles, parallelograms, triangles, trapezoids, and circles.

You were asked to explore the concept of volume by finding the volume and surface area of various block structures. You looked as the simple method of finding the volume and the more complex visual method of finding the surface area. By studying these block structures, you were able to then progress into finding the volume of rectangular prisms using a formula. Both finding the volume when given three measurements and finding a missing measurement when given the volume and other two measurements were applied. Finally, an introduction to polyhedra from polygons was discussed including construction of polyhedra.

## Strategies and Applications

The questions in this section will help you review and apply the important ideas and strategies developed in this chapter.

**Find the area of each shape.**

**1.**

3 in.

3 in.

**2.**
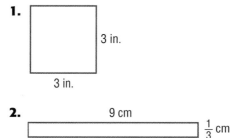

9 cm

$\frac{1}{3}$ cm

**Square each number.**

**3.** 16.4          **4.** 12          **5.** 0.5

### Finding and estimating areas

**6.** If two shapes have the same perimeter, must they also have the same area? Use words and drawings to help explain your answer.

**7.** Find the area of this parallelogram in centimeters. Explain the steps you followed.

8. In this chapter, you learned how to find the area of a triangle.

   **a.** Describe the base and height of a triangle.

   **b.** Explain how to find the area of a triangle if you know the lengths of the base and the height.

   **c.** How is finding the area of a triangle related to finding the area of a parallelogram?

9. A CD has a diameter of about 12 cm. The hole in the center of a CD has a diameter of about 1.5 cm. Find the area of a CD, not including the hole, to the nearest tenth of a square centimeter. Explain how you found your answer.

## Demonstrating Skills

**Find the perimeter and area of each figure.**

**10.**

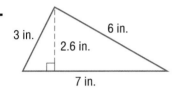

**11.**

**12.**

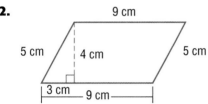

**13.**

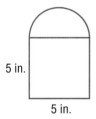

In Questions 14 and 15, find the value of *b*.

**14.**

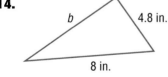

**15.**

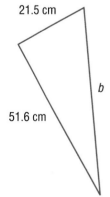

**Tell whether each is a perfect square.**

**16.** 289                **17.** 72                **18.** 1.69

**19.** Elena squared a number and the result was twice the number with which she started. What was her starting number?

**Find the area of each trapezoid.**

**20.**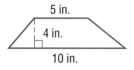

**21.**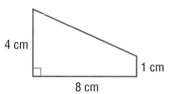

**22.** A trapezoid has an area of 240cm². Its two bases are 12 cm and 18 cm. What is its height?

**Calculate the area of each circle sector. Round to the nearest hundredth of a square unit.**

**23.**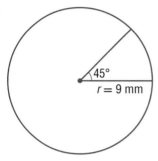

**24.**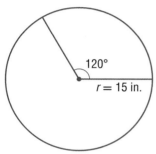

**Find the missing information.**

**25.**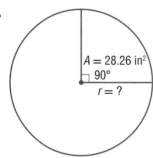

**26.**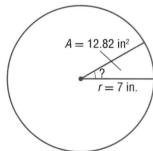

**Calculate the volume of each rectangular prism. Include the appropriate measurement units in your answer.**

**27.**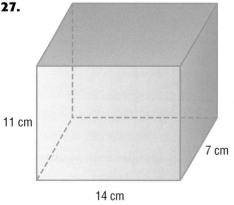

**28.**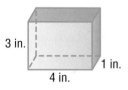

**29.**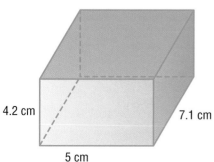

**30.** Jenelle's fish tank is 40 cm long, 15 cm wide, and 20 cm high. She has a bucket with a capacity of 19.8 liters.

**a.** What is the volume of the fish tank in cubic centimeters? Cubic meters?

**b.** About how many times would she need to fill the bucket and then transfer all the water to the fish tank in order to fill the tank?

**c.** Use your answer from Part b to estimate about how many liters of water the fish tank holds.

## Test-Taking Practice

### SHORT RESPONSE

**1** Shana is wrapping a package that has the shape of a right triangular prism, as shown to the right.

What is the minimum amount of paper that would be needed to cover all the surfaces of the package? The surface area of a right triangular prism = $wh + lw + lh + ls$.

**Show your work.**

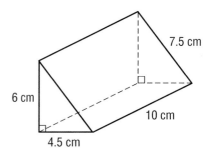

Answer _____

### MULTIPLE CHOICE

**2** Which of the following numbers is not a perfect square?

A  156
B  196
C  256
D  324

**3** What is the area of the triangle shown?

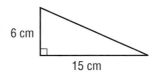

F  21 square centimeters
G  45 square centimeters
H  60 square centimeters
J  90 square centimeters

**4** A fish tank is 20 inches long, 10 inches wide, and 12 inches high. What is the volume of the tank?

A  240 cubic inches
B  1,120 cubic inches
C  1,200 cubic inches
D  2,400 cubic inches

**5** What is the area of a circle with a 15-cm diameter? Use 3.14 for $\pi$.

F  23.55 centimeters
G  47.1 centimeters
H  176.6 centimeters
J  706.5 centimeters

# Coordinate Plane

## Real-Life Math

**Let It Snow, Let It Snow, Let It Snow** Do you remember what you were doing on February 12, 2006? If you lived in New York City, it probably involved snow. The graph below shows the snowfall amounts in New York City for five months during the 2005–2006 season.

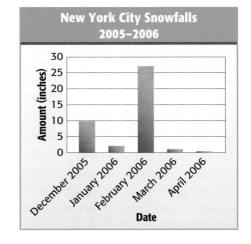

**Think About It** A graph is a useful tool for showing how quantities are related. You can tell at a glance that much more snow fell in February 2006 than in any other month. About how many times more snow fell in February 2006 than in December 2005?

**Math Online**

Take the **Chapter Readiness Quiz** at glencoe.com.

# Dear Family,

Graphs can be seen everywhere, on the sports page of the newspaper, in advertisements, and in your Science or Social Studies books. Most of these graphs are line graphs, bar graphs, or circle graphs. In this chapter, your student will learn about graphs that use points and lines to show patterns and relationships in data.

## Key Concept—Graphs

Here is an example. This graph has a horizontal axis and a vertical axis. It shows the prices of different bags of sugar. From this graph, facts like the following can be determined.

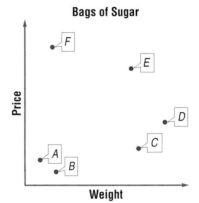

**Bags of Sugar**

- *D* is the heaviest.

- *B* and *F* are the same weight.

- *C*, *E*, and *D* are heavier than *B*.

- *E* and *F* cost more than *D*.

- *C* would give you a better value for the money than *B* because you get more sugar for only a little more cost.

In this chapter, your student will draw graphs for many types of situations. Some graphs, like the one above, will not have numerical values. Others will have scales, or sequences of numbers, along each axis.

## Chapter Vocabulary

| | |
|---|---|
| absolute value | opposites |
| axes | ordered pair |
| coordinates | origin |
| line graph | positive numbers |
| negative numbers | quadrants |

## Home Activities

- Help your student find examples of how graphs are used in everyday life by looking in newspapers and magazines.
- Encourage your student to determine what the graphs show, as well as the values for specific points in the graphs.

# Interpret Graphs

You can find graphs in many places, in your school books, on television, and in magazines and newspapers. Graphs are useful for displaying information so it can be understood at a glance. They are also a wonderful tool for making comparisons.

## Explore

Choose a topic in which you are interested, such as cars, dogs, or music. Think about some aspects of your topic that would be interesting to display in a graph. Describe what the graph might look like.

For example, if you choose cars for your topic, you might graph the number of cars of different sizes sold at a local dealership. You could list the sizes as compact, midsize, and full-size along the bottom and then draw bars to show how many cars of each size are sold in a day.

You have seen graphs that use bars, sections of circles, and pictures to display data. In this lesson, you will focus on graphs that use points and curves to show information.

## Investigation 1   Understand Graphs

**Vocabulary**

axes

origin

**Materials**

- drawing of the buildings
- graph of the buildings

Here are the front views of four buildings.

**SuperShop**     **Post Office**     **Westgate**     **Acme**

Some of the buildings are taller than others, and some are wider than others. In other words, the heights and widths of the buildings vary. As you may recall, quantities that vary, or change, are called variables.

Graphs are a convenient way to show information about two variables at the same time. This graph displays information about the heights and widths of the buildings shown on page 468. Each point represents one of the buildings.

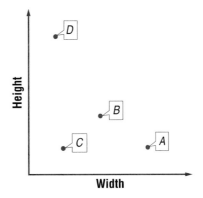

The horizontal line, representing the variable *width*, and the vertical line, representing the variable *height*, are called **axes**. The point where the axes meet is called the **origin**. The origin of a graph is usually the 0 point for each axis.

The arrow on each axis shows the direction in which the values of the variable are increasing. For example, points on the right side of the graph represent wider buildings than points on the left. Points near the top represent taller buildings than points near the bottom.

This graph does not have numbers along the axes, so it does not tell you the actual heights or widths of the buildings. However, it does show you how the heights and widths compare.

## Think & Discuss

Which point represents a wider building, *A* or *B*? Explain how you know.

Which building does Point *A* represent? How do you know?

## ✅ *Develop & Understand: A*

1. Identify the buildings represented by points *B*, *C*, and *D* on the graph of the buildings on page 469.

2. Which letters represent buildings that are less wide than the building represented by point *B*?

3. Which letters represent buildings that are shorter than the building represented by point *B*?

4. On a copy of the graph, add two more points, one to represent a skyscraper and the other to represent a doghouse. Explain how you decided where to put the points.

5. Now imagine a building with a different height and width from the four in the drawing.

   a. On a copy of the drawing, make a sketch of the building you are imagining.

   b. Add a point to the graph to represent your building.

   c. Show your graph to a partner. Ask him or her to sketch or describe your building.

In the graph you used in Exercises 1–5, the variable *width* is represented by the horizontal axis. The variable *height* is represented by the vertical axis. You could instead label the axes the other way.

## ✅ *Develop & Understand: B*

6. In this graph, the horizontal axis shows height. The vertical axis shows width.

   a. Plot a point for each of the four buildings on a copy of this set of axes.

   b. How does your graph compare to the graph on the previous page?

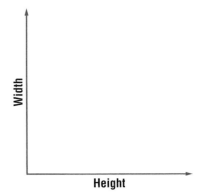

**Real-World Link**

One of the world's largest hotels is the MGM Grand in Las Vegas, Nevada. The hotel has 5,044 guest rooms and covers 112 acres.

7. The points on the graph below represent the height and weight of the donkey, dog, crocodile, and ostrich shown in the drawing.

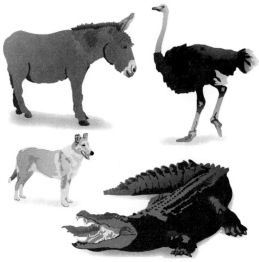

a. What are the two variables represented in the graph?

b. Tell which point represents each animal. Explain how you decided.

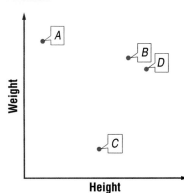

8. The two graphs below compare car A and car B. The left graph shows the relationship between age and value. The right graph shows the relationship between size and maximum speed.

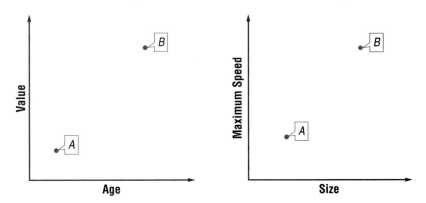

Use the graphs to determine whether each statement is true or false. Explain how you know.

a. The older car is less valuable.

b. The faster car is larger.

c. The larger car is older.

d. The faster car is older.

e. The more valuable car is slower.

## Share & Summarize

This graph shows the heights and widths of four chimneys. Sketch four chimneys that the points could represent. On your sketch, label each chimney with the correct letter.

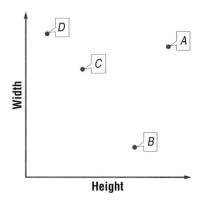

# Investigation 2 Interpret Points

It takes practice to become skillful at reading graphs. When you look at a graph, you need to think carefully about what the variables are and where each point is located.

## Example

What information does this graph show?

- The two variables are speed and age.
- *Y* is faster than *Z*. *Z* is older than *Y*.

What might *Y* and *Z* represent?

- *Y* and *Z* could be computers since a newer computer usually processes information faster than an older computer.
- *Y* could represent a boy. *Z* could represent his grandfather. The grandfather is older than the boy. The boy runs faster than the grandfather.

# ✅ Develop & Understand: A

Complete Parts a and b for each graph.

    **a.** Tell what two variables are shown on the graph.

    **b.** Describe what the graph tells you about the variables represented by the points. Then state what the points could represent.

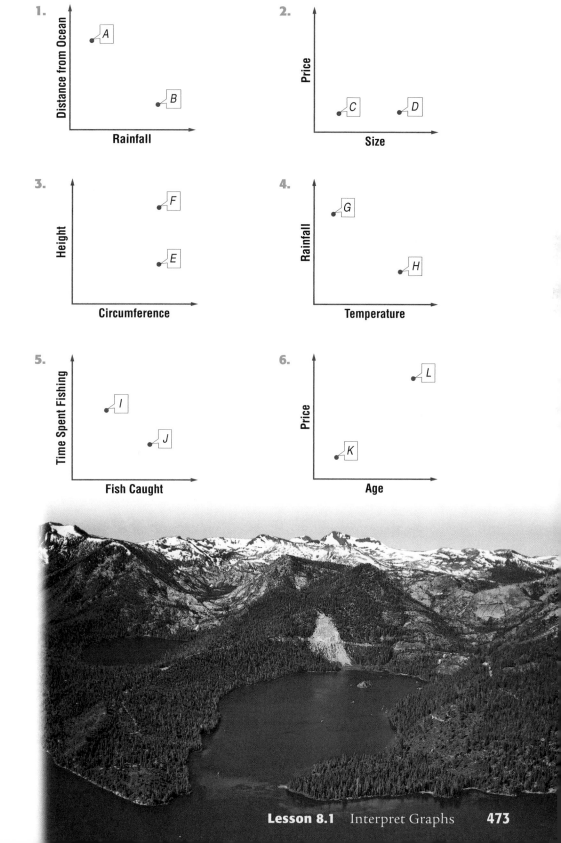

**1.**
Distance from Ocean (vertical axis)
Rainfall (horizontal axis)
A
B

**2.**
Price (vertical axis)
Size (horizontal axis)
C
D

**3.**
Height (vertical axis)
Circumference (horizontal axis)
F
E

**4.**
Rainfall (vertical axis)
Temperature (horizontal axis)
G
H

**5.**
Time Spent Fishing (vertical axis)
Fish Caught (horizontal axis)
I
J

**6.**
Price (vertical axis)
Age (horizontal axis)
L
K

**7.** Each point on this graph represents one bag of sugar.

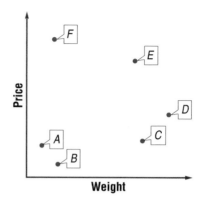

a. Which bag is heaviest? Which bag is lightest?

b. Which bags are the same weight? Which bags are heavier than *B*?

c. Which bags are the same price? Which bags cost more than *D*?

d. Which bag is heavier than *B* and costs more than *D*?

e. Assuming all of the bags contain sugar of the same quality, is *D* or *F* a better value? How can you tell?

**8.** Gina is careless with her pens, often losing some and then finding them again. This graph shows how many pens were in her case at noon each day last week. Use the information in the graph to write a story about Gina and her pens over the course of the week.

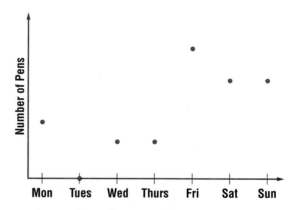

The graphs you have been exploring have had only a few points. Graphs of real data often contain many points. Although looking at individual points gives you information about the data, it is also important to consider the overall *pattern* of points.

## ✅ *Develop & Understand: C*

Ms. Dimas surveyed two of her classes to find out how much time each student spent watching television and reading last weekend. She made this graph of her results.

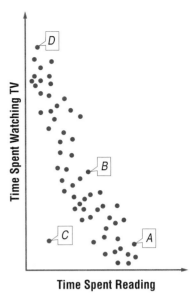

9. Choose one of the four students represented by points *A*, *B*, *C*, and *D*. Write a sentence or two describing the time that student spent reading and watching television. Do not mention the student's letter in your description.

10. Exchange descriptions with your partner. Try to figure out which student your partner described.

11. Now think about all the points in the graph, not just the four labeled points. Which of these statements best fits the graph?

    a. There is not much connection between how much time the students spent reading and how much time they spent watching TV.

    b. Most students spent about the same amount of time reading as they did watching TV.

    c. The more time students spent reading, the less time they spent watching TV.

### Share & Summarize

You have seen several uses for graphs.

- Graphs can help *tell a story*. For example, the graph on page 474 shows how the number of pens Gina had changed throughout the week.

- Graphs can be used to *make comparisons*. For example, the graph on page 474 shows weights and prices of bags of sugar.

- Graphs can *show an overall relationship*. For example, the graph on page 475 shows how time spent watching TV is related to time spent reading.

Create your own graph. Explain its use.

## Investigation 3  Interpret Lines and Curves

**Vocabulary**

line graph

**Materials**

- Megan's height graph

The graphs you have seen so far have been made up of separate points. When a graph shows a line or a curve, each point on the line or curve is part of the graph. The skills you developed for interpreting graphs with individual points can help you understand information given by a line or a curve.

This graph shows the noise level in Ms. Whitmore's classroom one Tuesday morning between 9 A.M. and 10 A.M.

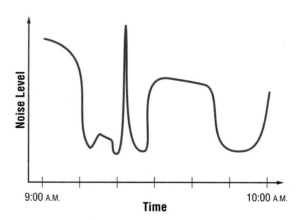

The variable on the horizontal axis is *time*. If you read this graph from left to right, it tells a story about how the noise level in the classroom changed over time.

## Real-World Link

Sound levels are measured in decibels (dB). Sound levels above 80 dB for extended periods can be harmful. Here are decibel measures for some common sounds.

| | |
|---|---|
| whisper | 25 dB |
| conversation | 60 dB |
| lawn mower | 90 dB |
| chain saw | 100 dB |
| rock concert | 110 dB |

## Think & Discuss

• At about what time did the room first get suddenly quiet? How is this shown on the graph?

• At one point during the hour, the class was interrupted for a very short announcement on the public address system. At about what time did this happen? How do you know?

• When were the students the noisiest? Ignore the PA announcement.

• During part of the hour, students worked on an exercise in small groups. When do you think this happened? Explain why you think so.

• Ms. Whitmore stopped the group activity to talk about the next day's homework. When did this happen? Explain how you know.

## Develop & Understand: A

1. This graph shows the audience noise in a school auditorium on the evening of a school play. The graph shows only the noise made by the audience, not the noise created by the actors on stage.

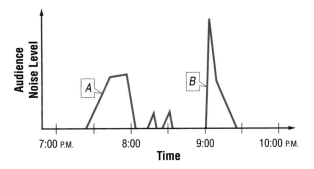

a. At 7:00 P.M., the auditorium was empty. At about what time do you think people started entering the auditorium? Explain why you think so.

b. What time do you think the play started? How is this shown on the graph?

c. At the end of the performance, the audience burst into applause. At what time do you think this happened? Why?

d. There are some small "bumps" in the graph between 8:00 P.M. and 9:00 P.M. What might have caused these bumps?

e. A and B mark sections that show an increase in noise level. In which section does the noise level increase more quickly? How can you tell?

**2.** This graph shows how Rita's hunger level changed over one Saturday. Write a story about Rita's day that fits the information in the graph. Your story should account for all the increases and decreases in her hunger level.

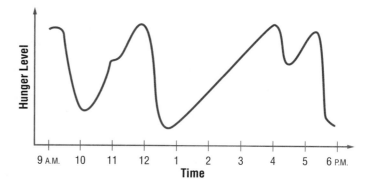

The graphs in this investigation have shown how the values of a variable change over a period of time. In some cases, you will have information about a variable only at certain times during a time period. You can often use what information you *do* have to estimate what happens between those times.

## ✅ Develop & Understand: B

As Megan was growing up, her mother measured her height on each of her birthdays. She recorded the results in a graph in the family scrapbook.

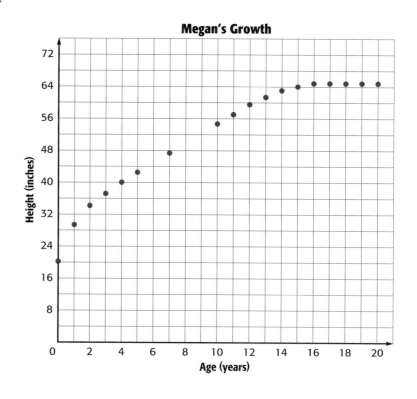

On Megan's sixth birthday, her family was on vacation, and her mother forgot to measure her. Megan's family moved right before her eighth birthday, and the scrapbook was misplaced. A few weeks before Megan turned ten, her mother found the scrapbook and started recording again.

Megan wonders what her height was in the missing years.

3. From the graph, what do you know about Megan's height when she was six years old?

Megan thought connecting the points might help her estimate her height in the missing years.

4. On a copy of Megan's graph, draw line segments to connect the points in order. Use the segments to estimate Megan's heights at ages 6, 8, and 9.

5. Do you think the values you found in Exercise 4 give Megan's exact heights at ages 6, 8, and 9? Explain.

6. Estimate how tall Megan was at age $1\frac{1}{2}$.

7. At what age did Megan's height begin to level off?

Connecting the points in Megan's height graph allowed you to make predictions. It also helped you to see *trends*, or patterns, in the data. Graphs in which points are connected with line segments are called **line graphs**.

In Megan's height graph, it makes sense to connect the points because she continues to grow between birthdays. In other words, there are values between those plotted on the graph.

For some graphs, there are no values between the plotted values. For example, Megan's father keeps a graph of how many fish he catches each year on his annual fishing trip.

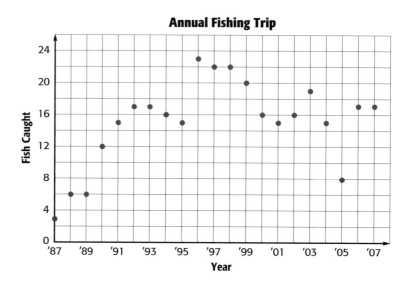

**Annual Fishing Trip**

Since the number of fish caught does not change between fishing trips, there are no values between the values shown in the graph. In this case, connecting the points does not make sense.

## ✅ *Develop & Understand: C*

Tell whether it would make sense to connect the points in each graph described below. Explain why you think so.

8. a graph showing the number of tickets sold for each football game during the season

9. a graph showing the speed of a race car every ten minutes during a race

10. a graph of the sun's perceived height at each hour during the day

11. a graph of Clara's weekly paycheck amount for ten weeks

## Share & Summarize

These graphs show how something changes over time. For each graph, write a sentence or two describing the change. Then try to think of something that might change in this way.

For example, this graph shows something increasing slowly at first and then more quickly. It could represent the speed of someone running a long-distance race. The runner starts out slowly and then moves faster and faster as she sprints toward the finish line.

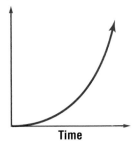

**1.**

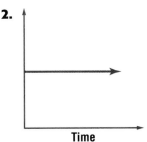

**2.**

**3.**

**4.**

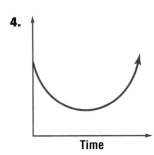

**5.**

**6.**

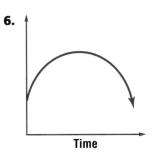

**7.** Describe a graph involving change over time for which it would make sense to connect the points.

**8.** Describe a graph involving change over time for which it would not make sense to connect the points.

**Practice** & **Apply**

**Real-World Link**

One of the world's tallest apartment buildings is the John Hancock Center in Chicago, Illinois. The building is 1,127 feet tall and has 100 stories.

**1.** This drawing shows the front view of several buildings.

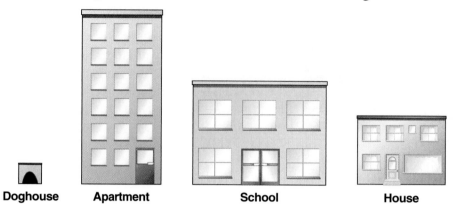

Doghouse    Apartment        School        House

**a.** Copy this set of axes. Plot a point to represent the height and width of each building.

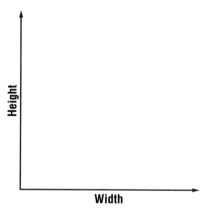

**b.** Add a new point to your graph. Write a brief description of the building it represents. Describe how the height and width of your building compare to the height and width of at least two of the other buildings.

**2.** This graph shows the relationship between effort and test results for five students.

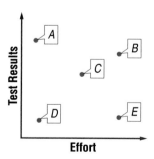

The teacher wrote the following comments on the report cards for these students.

- Allen's poor attendance this term has resulted in an extremely poor test performance.

- Nicola is a very able pupil, as her test mark clearly shows. But her concentration and behavior in class are poor. With more effort, she could do even better.

- Hoang has worked very well and deserves his marvelous test results. Well done!

- Adrienne has worked reasonably well this term. She has achieved a satisfactory test mark.

   **a.** Match each student's performance to a point on the graph.

   **b.** Write a comment about the student represented by the point that you did not mention in Part a.

**3.** Complete Parts a and b for each graph.

   **a.** Tell what two variables the graph shows.

   **b.** Describe what the graph tells you about the subjects represented by the points. Then try to think of an idea about what the points could represent.

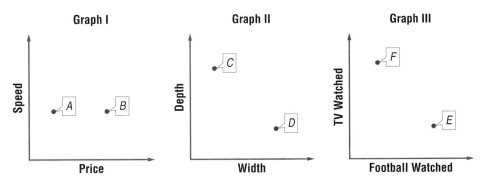

**4.** The age and height of each person in the photograph are represented by a point on the graph. Going from left to right, match each person to a point.

**5.** This graph shows the height of a hay crop over a summer.

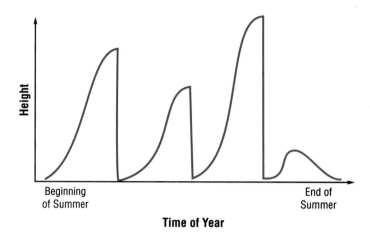

**a.** How many times was the crop harvested during the summer? How can you know this by looking at the graph?

**b.** Before which harvest was the hay tallest?

**c.** Describe the change in the hay's height after the third harvest. Why do you think the height changed this way?

**6.** This graph shows Tyson's mood during one Saturday.

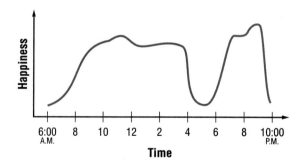

**a.** What two variables does this graph show?

**b.** Write a short story about what might have happened during Tyson's day. Your story should account for all his mood changes.

**7.** The De Marte family went on a picnic last Sunday. This graph shows how far the family was from home at various times of the day.

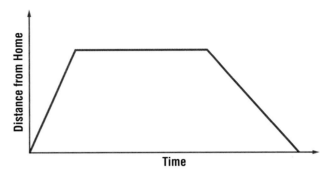

Mitchell and Dario each told a story about the graph. Which story best fits the graph? What is wrong with the other story?

**Mitchell:** "The family drove up a tall mountain and stayed on a level area for several hours. Then the family came down the mountain on a road that was less steep than the first."

**Dario:** "The family drove fairly quickly to the picnic spot and stayed there for most of the day. The family drove home more slowly."

**Connect & Extend**

**8.** In Exercise 8 on page 471, you used these graphs to compare two cars.

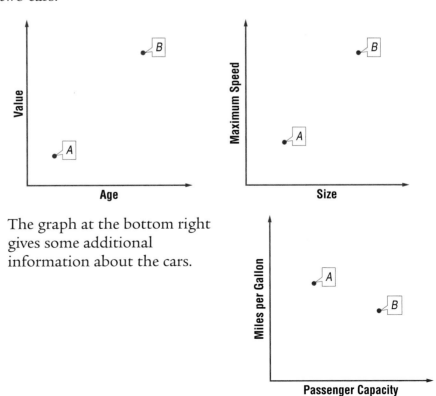

The graph at the bottom right gives some additional information about the cars.

**a.** *True or false?* The car that holds more passengers gets fewer miles per gallon.

**b.** Copy each set of axes below. Mark and label points to represent car A and car B.

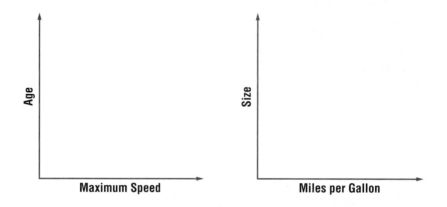

9. Complete Parts a and b for each graph below.

   **a.** Copy the graph. Make up a variable for each axis.

   **b.** Describe what the graph tells you about the things represented by the points. Then try to think of an idea about what the points could represent.

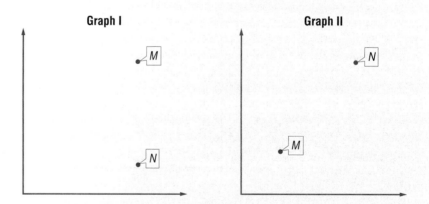

10. **In Your Own Words** Make a graph that shows how something changes over time. Write a story to go with your graph.

**11.** In an experiment, the heights of 192 mothers and their adult daughters were measured.

Based on this graph, does there appear to be a connection between the heights of the mothers and the heights of their daughters? Explain your answer.

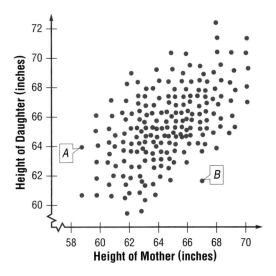

**12. Science** This graph was created to show how the amount of air in a balloon changed as it was blown up and then deflated. What is wrong with this graph?

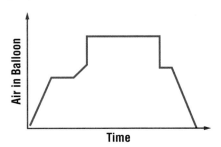

**13.** A firefighter walked from the fire truck up the stairs to the second floor of the firehouse. Several minutes later, the alarm rang. The firefighter slid down the pole and returned to the fire truck.

Sketch a graph to show the firefighter's height above the ground from the time she left the fire truck to the time she returned to it.

*Mixed Review*

**Economics** Tell how much change you would get if you paid for each item with a $5 bill.

**14.** a small salad for $1.74

**15.** a pack of mints for $.64

**16.** a magazine for $3.98

**17.** a pack of collector cards for $2.23

**Measurement** Express each measurement in meters.

**18.** 13 mm          **19.** 123 cm

**20.** 0.05 cm          **21.** 430 mm

**Find each sum or difference.**

**22.** $5\frac{6}{7} + 1\frac{5}{14}$          **23.** $\frac{11}{12} - \frac{5}{18}$

**24.** $\frac{5}{9} + \frac{13}{7}$          **25.** $3 - 1\frac{31}{72}$

**26.** $2\frac{5}{8} - 1\frac{3}{4}$          **27.** $\frac{7}{8} + \frac{7}{12} + \frac{7}{16}$

**Complete each table.**

**28.**

| Fraction | Decimal | Percent |
|---|---|---|
| $\frac{1}{2}$ | 0.5 | 50% |
| $\frac{3}{4}$ | | |
| | | 30% |
| | 0.25 | |

**29.**

| Fraction | Decimal | Percent |
|---|---|---|
| $\frac{1}{2}$ | 0.5 | 50% |
| $\frac{1}{8}$ | | |
| | 0.8 | |
| | | 5% |

**Write each fraction or decimal as a percent. Round to the nearest tenth of a percent.**

**30.** $\frac{1}{3}$          **31.** 0.06

**32.** $\frac{1}{2}$          **33.** $\frac{3}{10}$

**34.** 0.561          **35.** $\frac{2}{3}$

**36.** 1.25          **37.** $\frac{11}{10}$

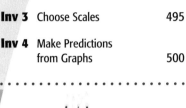

### Real-World Link

When an Argentinean child has a birthday, it is customary to tug his or her earlobe once for each year of age. In Israel, the birthday child sits in a chair while adults raise and lower it once for each year, plus once for good luck.

# LESSON 8.2

# Draw and Label Graphs

You have already had experience describing the information shown by graphs. In this investigation, you will practice making your own graphs.

**Explore**

Ella and Chase wanted to make a graph to show how the noise level at a typical birthday party is related to time. Here is the first graph that they drew.

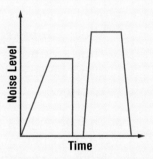

Ella said the graph was not quite right since it is not usually completely quiet at the start of a party. Chase said that most people arrive on time for a party, and the noise level would go up more quickly. They also realized that the party would probably never be completely silent. They drew a new graph.

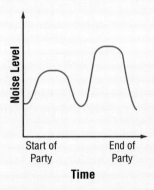

Think about birthday parties that you have attended. What would you change to improve Ella and Chase's graph?

Another pair of variables that might be useful for describing what happens at a birthday party is the amount of food and the time. Draw a graph showing how these variables might be related. Explain what your graph represents.

# Investigation (1) Draw Graphs

In this investigation, you will sketch graphs to fit various situations.

## ✅ Develop & Understand: A

For Exercises 1–6, create your own graph. Then compare and discuss graphs with your partner. Work together to create a final version of your graph. Be sure to label the axes.

1. At a track meet, the home team won a relay race, and the crowd roared with excitement. Make a graph to show how the noise level might have changed from just before the win to a few minutes after the win.

2. A child climbs to the top of a slide, sits down, and then starts to slide, gaining speed as he goes. At the bottom, he gets up quickly, runs around to the ladder, and climbs up again. Make a graph to show how the height of the child's feet above the ground is related to time.

3. At Computer Cafe, customers are charged a fixed price plus a certain amount per minute for using a computer. Create a graph that shows how the amount a customer is charged is related to the time he or she spends using a computer.

4. Smallville is a town surrounded by farms. The number of people in Smallville changes a lot during a typical school day. During the day, many children come to town for school, and adults drive in for business and shopping. In the evenings, people come to town to eat dinner or attend social events. Draw a graph to show how the number of people in town might be related to time on a typical school day.

5. Make a graph showing how the number of hours of daylight is related to the time of year. Assume the time axis starts in January and goes through December.

6. Make a graph showing approximately how the temperature outside has changed over the past three days.

## Share & Summarize

Write a description of something that changes over time. Then sketch three graphs, one that matches your description and two that do not. See if your partner can guess which graph is correct.

# Investigation 2 Plot Points

## Vocabulary

**coordinates**

**ordered pair**

## Materials

- copy of the map
- copy of the grid
- grid paper

When you made your graphs in the last investigation, you had to think about the overall shape of the graph, not about exact values. To draw a graph that shows exact values, you need to plot points.

This graph shows a map of an island just off the coast of a continent. The point labeled S represents a major city on the coast. The distance between grid lines represents one mile.

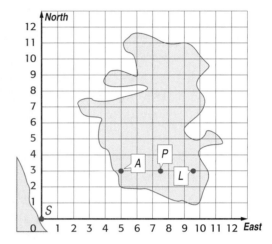

Point A represents a resort that is located five miles east and three miles north of point S. The values 5 and 3 are the **coordinates** of point A. The coordinates can be given as the **ordered pair** (5, 3), where 5 is the horizontal coordinate and 3 is the vertical coordinate.

As you might guess, the *order* of the numbers in an ordered pair is important. The first number is always the horizontal coordinate. The second number is always the vertical coordinate.

## ✅ Develop & Understand: A

1. On a copy of the map on page 491, mark the point that is three miles east and five miles north of point *S*. Label it *B*. Is point *B* in the water or on the island? Is point *B* in the same place as point *A*?

2. Mark the point that is seven miles east and five miles north of point *S*. Label it *C*. Then mark the point that is five miles east and seven miles north of point *S*. Label it *D*. Are points *C* and *D* in the same place? Give the coordinates of points *C* and *D*.

3. Which point is in the water, (2, 7) or (7, 2)? Mark the point on your map. Label it *E*.

4. Developers want to build another resort on the island. Which would be the better location, (6, 11) or (11, 6)? Why?

5. Give the coordinates of two points on the island that are exactly two miles from point *A*.

6. Coordinates are not always whole numbers. For example, point *L*, the island lighthouse, has coordinates (9.5, 3). Point *P* represents the swimming pool. What are the coordinates of point *P*? How far is the lighthouse from the pool?

7. Give the coordinates of the point that is halfway between points *L* and *P*.

8. Give the coordinates of the point that is halfway between points *A* and *P*.

9. List three points on the island with a first coordinate greater than 8.

10. List three points on the island with a second coordinate equal to 8.

11. List three points on the island with a second coordinate less than 4.

In Exercises 12–17, you will use what you know about plotting points to make a graph.

# ✅ Develop & Understand: B

Roberta wants to use some of the money she earns from her paper route to sponsor a child in a developing country. Sponsors donate money each month to help pay for food, clothing, and education for a child in need.

Roberta learned that sponsoring a child costs $48 a month. This is more than she can afford, so she wants to ask some of her friends to share the cost.

12. If two people divide the monthly sponsorship cost, how much will each person pay? If three people divide the cost, how much will each pay?

13. Copy and complete the table to show how much each person would pay if the given number of people split the cost.

| Number of People | 1 | 2 | 3 | 4 | 6 | 8 | 12 | 16 | 24 | 48 |
|---|---|---|---|---|---|---|---|---|---|---|
| Cost per Person (dollars) | 48 | | | | | | | 3 | | |

14. On a copy of the grid below, plot and label points for the values from your table. For example, for the first entry, plot the point with coordinates (1, 48). Two of the points have been plotted for you.

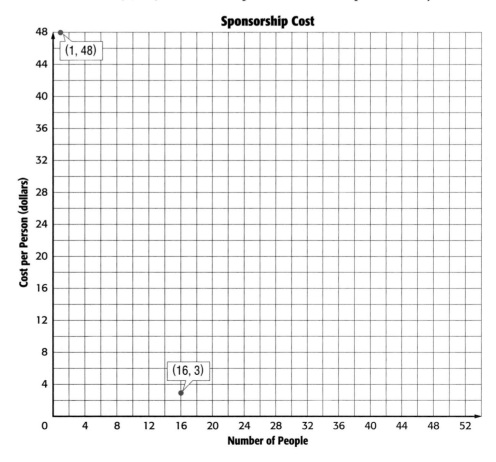

**Sponsorship Cost**

**15.** Describe the overall pattern of points in your graph.

**16.** Jake wanted to connect the points on his graph with line segments. Lamar said, "That wouldn't make sense. How can $1\frac{1}{2}$ people share the cost?" What do you think?

**17.** As the number of people increases, the amount each person pays decreases.

    **a.** If you continued to plot points for more and more people, what would the graph look like?

    **b.** Could there ever be so many people that each person would pay nothing? Explain.

## Share & Summarize

**1.** Give the coordinates of the point halfway between (2, 7) and (2, 3). Then make a graph showing all three points. Label the points with their coordinates.

**2.** Give the coordinates of two more points on the same vertical line. Plot and label the points.

**3.** Give the coordinates of two points that, along with (2, 7) and (2, 3), form the vertices of a rectangle.

**4.** Is your answer for Question 3 the only one possible? Explain.

# Investigation 3 Choose Scales

## Materials

- 10-by-10 grids

### Real-World Link

On January 15, 1919, in Boston, Massachusetts, a tank of molasses burst, releasing 2.2 million gallons of the sticky substance. The giant wall of molasses moved at speeds up to 35 mph, killing 21 people and injuring 150 others.

When you created a graph showing the relationship between the number of sponsors and the amount each sponsor would have to pay, you were given a set of axes labeled with *scale values*. Often when you make a graph, you have to choose the scales yourself.

### Think & Discuss

Sara and Nina are baking ginger snaps for the school bake sale. They need half a cup of molasses for each batch of cookies.

Sketch a rough graph showing the relationship between the number of batches and the number of cups of molasses. Your graph does not need to show precise points.

Sara and Nina made a table to show how much molasses they would need for different numbers of batches.

| Batches | Molasses (cups) |
|---------|-----------------|
| 1 | $\frac{1}{2}$ |
| 12 | 6 |
| 20 | 10 |
| 3 | $1\frac{1}{2}$ |
| 8 | 4 |
| 40 | 20 |

They each decided to graph the data. Here are their graphs.

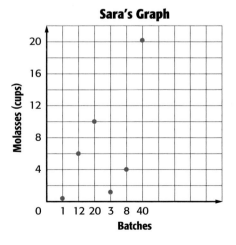

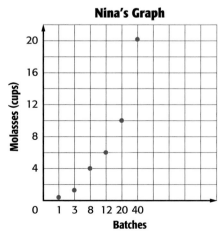

Discuss each graph. Do you think the graph is correct? If not, explain what is incorrect. Tell how you would fix it.

When you draw a graph, you need to think about the greatest value that you want to show on each axis. For the cookie data, you need to show values up to 40 on the "Batches" axis and up to 20 on the "Molasses (cups)" axis.

You also need to consider the scale to use on each axis. The *scale* is the number of units each equal interval on the grid represents. You want to choose a scale that will make your graph easy to read but will not make it so large that it will not fit on your paper. Here are two possibilities for the cookie data.

- Let each interval represent four batches on the horizontal axis and two cups on the vertical axis. Then the graph would fit on a 10-by-10 grid.

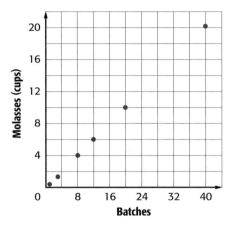

- Let each interval represent two batches on the horizontal axis and two cups on the vertical axis. Then the graph would fit on a 20-by-10 grid.

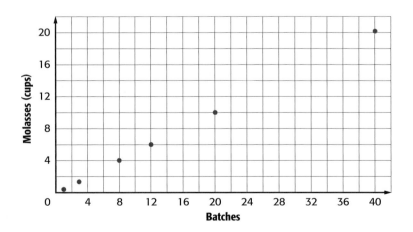

# ✅ Develop & Understand: A

Pablo wants to buy a birthday present for his sister. He has $10 to spend on the present and the gift wrap.

**Real-World Link**
The birthday cake with lit candles that we know today was a tradition started by Germans in the Middle Ages who adapted it from an ancient Greek custom.

1. Copy and complete this table to show some of the ways the $10 Pablo has to spend can be distributed between the cost of the present and the cost of the gift wrap.

| Cost of Present (dollars) | 10 | 9 | 8 | 7 | 6 | | | 3 |
|---|---|---|---|---|---|---|---|---|
| Cost of Gift Wrap (dollars) | 0 | 1 | 2 | | | 5 | 6 | |

2. Suppose you want to graph these data with the cost of the present on the horizontal axis and the cost of the gift wrap on the vertical axis. What is the greatest value you need to show on each axis?

3. What scale would you use on each axis? In other words, how many dollars would you let each interval on the grid represent?

4. Make a graph of the data. Be sure to do the following.
   - Label the axes with the names of the variables.
   - Add scale labels to the axes.
   - Plot a point for each pair of values in the table.

5. Describe the pattern of points on your graph.

6. In this situation, either variable could have a value that is not a whole number. For example, Pablo could spend $6.50 on the present and $3.50 on the wrapping. Plot the point (6.5, 3.5) on your graph. Does this point follow the same pattern as the others?

7. Find two more pairs of non-whole-number values that fit this situation. Plot points for these values. Check that the points follow the same pattern as the others.

8. You can use your graph to make predictions.
   a. Connect the points with line segments.
   b. Choose a point on the graph that is not one of the points that you plotted. Use the coordinates of the point to predict a pair of values for the present cost and the wrapping cost. Then check your prediction by verifying that the values add to $10.

When you make the graphs in Exercises 9 and 10, you will need to choose a scale that gives a good view of the overall pattern in the data.

### ☑ Develop & Understand: B

Make each graph on a 10-by-10 grid in Exercises 9 and 10.

9. James is planning a party at a local pizza parlor. The party will cost $6 per person.

   a. Complete the table to show the cost for various numbers of people.

   | People | 1 | 2 | 3 | 4 | 5 | 6 | 7 | 8 |
   |--------|---|---|---|---|---|---|---|---|
   | Cost (dollars) | 6 | 12 | | | | | | |

   b. James wants to graph these data on a 10-by-10 grid with the number of people on the horizontal axis. What scale do you think he should use on each axis?

   c. Graph the values in the table using an appropriate scale.

   d. Do the points on your graph form a pattern? If so, describe it.

   e. Use your graph to predict the cost of the pizza for nine people. Check that your prediction is correct.

   f. Would it make sense to connect the points on this graph? Explain.

**10.** The social committee needs streamers for decorations. Material for the streamers costs 20¢ per yard.

a. Complete the table to show the cost of various lengths of material.

| Length (yards) | 1 | 2 | 3 | 4 | 5 | 6 | 7 | 8 | 9 |
|---|---|---|---|---|---|---|---|---|---|
| Cost (cents) | 20 | 40 | | | | | | | |

b. Suppose you graphed these data with the length on the horizontal axis and the cost on the vertical axis. What would happen if you let each interval on the vertical axis represent 1¢?

c. What scale would be appropriate for each axis?

d. Make a graph of the data with length on the horizontal axis.

e. Do the points on your graph form a pattern? If so, describe it.

f. Would it make sense to connect the points on this graph? Explain.

g. Use your graph to predict the cost for $4\frac{1}{2}$ yards. Check your prediction by multiplying the number of yards by the cost per yard.

h. Draw another graph using the same data. But this time, put the cost on the horizontal axis.

i. Describe how your two graphs are alike and different.

## Share & Summarize

1. When you make a graph, how do you decide on the scale for each axis?

2. After you have plotted points from a table, how do you decide whether to connect the points?

# Investigation  Make Predictions from Graphs

## Materials

- graph paper
- copy of the graph

You have seen that when the points in a graph form a pattern, you can use the graph to make predictions. Connecting the points on a graph with line segments can help you find in-between values. Extending the pattern can help you make predictions about values beyond those shown in the graph.

### ✓ Develop & Understand: A

This graph shows the cost of various numbers of decks of playing cards.

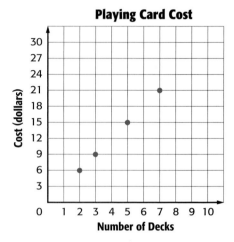

**Playing Card Cost**

1. How much do five decks of cards cost?

2. How many decks of cards can you buy for $9?

3. Can you use the graph to find the cost for four decks of cards? For ten decks of cards? Why or why not?

Here is how Hannah thought about Exercise 3.

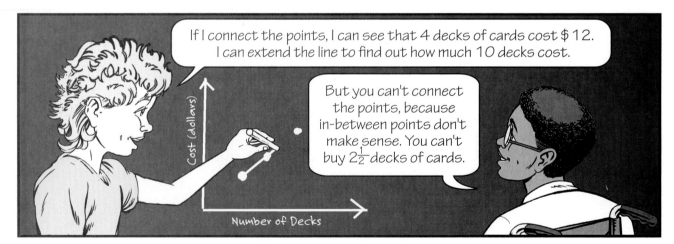

### Think & Discuss

What do you think about Hannah's method? Does it make sense? Is Jahmal correct?

Sometimes connecting the points in a graph can help you find information or to see a pattern, even when all the in-between points do not make sense. In cases like this, people often use dashed segments to connect the points.

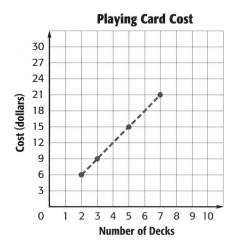

**Playing Card Cost**

The dashed segments in the graph above make it easy to find information and to see that all the points fall on a straight line. The dashes also indicate that not every point on the line makes sense.

## ✅ Develop & Understand: B

4. This table shows the cost for different numbers of packs of fruit snacks.

| Packs | 3 | 5 | 8 | 15 |
|---|---|---|---|---|
| Cost | $1.50 | $2.50 | $4.00 | $7.50 |

   a. Make a graph of the data. Be sure to choose appropriate scales and to label the axes with the variable names.

   b. Use your graph to find the cost of six packs of fruit snacks and the cost of nine packs of fruit snacks. Explain how you found your answers.

   c. Use your graph to find the cost of 19 packs of fruit snacks. Explain how you found your answer.

   d. Use your graph to find how many packs of fruit snacks you can buy for $6. Explain how you found your answer.

**5.** This graph shows the masses of different lengths of a certain type of copper pipe.

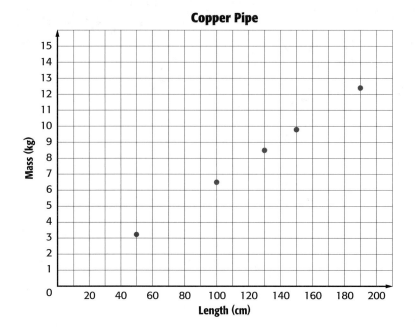

**Copper Pipe**

**a.** What is the mass of 100 centimeters of copper pipe?

**b.** Would you connect the points on this graph? If so, would you use dashed or solid segments? Explain your answers.

**c.** Estimate the mass of a copper pipe with length 180 centimeters. Explain how you found your answer.

**d.** Estimate the length of a copper pipe with a mass of 5 kilograms. Explain how you found your answer.

## Share & Summarize

**1.** If you were to make a graph of these data, would you connect the points? If so, would you use dashed or solid segments? Explain your answers.

| Tickets | Price ($) |
|---------|-----------|
| 1 | 3 |
| 5 | 15 |
| 8 | 24 |
| 15 | 45 |
| 21 | 63 |

**2.** If you were to make a graph of these data, would you connect the points? If so, would you use dashed or solid segments? Explain your answers.

| Time | Temp. (°F) |
|------|------------|
| 6 A.M. | 17 |
| 8 A.M. | 20 |
| 10 A.M. | 25 |
| noon | 32 |
| 2 P.M. | 30 |
| 4 P.M. | 28 |
| 6 P.M. | 15 |

**Practice & Apply**

1. A weight lifter grips a barbell and struggles with it for several seconds. She suddenly lifts it part of the way and then steadily raises it until it is fully above her head. She holds it for a few seconds and then drops it.

   Sketch a graph to show how the height of the barbell changes from the time the weight lifter first grips the barbell until just after she drops it.

2. Goin' Nuts sells mixed nuts by the pound. Draw a graph that shows how the cost for cashews is related to the number of pounds purchased.

3. For this exercise, use a grid with horizontal and vertical axes from 0 to 14. Plot the points below in order, reading across the rows. Connect the points as you go with straight line segments. The line segments should form a picture.

   | | | | | | | |
   |---|---|---|---|---|---|---|
   | (7, 3) | (13, 3) | (10, 1) | (2, 1) | (1, 2) | (1, 3) | (7, 3) |
   | (7, 12) | (13, 4) | (7, 4) | (8, 5) | (8, 10) | (7, 12) | (6, 10) |
   | (6, 5) | (7, 4) | (0, 4) | (7, 12) | (5, 13) | (7, 13) | (7, 12) |

4. **Geometry** The area of a square is the product of the lengths of two sides.

   **a.** What is the area of a square with side length 1? What is the area of a square with side length 2?

   **b.** Complete the table to show areas of squares with the given side lengths.

   | Side Length | 1 | 2 | 3 | 4 | 5 | 6 |
   |---|---|---|---|---|---|---|
   | Area | 1 | | | | 25 | |

   **c.** On a copy of the axes at right, plot and label points for the values from your table. For example, for a square with side length 5, plot the point (5, 25).

   **d.** Describe the overall pattern of points in your graph.

   **e.** Would it make sense to connect the points on this graph? Explain.

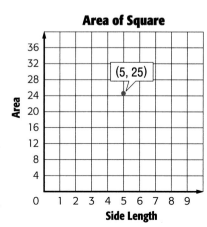

**5.** Calvin has only 30 minutes to finish his homework, which includes practicing the violin and studying for a history quiz.

**a.** Make a table showing at least six ways Calvin can split up the time between the two activities.

**b.** Suppose you want to graph your data with the time spent practicing the violin on the horizontal axis. What is the greatest value you need to show on each axis?

**c.** What scale would you use on each axis?

**d.** Make a graph of your data. Be sure to add labels and scale values to the axes.

**e.** Describe the general pattern of points on your graph.

**f.** Connect the points on your graph with line segments. Then choose a point on the graph that is not one of the points you plotted. Use the coordinates of the point to predict a pair of values for the violin time and the study time.

**6. Economics** Sparks Internet Service charges $5 per month plus $0.02 for each minute a customer is online.

| Time (hours) | 0 | 1 | 2 | 3 | 4 | 5 | 6 | 7 | 9 | 10 |
|---|---|---|---|---|---|---|---|---|---|---|
| Cost (dollars) | 5 | 6.20 | | | | | | | | |

**a.** Complete the table to show the cost for various amounts of time online.

**b.** Graph the data from the table on a 10-by-10 grid. Show time on the horizontal axis. Use appropriate scales.

**c.** Do the points on your graph make a pattern? If so, describe it.

**d.** Does it make sense to connect the points on this graph? Explain.

**e.** Use your graph to predict the cost for eight hours. Check that your prediction is correct.

**7. Geometry** *Circumference* is the distance around a circle. The table lists the approximate circumferences of circles with given diameters.

| Diameter (cm) | Circumference (cm) |
|:---:|:---:|
| 0.5 | 1.6 |
| 1.5 | 4.7 |
| 2 | 6.3 |
| 3 | 9.4 |
| 3.5 | 11.0 |
| 5 | 15.7 |

**a.** Make a graph of the data. Be sure to choose appropriate scales and to label the axes with the variable names.

**b.** Use your graph to estimate the circumference of a circle with diameter 4 cm. Explain how you found your answer.

**c.** Use your graph to estimate the diameter of a circle with circumference 8 cm. Explain how you found your answer.

**Connect & Extend**

**8.** A family is driving to visit some relatives a few hundred miles away. The drive begins at a moderate pace along back roads and moves to several hours on a major highway. When the family gets to the city where the relatives live, travel is slowed by heavy traffic and lights at intersections.

**a.** Sketch a graph showing the time since the trip began on the horizontal axis and the family's speed on the vertical axis.

**b.** Sketch a graph showing the time on the horizontal axis and the family's distance from the starting point on the vertical axis.

**c.** Sketch a graph showing the time on the horizontal axis and the distance from the destination on the vertical axis.

**9.** Annie said, "When my schoolwork is much too easy, I don't learn very much. But I also don't learn very much when it is much too hard. I learn the most when the difficulty level is somewhere between 'too easy' and 'too hard.'" Draw a graph to illustrate Annie's ideas.

**10.** You could play tic-tac-toe on a grid like this. Instead of writing X's and O's in the squares, you would write them at points where the grid lines meet. For example, you could mark (0, 0) or (2, 1).

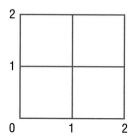

**a.** How can you tell from just the coordinates whether three points are on the same horizontal line?

**b.** How can you tell from just the coordinates whether three points are on the same vertical line?

**c.** How can you tell from just the coordinates whether three points are on the same diagonal line?

**11.** Consider the toothpick pattern below.

**Stage 1**     **Stage 2**          **Stage 3**               **Stage 4**

**a.** Imagine that this pattern continues. Complete the table to show the number of toothpicks in the first six stages.

| Stage | 1 | 2 | 3 | 4 | 5 | 6 |
|---|---|---|---|---|---|---|
| Toothpicks | 4 | | | 13 | | |

**b.** Make a graph with the stage number on the horizontal axis and the number of toothpicks on the vertical axis. Make the horizontal axis from 0 to 10 and the vertical axis from 0 to 30.

**c.** Describe the pattern of points in your graph.

**d.** Use the pattern in your graph to predict the number of toothpicks in stages 7 and 8. Check your answers by drawing these stages.

**e.** Would it make sense to connect the points on this graph? Explain.

**12.** Consider this input/output table.

| Input | 1 | 2 | 4 | 5 | 7 |
|---|---|---|---|---|---|
| Output | 2 | 5 | 11 | 14 | 20 |

**a.** Describe a rule relating the input and output values.

**b.** Graph the values from the table. Make the "Input" axis from 0 to 8 and the "Output" axis from 0 to 24.

**c.** Use your graph to predict the outputs for inputs of 3, 6, and 8. Use your rule to check your predictions.

**Real-World Link**

For two weeks each spring, the city of Washington, D.C., hosts the Cherry Blossom Festival. This event commemorates the gift of 3,000 cherry trees received in 1912 by the United States from Tokyo, Japan.

**13. Earth Science** This table represents the normal monthly temperatures for Washington, D.C.

**a.** Graph the data with the month on the horizontal axis and the temperature on the vertical axis.

**b.** In which month is the temperature highest? In which month is the temperature lowest?

**c.** Would you connect the points on your graph? If so, would you use dashed or solid segments? Explain your answers.

**d.** Use your graph to estimate the temperatures for April and November. Explain how you made your estimates.

**Washington, D.C.
Normal Monthly Temperatures**

| Month | Temperature (°F) |
|---|---|
| January | 31 |
| February | 34 |
| March | 43 |
| April | |
| May | 62 |
| June | 71 |
| July | 76 |
| August | 74 |
| September | 67 |
| October | 55 |
| November | |
| December | 35 |

**Source:** *World Almanac and Book of Facts 2000.*
Copyright © 1999 Primedia Reference Inc.

**14. In Your Own Words** Sketch a graph that shows how your height has changed since you were born. Use as much exact information as you can recall or that you can obtain from family members. Extend your sketch to show your prediction for your full adult height and the age when you will reach it. Write an explanation of what your graph represents about your growth.

**15.** Here are the first seven rows of Pascal's triangle.

```
                1                      Row 0
              1   1                    Row 1
            1   2   1                  Row 2
          1   3   3   1                Row 3
        1   4   6   4   1              Row 4
      1   5  10  10   5   1            Row 5
    1   6  15  20  15   6   1          Row 6
```

**a.** Complete a table to show the sum of the numbers in each row.

| Row | 0 | 1 | 2 | 3 | 4 | 5 | 6 |
|-----|---|---|---|---|---|---|---|
| Sum |   |   |   |   |   |   |   |

**b.** Make a graph of these values. Put the row number on the horizontal axis and the sum on the vertical axis.

**c.** Describe the pattern of points in your graph.

**Mixed Review**

**Find each product or quotient.**

**16.** $\frac{4}{5} \cdot \frac{5}{7}$

**17.** $\frac{10}{13} \div \frac{5}{26}$

**18.** $4\frac{2}{3} \div 1\frac{5}{6}$

**Find each sum or difference.**

**19.** $3\frac{1}{2} - 1\frac{5}{8}$

**20.** $1\frac{7}{12} + 4\frac{2}{3}$

**21.** $12\frac{6}{7} + 5\frac{5}{6}$

**22. Economics** The Book Bin is having a clearance sale.

**a.** All dictionaries are marked $33\frac{1}{3}$% off. Ramesh bought a French dictionary with a sale price of $18. What was the dictionary's original price?

**b.** Novels are on sale for 20% off. Diana bought a novel with an original price of $11.95. What was the sale price?

**c.** Travel books are all marked down by a certain percent. Nestor bought a book about African safaris. The book was originally priced at $27.50, but he paid only $16.50. What percent did Nestor save?

**For each square, estimate the percent of the area that is shaded.**

**23.**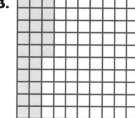

**24.**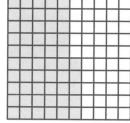

# Graph in Four Quadrants

As you have seen, making a graph is a useful way to represent the relationship between two quantities. For example, during a two-week snorkeling vacation, Deane timed how long he could hold his breath under water. The graph shows his maximum breath-holding time each day.

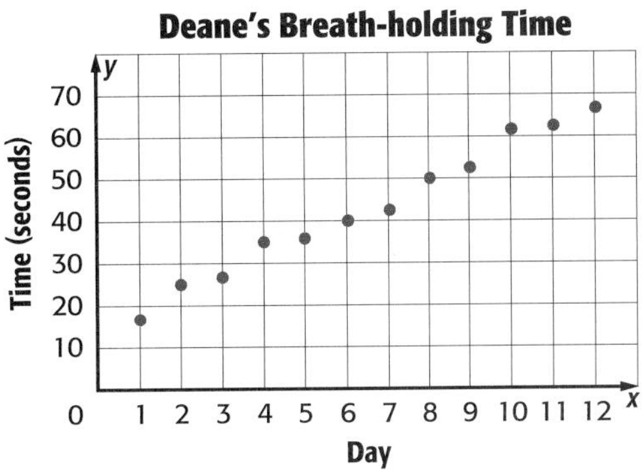

## Vocabulary

**negative numbers**

**positive numbers**

**Deane's Breath-holding Time**

### Think & Discuss

Choose two points on the graph. Give their coordinates. Explain what the coordinates tell you about Deane's breath-holding time.

How long could Deane hold his breath at the end of his sixth day of practicing? How did you find your answer from the graph?

In the graph above, both quantities, day and time, are always positive. What if one or both of the quantities you want to graph are sometimes negative?

You have had a lot of experience working with **positive numbers**, numbers that are greater than 0. You are familiar with positive whole numbers, positive decimals, and positive fractions. You will now turn your attention to **negative numbers**, which are numbers that are less than 0.

# Investigation 1 Understand Integers

## Vocabulary

opposites

Below, three students describe the current temperatures in the cities where they live. Use the thermometer to help figure out the temperature in each city.

- Carlita lives in Cincinnati, Ohio. She says, "It's not very cold here. If the temperature rises 5 degrees, it will be 47.5°F."
- Trey lives in Niagara Falls, New York. He says, "You think that's cold? If the temperature rises 30 degrees here, it will be only 30°F!"
- Jean lives in Juneau, Alaska. She says, "We can top all of you up here! If our temperature went up 30 degrees, it would be only 10°F!"

Create a temperature puzzle, like those above, for which the answer is negative. Exchange puzzles with your partner. Solve your partner's puzzle.

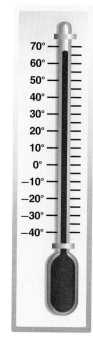

In colder climates, it is not unusual for the temperature to drop to −20°F, meaning 20 degrees *below* 0. There are many other contexts in which the number −20 might be used. For example, if you were standing at a location 20 feet below sea level, your elevation would be −20 feet. If you wrote checks for $20 more than you had in your bank account, your account balance would be −$20.

Here are a few facts about the number −20.

- −20 is read as "negative twenty" or "the opposite of twenty."
- −20 is located 20 units to the left of 0 on a horizontal number line.
- −20 is located 20 units below 0 on a vertical number line or thermometer.
- −20 is located halfway between −21 and −19 on a number line.

### Real-World Link

The lowest temperature ever recorded on Earth is −128.6°F. This temperature occurred on July 21, 1983 at Vostok, a Russian station in Antarctica.

The numbers −20 and 20 are *opposites*. Two numbers are **opposites** if they are the same distance from 0 on the number line but on different sides of 0. What is the result when you add two opposites?

As you know, the number 20 is positive. You will occasionally see 20 written as +20. The notations 20 and +20 have the same meaning. Both can be read as "positive twenty" or just "twenty."

## ✅ Develop & Understand: A

**Real-World Link**

Accountants use red ink when recording negative numbers and black when recording positive numbers. A company that is "in the red" is losing money. One that is "in the black" is showing profit.

1. Copy the number line below. Plot the following points to show the locations of the numbers listed, and label each point with the corresponding number.

$$1.25, -2, -\frac{1}{3}, 3.7, -4\frac{3}{4}, -1\frac{3}{8}$$

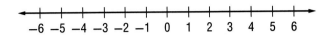

2. On your number line from Exercise 1, plot and label points for three more negative mixed numbers.

3. Give a number that describes the approximate location of each labeled point.

4. Find the opposite of each number.

   a. 3.2        b. $-\frac{3}{4}$        c. $-2$        d. 317

5. Is the *opposite* of a number always negative? Explain.

> ### Think & Discuss
>
> How can you tell which of two numbers is greater by looking at their locations on a horizontal number line?
>
> Which is the warmer temperature, $-20°F$ or $-15°F$? How do you know?

## ✅ Develop & Understand: B

**Real-World Link**

Antarctica is covered with 7 million cubic miles of ice with an average thickness of 1.5 miles. The ice is so heavy that it deforms the South Pole, giving Earth a slight pear shape.

6. The table shows record low temperatures for each continent.

| Continent | Location | Temperature |
|---|---|---|
| Africa | Ifrane, Morocco | $-11°F$ |
| Antarctica | Vostok Station | $-129°F$ |
| Asia | Oimekon and Verkhoyansk, Russia | $-90°F$ |
| Australia | Charlotte Pass, New South Wales | $-9.4°F$ |
| Europe | Ust'Shchugor, Russia | $-67°F$ |
| North America | Snag, Yukon, Canada | $-81.4°F$ |
| South America | Sarmiento, Argentina | $-27°F$ |

**Source:** *World Almanac and Book of Facts 2000.* Copyright © 1999 Primedia Reference Inc.

**a.** Order the temperatures from coldest to warmest.

**b.** When temperatures drop below −30°F, people often experience frostbite. For which locations in the table is the record low temperature below −30°F?

**c.** In the table, how many degrees below the warmest temperature is the coldest temperature?

**7.** Ten students measured the outside temperature at different times on the same winter day. Their results are shown in the table.

| Student | Temperature |
|---------|-------------|
| Gabe | −0.33°F |
| Jill | 1.30°F |
| Fabiana | −0.80°F |
| Marco | 0.33°F |
| Micheala | −1.30°F |
| Brad | −1.80°F |
| Phil | 1.08°F |
| Jasmine | −1.75°F |
| Kurt | −1.00°F |
| Sophia | −0.80°F |

**a.** List the temperatures in order from coldest to warmest.

**b.** On the night the students recorded the temperatures, a weather reporter said, "The average temperature today was a chilly 0°F." Which students recorded temperatures that were closest to the average temperature?

## Share & Summarize

**1.** Explain what the opposite of a number is.

**2.** Name three negative numbers between −3 and −2.

# Investigation 2 Absolute Value

**Vocabulary**

absolute value

For a Science project, Trina studied how the temperature varies from day to day. She recorded the temperature at noon every day for a week.

| Day | Monday | Tuesday | Wednesday | Thursday | Friday | Saturday | Sunday |
|-----|--------|---------|-----------|----------|--------|----------|--------|
| Temperature | 60°F | 51°F | 61°F | 67°F | 45°F | 50°F | 42°F |

## Develop & Understand: A

Trina calculated the change in temperature between each day and the next. If the temperature rose, she recorded the change as a positive number. If the temperature fell, she recorded the change as a negative number.

1. Calculate each temperature change. Complete the table.

| Days | Monday–Tuesday | Tuesday–Wednesday | Wednesday–Thursday | Thursday–Friday | Friday–Saturday | Saturday–Sunday |
|---|---|---|---|---|---|---|
| Change in Temperature | −9°F | | | | | |

2. Between which days did the greatest increase in temperature occur?

3. Between which days did the greatest decrease in temperature occur?

Trina saw that sometimes the temperature changed a lot from one day to the next, and sometimes it changed very little. She wanted to compare the differences in the changes. She made a new table. This time, she listed just the size of each change, without showing whether it was an increase or a decrease.

4. Complete the table.

| Days | Monday–Tuesday | Tuesday–Wednesday | Wednesday–Thursday | Thursday–Friday | Friday–Saturday | Saturday–Sunday |
|---|---|---|---|---|---|---|
| Size of Change | 9°F | 10°F | | | | |

5. When was the greatest temperature change?

6. When was the least temperature change?

7. Trina found that during the next week, the temperature change from Monday to Tuesday was 8°. The temperature on Monday was 63°. Give two possible temperatures for Tuesday.

### Think & Discuss

Compare the two tables of temperature change.

Can you get the same information from each of them?

Could Trina have used the first table to figure out what the greatest and least changes were?

The **absolute value** of a number is its distance from zero on the number line.

On the number line, 5 and −5 are both 5 units away from 0. They both have the same absolute value, 5. *Absolute value* is indicated by vertical bars around a number, $|-5|$ or $|5|$.

## ✅ Develop & Understand: B

$$\begin{array}{cccccccccccc} \text{−5} & \text{−4} & \text{−3} & \text{−2} & \text{−1} & 0 & 1 & 2 & 3 & 4 & 5 \end{array}$$

**Find the distance of each number from 0.**

8. −4

9. −3.5

10. $4\frac{1}{2}$

**Which is further from 0?**

11. 4 or −3?

12. −2.5 or −4.5?

13. $-\frac{1}{2}$ or $-\frac{1}{3}$?

**Find the absolute value of each expression.**

14. $|-6|$

15. $\left|\frac{3}{5}\right|$

16. $|-3^3|$

17. $|57.01|$

18. Quin wrote down a number whose absolute value is 21.5. What could be Quin's number?

19. The Bulldogs are playing the Rangers in football. The Bulldogs have 30 yards to go to score a touchdown. On the next play, the Rangers push them back to the 42 yard line. Write an equation to show what happened. How many yards did the Bulldogs lose?

You can take the absolute value of any expression. $|4 + 2|$, $|-25 \div 3|$, and $|-7 \cdot 40|$ all mean "evaluate the expression between the absolute value bars, and then find the absolute value of the result."

> **Think & Discuss**
>
> Which is greater, $|-5| + |4|$ or $|-5 + 4|$?

## ✅ Develop & Understand: C

**Evaluate each expression.**

20. $|50 - 25|$

21. $|-3| + |-15|$

22. $|-17| - |-17|$

23. $\left|\frac{2}{3}\right| - \left|-\frac{1}{5}\right|$

> **Share & Summarize**
>
> What happens to a negative number when you take its absolute value?
>
> What happens to a positive number when you take its absolute value?
>
> What is $|0|$?

# Investigation ③ Plot Points with Negative Coordinates

## Materials

- graph paper
- coordinate grids with x-axis and y-axis from −3 to 3

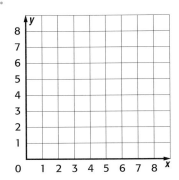

You know how to plot points on a coordinate grid that looks like the one at left. The x-axis of the grid is a horizontal number line. The y-axis is a vertical number line.

In the graphs with which you have previously worked, the number lines included only numbers greater than or equal to 0. But if they are extended to include negative numbers, the coordinate grid will look something like this.

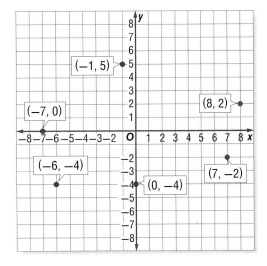

Using a grid like this, you can plot points with negative coordinates.

### Think & Discuss

Rashelle plotted six points on the grid above. See if you can discover the procedure she used to plot the points.

### ✓ Develop & Understand: A

1. Plot points A–F on the same coordinate grid. Label each point with its letter.

Point A: (6, −1)      Point B: (−2, −2)      Point C: (−1, −3)

Point D: (−1, 0)      Point E: (−2, 3.5)      Point F: $(-\frac{1}{3}, 4)$

**2.** Give the coordinates of each point plotted on this grid.

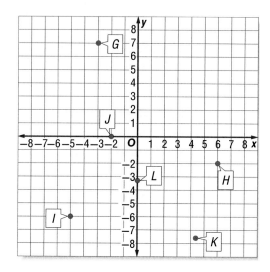

**3.** The graph shows daily average temperatures at the Gulkana Glacier basin in Alaska from September 9 to September 18 in a recent year.

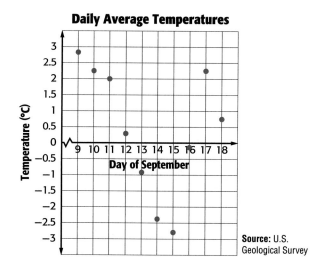

**Source:** U.S. Geological Survey

   **a.** What was the lowest of these temperatures?

   **b.** On which day was the temperature lowest?

   **c.** What was the highest of these temperatures?

   **d.** On which day was the temperature highest?

The game you will now play will give you practice locating points on a coordinate grid.

## ✅ Develop & Understand: B

In the *Undersea Search* game, you and your partner will hide items from each other on a coordinate grid. To win the game, you need to find your partner's hidden items before he or she finds yours.

Each player will need two coordinate grids with *x*- and *y*-axes that range from −4 to 4. Think of each grid as a map of part of the ocean floor. During the game, you will be hiding a buried treasure and a coral reef on one of your grids.

......................................

**Math Link**

The area of each shape is the number of square units that fit inside it.

......................................

The grid below shows one way you could hide the items.

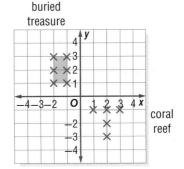

You can bury the items anywhere on the grid, but they must have the shapes shown above.

- The buried treasure must be a rectangle with an area of 2 square units. It must be drawn using two Xs along one side and three along the other.
- The coral reef must be a T-shape made from five Xs.

The Xs must all be placed where grid lines intersect. The buried treasure and the reef cannot overlap, so they cannot share points on your map.

Here is how you play the game.

- *Hide the buried treasure and the coral reef.* Start with one of your grids. Without showing your partner, use Xs to mark the places you want to hide the buried treasure and the coral reef. Make sure you put all your Xs where grid lines intersect.

- *Search the sea.* You and your partner take turns calling out the coordinates of points, trying to guess where the other has hidden the items.

  If your partner calls out a point where you have hidden something, say "X marks the spot." If your partner calls out any other point, say "Sorry, nothing there."

  Use your blank grid to keep track of your guesses. If you guess a point where your partner has hidden something, put an X on that point. If you guess a point where nothing is hidden, circle the point so you know not to guess it again.

- *Victory at sea.* The first person to guess all the points for both hidden items wins.

Play *Undersea Search* with your partner at least once. Then answer the questions.

4. Suppose your partner said "X marks the spot" when you guessed the points (−3, 1), (−3, 2), and (−2, 2). Can you tell whether you have found the buried treasure or the coral reef? Why or why not?

5. Suppose your partner said "X marks the spot" when you guessed the points (−2, −2), (−3, 0), and (−1, 0). Can you tell whether you have found the buried treasure or the coral reef? Why or why not?

6. Suppose you have already found the coral reef, and you know that part of the buried treasure is at the points (1, −2), (0, −2), and (−1, −2). What could be the coordinates of the other three points that make up the buried treasure? Name as many possibilities as you can.

7. Suppose you are playing *Undersea Search* with your younger cousin. You want to alter the rules of the game so it will be easier to locate the buried treasure. How could you change the area of the treasure?

## Share & Summarize

Write a letter to a student a grade below you explaining how to plot points with negative coordinates on a coordinate grid.

# Investigation  Parts of the Coordinate Plane

**Vocabulary**

**quadrants**

The *x*- and *y*-axes divide the coordinate plane into four sections called **quadrants**. The quadrants are numbered with roman numerals as shown below. Points on the axes are not in any of the quadrants.

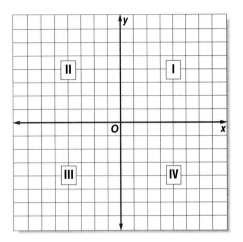

## ✓ *Develop & Understand: A*

Points *A* through *R* are plotted on the grid.

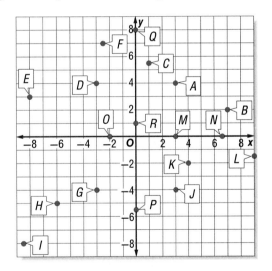

1. Look at the points in Quadrant I.

   a. Record the coordinates of each point in Quadrant I.

   b. What do you notice about the signs of the coordinates of each point?

   c. If someone gives you the coordinates of a point, how can you tell whether it is in Quadrant I without plotting the point?

2. Look at the points in Quadrant II.

   a. Record the coordinates of each point.

   b. What do you notice about the signs of the coordinates of each point?

   c. If someone gives you the coordinates of a point, how can you tell whether it is in Quadrant II without plotting the point?

3. Look at the points in Quadrant III.

   a. Record the coordinates of each point.

   b. What do you notice about the signs of these coordinates?

   c. If someone gives you the coordinates of a point, how can you tell whether it is in Quadrant III without plotting the point?

4. Look at the points in Quadrant IV.

   a. Record the coordinates of each point.

   b. What do you notice about the signs of these coordinates?

   c. If someone gives you the coordinates of a point, how can you tell whether it is in Quadrant IV without plotting the point?

5. Look at the points on the $x$-axis.

   a. Record the coordinates of each point.

   b. What do these coordinates have in common?

   c. If someone gives you the coordinates of a point, how can you tell whether it is on the $x$-axis without plotting the point?

6. Look at the points on the $y$-axis.

   a. Record the coordinates of each point.

   b. What do these coordinates have in common?

   c. If someone gives you the coordinates of a point, how can you tell whether it is on the $y$-axis without plotting the point?

## ✓ Develop & Understand: B

In Chapter 7 you found the area of several figures using formulas. You can also find the area of a figure when placed on a coordinate plane.

7. On a sheet of graph paper, draw a coordinate plane showing the four quadrants. Make the $x$- and $y$-axis range from $-10$ to $10$.

   a. Plot the following ordered pairs. Connect consecutive points with straight lines.

   $(4, -3), (4, 2), (2, 2), (2, 4), (8, 4), (8, 2), (6, 2), (6, -3), (4, -3)$

**b.** What kind of figure did you create?

**c.** In which Quadrants does this figure lie?

**d.** Find the area of this figure. Explain how you found it.

8. On the same sheet of graph paper, plot four points to create a rectangle that lies in all four quadrants.

   **a.** Record the coordinates of each point.

   **b.** What is the area of the rectangle?

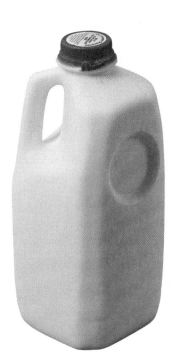

## *Share* & *Summarize*

1. Without plotting each point, determine in which quadrant or on which axis or axes it lies.

   **a.** $(-5, -2)$

   **b.** $(0, 0)$

   **c.** $\left(3, -\frac{2}{7}\right)$

   **d.** $(-35, 0)$

2. In general, if you are given the coordinates of a point, how can you tell which part of the coordinate plane the point is in without plotting it? You might organize your ideas in a chart like the one below.

| x-coordinate | y-coordinate | Part of Coordinate Plane |
|---|---|---|
|  |  |  |
|  |  |  |
|  |  |  |
|  |  |  |
|  |  |  |
|  |  |  |
|  |  |  |
|  |  |  |

## *Inquiry*

## Investigation ⑤ Travel on a Grid

### Materials

• grid paper

The Robinson family is planning a weekend trip. During the trip, the family plans to visit a zoo, an amusement park, and a water park. There is a strict gasoline budget for the trip. Destinations must be reached using the shortest distance possible.

There is not a map available, so you will have to help guide the Robinson family. Each line on the grid represents a road, and the Robinsons must stay on the road during the trip. Your directions can have them move only up or down spaces or to the left and right. Remember, they cannot cut *through* spaces.

### Try It Out

1. Draw a Cartesian plane. Be sure to include the four quadrants. Give the x-axis a scale of $-10$ to 10. Give the y-axis a scale from $-10$ to 10.

2. The Robinson's house is located at $(-2, 4)$. Plot and label a point representing the Robinson's house. In what quadrant do the Robinsons live?

3. The family plans to visit the zoo, which is located at $(-7, -1)$. There is only enough gas to move 35 spaces. Each line on the grid represents one space.

   a. Can the Robinson family make it to the zoo on that tank of gas?

   b. In what quadrant is the zoo located?

   c. Plot and label the point representing the zoo. Trace the lines from the Robinson's house to the zoo to show the route.

   d. How many spaces did the family move to get to the zoo?

4. The next destination is the amusement park. It is located in Quadrant I. The family planned 15 moves to get to the park. The amusement park is located nine spaces to the right and up three spaces from the zoo. Draw your route and place a point representing the amusement park. Label the point.

   a. What are the coordinates of this point?

   b. Did the family follow its plan? How do you know?

5. If you recall, there was enough gas to move 35 spaces for the entire weekend trip. The family has already visited the zoo and the amusement park. How many spaces are left to visit the water park and to return home?

**6.** From the amusement park, the family is ready to travel to the water park to finish out the weekend trip. The water park is located in Quadrant IV.

   **a.** Is the *x*-coordinate of the water park a positive or negative number? Explain.

   **b.** Is the *y*-coordinate of the water park a positive or a negative number? Explain.

**7.** The family leaves the amusement park and travels vertically three spaces and to the left one space. Name the ordered pair in which the Robinsons have arrived. Place a point on this ordered pair and label it.

**8.** After a day of swimming, the family is ready to make the trip home. On your graph, trace a route from the water park back to the Robinson's home. How many spaces did it take to go from the water park to the family's home?

**9.** Was the family able to stay within its allotted gasoline budget of 35 spaces? Explain.

## Try It Again

**10.** The following weekend, the Robinsons want to visit relatives. For this trip, there is a way that allows the family to "cut" through spaces. The Robinsons travel vertically for 9.5 spaces and horizontally for 6 spaces. The relatives live in Quadrant III. Give the coordinates of the relatives' house.

## What Did You Learn?

**11.** Give the coordinates of the point on your grid that is halfway between the zoo and the water park.

**12.** List three coordinates on the grid with the first coordinate greater than 3.

**13.** List three coordinates on the grid with a second coordinate less than 2.

**14.** List three coordinates in Quadrant IV.

# On Your Own Exercises

**Lesson 8.3**

**Practice & Apply**

1. If the temperature rises 10°F, it will be −25°F. What is the temperature now?

2. If the temperature goes down 33°F, it will be −10°F. What is the temperature now?

3. If the temperature goes up $1\frac{1}{4}$°F, it will be 0°F. What is the temperature now?

4. Copy the number line. Plot points to show the locations of the numbers listed. Label each point with the corresponding number.

$$-2.5 \quad -1.75 \quad 2\frac{1}{3} \quad -3 \quad 0.4 \quad 4\frac{1}{2} \quad -0.8$$

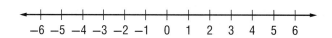

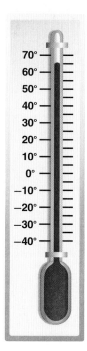

In Exercises 5–8, tell which temperature is warmer.

5. 5°F or −15°F

6. −35°F or −25°F

7. $-5\frac{1}{2}$°F or $-5\frac{3}{4}$°F

8. −100.9°F or −100.5°F

9. Which is greater, $|-5|$ or $|3|$?

10. Which is greater, $-8$ or $|-8|$?

**Evaluate each expression.**

11. $|5.8|$

12. $\frac{1}{3} \cdot |-21|$

13. $|-4.5| + |-3.25|$

14. $\left|5\frac{1}{3}\right| - 8$

15. If $|x| = 7.1$, what is $x$ equal to?

16. Plot these points on a coordinate plane. Label each point with its letter.

   a. $(3\frac{2}{5}, 2)$   b. $(-2, 6)$   c. $(0.4, -4.4)$   d. $(-5, -2)$

**17.** Find the coordinates of points *J* through *O*.

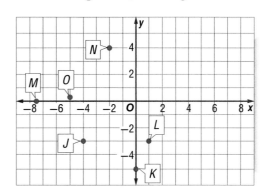

**18.** Without plotting each point, tell in which quadrant or on which axis it lies.

    **a.** $(2, 0)$          **b.** $(0, -24)$          **c.** $(35, -23)$

    **d.** $(3, 5)$          **e.** $(-2, -2)$          **f.** $(-52, 5)$

**Earth Science** Elevations are measured from sea level, which has an elevation of 0 feet. Elevations above sea level are positive. Elevations below sea level are negative. In Exercises 19–23, use this table, which shows the elevation of the lowest point on each continent.

| Continent | Location of Lowest Point | Elevation |
|---|---|---|
| North America | Death Valley | −282 ft |
| South America | Valdes Peninsula | −131 ft |
| Europe | Caspian Sea | −92 ft |
| Asia | Dead Sea | −1,312 ft |
| Africa | Lake Assal | −512 ft |
| Australia | Lake Eyre | −52 ft |
| Antarctica | Bentley Subglacial Trench | −8,327 ft |

**Source:** *World Almanac and Book of Facts 2000.* Copyright © 1999 Primedia Reference Inc.

**19.** Order the elevations in the table from lowest to highest.

**20.** Draw a number line. Plot each elevation given in the table. Label the point with the name of the continent.

**21.** How much lower is the Dead Sea than the Caspian Sea?

**22.** One of the elevations in the table is significantly lower than the others.

    **a.** On which continent is this low point located?

    **b.** How much lower is this point than the next lowest point?

23. **Challenge** The highest point in North America is the summit of Denali, a mountain in Alaska with an elevation of 20,320 feet. How many feet higher than Death Valley is the Denali summit?

Graph each of the following coordinates. Answer Parts a and b for Exercises 24–27.

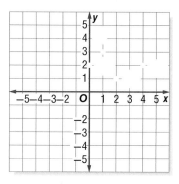

    **a.** In which quadrant did you graph the point?

    **b.** Would the point have been in the same location if you did not take the absolute value?

24. *A:* $(|0|, |2|)$

25. *B:* $(|-2|, |1|)$

26. *C:* $(|-1|, |-3|)$

27. *D:* $(|4|, |-2|)$

28. Is there a pattern to your answers to Exercises 24—27? Explain.

29. **Astronomy** The average surface temperatures of the planets in the solar system are related to their average distances from the Sun.

**Planets in the Solar System**

| Planet | Distance from Sun (millions of miles) | Surface Temperature (°F) |
|---|---|---|
| Mercury | 36 | 662 |
| Venus | 67 | 860 |
| Earth | 93 | 68 |
| Mars | 142 | −9 |
| Jupiter | 483 | −184 |
| Saturn | 888 | −292 |
| Uranus | 1,784 | −346 |
| Neptune | 2,799 | −364 |

    **a.** Create a graph with distance from the Sun on the *x*-axis and average surface temperature on the *y*-axis. Plot the eight points listed in the table.

    **b.** Generally speaking, how does temperature change as you move further from the Sun?

    **c.** Why do you think the relationship you noticed in Part b happens?

    **d.** Which planet or planets do not fit the general pattern? Why might a planet not follow the pattern?

**30.** If you plotted all points for which the *y*-coordinate is the square of the *x*-coordinate, in which quadrants or on which axes would the points lie? Explain how you know your answer is correct.

**31. In Your Own Words** Explain how a number line can help you order a set of positive and negative numbers.

*Mixed Review*

**Find the value of each expression in simplest form.**

**32.** $\frac{5}{6} + \frac{4}{9}$

**33.** $\frac{19}{26} - \frac{17}{39}$

**34.** $\frac{45}{56} \cdot \frac{32}{35}$

**35.** $\frac{14}{15} \div \frac{2}{5}$

**36.** $11\frac{19}{21} + 6\frac{1}{7}$

**37.** $5\frac{1}{4} - 2\frac{5}{12}$

**38.** $3\frac{5}{8} \cdot 1\frac{3}{4}$

**39.** $9\frac{1}{3} \div \frac{5}{9}$

**40.** Mrs. Heflin conducted a survey of students in her class to determine their opinions about the amount of homework they were given in class. Write each fraction as a decimal and a percent. Write each given percent as a decimal and as a fraction.

   **a.** She found that $\frac{9}{10}$ of the students did their homework every night.

   **b.** She found that $\frac{3}{4}$ of the students would prefer less homework.

   **c.** She learned that 1% would prefer to have no homework.

**41.** Which is greater, 0.25 or $\frac{1}{8}$?

**42.** Which is greater, 1.2 or 120%?

**43.** List in order from least to greatest: $\frac{2}{3}$, 60%, 0.66.

# Review & Self-Assessment

## Vocabulary

absolute value

axes

coordinates

line graph

negative numbers

opposites

ordered pair

origin

positive numbers

quadrants

## Chapter Summary

In this chapter, you interpreted and created graphs. You started by looking at graphs with only a few points and discovering what information was revealed by the positions of the points. Then you looked at graphs with lines and curves. You saw that when the variable on the horizontal axis is time, the line or curve tells a story about how the other variable changes.

You then made your own graphs. For some graphs, you drew a line or curve to fit a story or description without worrying about specific values. For other graphs, you made a table of values, chose scales for the axes, and plotted points. You saw that sometimes it makes sense to connect the points on a graph and that connecting points can help you see patterns and make predictions.

You also learned that plotting collected data and looking for trends, or patterns, can help you determine whether the variables are related. When the plotted points do show a pattern, you can make predictions about values not on the graph.

Finally, you can use a graph to show a relationship between quantities. On a single horizontal number line, negative integers are graphed to the left of zero and positive integers are graphed to the right of zero. The distance a number is from zero is called the absolute value. The further a number is to the left, the smaller it is.

On a coordinate grid, the number lines divide the plane into four sections labeled quadrants. Points may either lie in one of the four quadrants or on the *x*- or *y*-axis. You learned that a point in Quadrant I will have positive *x*- and *y*-coordinates. A point in Quadrant II will have a negative *x*-coordinate and a positive *y*-coordinate. A point in Quadrant III will have both negative *x*- and *y*-coordinates. A point in Quadrant IV will have a positive *x*-coordinate and a negative *y*-coordinate.

## Strategies and Applications

The questions in this section will help you review and apply the important ideas and strategies developed in this chapter.

## Interpreting graphs

**1.** These graphs give information about Lydia's dogs. Use them to determine whether each statement that follows is true or false. Explain how you decided.

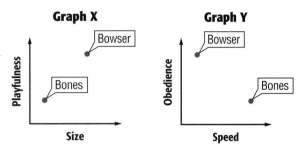

**Graph X**

Bowser

Bones

Playfulness / Size

**Graph Y**

Bowser

Bones

Obedience / Speed

  **a.** The smaller dog is less playful.

  **b.** The larger dog is more obedient.

  **c.** The faster dog is more playful.

  **d.** The slower dog is smaller.

**2.** This graph shows how Jackie's distance from home changed one Saturday morning. Write a story about Jackie's morning. Your story should account for all the changes in the graph.

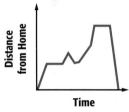

Distance from Home / Time

**3.** This graph shows the percent of the U.S. labor force that was unemployed in even-numbered years from 1980 to 2000.

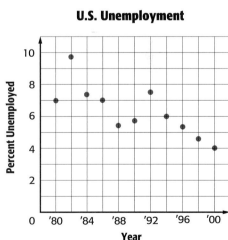

**U.S. Unemployment**

Percent Unemployed / Year

Source: *World Almanac and Book of Facts 2003.* Copyright © 2003 Primedia Reference Inc.

  **a.** In which year was the unemployment rate highest? In which year was it lowest?

  **b.** Over which two-year period did the unemployment rate decrease most? By about how much did it decrease?

  **c.** Over which two-year period did the unemployment rate increase most? By about how much did it increase?

  **d.** Over which two-year period did the unemployment rate change least?

  **e.** Predict what the unemployment rate was in 1993.

## Creating graphs

**4.** Frank went diving to explore a sunken ship. He descended quickly at first and then more slowly. As he was descending, he noticed an interesting fish. Frank ascended a little and stayed at a constant depth for awhile as he watched the fish. Then Frank slowly descended until he reached the sunken ship. He explored the ship for several minutes and then returned to the surface of the water at a slow, steady pace. Draw a graph that shows how Frank's depth changed during his dive.

**5.** A plumber charges $40 for a house call plus $15 for each half hour he works.

**a.** Copy and complete the table to show how much the plumber charges if he works the given numbers of hours.

| Time (hours) | 0 | 0.5 | 1 | 1.5 | 2 | 2.5 | 3 | 3.5 | 4 | 4.5 | 5 |
|---|---|---|---|---|---|---|---|---|---|---|---|
| Charge (dollars) | 40 | 55 | | | | | | | | | |

**b.** Choose an appropriate scale. Graph the data.

**c.** Does it make sense to connect the points on your graph? Explain.

**d.** Suppose the plumber charges by fractions of a half hour. Use your graph to estimate how much the plumber would charge if he worked 3 hours 15 minutes. Explain how you found your answer.

**Find the distance of each number from 0.**

**6.** 3.5

**7.** $-1\frac{1}{2}$

**Which is further from 0?**

**8.** 3.5 or $-2.5$

**9.** $-4.5$ or $-3.5$

**Find the value of each expression.**

**10.** $|-4|$

**11.** $-|-2.5|$

**12.** $-|1.5|$

**13.** $|-2^2|$

**The table shows the change in the number of people at the movies for each day of the week.**

| Day of the Week | Monday | Tuesday | Wednesday | Thursday | Friday | Saturday | Sunday |
|---|---|---|---|---|---|---|---|
| Size of change | $-75$ | $-10$ | 5 | 10 | 25 | 60 | $-25$ |

**14.** When was the greatest change in the number of people at the movies?

**15.** When was the smallest change in the number of people at the movies?

## Demonstrating Skills

**16.** Plot each point on one set of axes. Label each point with its coordinates.

**a.** $(0, 5)$      **b.** $\left(1\frac{1}{2}, 5\frac{1}{2}\right)$      **c.** $(9.5, 8)$

**d.** $(7, 7)$      **e.** $(7, 0)$      **f.** $(3.3, 6.5)$

**17.** Give the coordinates of the point halfway between $(3, 5)$ and $(7, 5)$.

**18.** Plot the points $(1, 4)$ and $(5, 4)$. Plot two more points so that the four points can be connected to form a square. Give their coordinates.

**Identify the quadrant where each point is located.**

**19.** $(-2, 3)$          **20.** $(-1.5, -6.5)$

**21.** $(4, 6)$          **22.** $\left(3\frac{1}{2}, -7\right)$

**23.** What is the sign of the $x$-value of a point in Quadrant II?

**24.** What is the sign of the $y$-value of a point in Quadrant I?

**25.** What is the $x$-value of a point on the $y$-axis?

**26.** What is the $y$-value of a point on the $x$-axis?

**Order the numbers from least to greatest.**

**27.** $-3, 2, -9, 5, 6, -10, 3, -6$

**28.** $-35, -15, -45, -25, 0, -5, -20, 5, 10, -15$

## Test-Taking Practice

**New York Population**

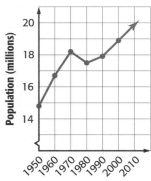

*SHORT RESPONSE*

**1** Use the data in the graph to describe the pattern or trend of the New York population.

What do you predict the population to be in 2010?

*Show your work.*

*Answer* _____

*MULTIPLE CHOICE*

**2** Which point lies on the $y$-axis in the coordinate plane?

**A** $(0, -4)$

**B** $(-6, 0)$

**C** $(-3, 3)$

**D** $(7, 0)$

**3** Which expression is equivalent to $|-16| + |4| + |-8|$?

**F** $-16 + 4 - 8$

**G** $-16 + 4 + 8$

**H** $16 + 4 + 8$

**J** $16 + 4 - 8$

# Equations

## Real-Life Math

**It's for the Birds** Amy makes and sells birdhouses at her town's craft fair. It costs $10 to rent a table. Amy spends an average of $5 for supplies to make each birdhouse.

The equation $c = \$10 + \$5n$ represents the total cost of making the birdhouses. In the equation, c represents the total cost, and $n$ represents the number of birdhouses. Suppose Amy has a total of $75 to spend on rent and supplies. The equation $\$75 = \$10 + \$5n$ represents the number of birdhouses that she can make.

**Think About It** Amy plans to sell her birdhouses for $8 each. The equation $m = \$8n$ represents the amount of money she will earn. In the equation, $m$ represents the amount of money she will earn, and $n$ represents the number of birdhouses. She hopes to earn $120. What equation should she solve?

**Math Online**
Take the **Chapter Readiness Quiz** at glencoe.com.

# Dear Family,

The next chapter is about solving equations. An equation is a number sentence that includes an equals sign, which means that two expressions have the same value. Here are three examples.

$$9 + 6 = 15 \qquad 9 + 6 = 5 \cdot 3 \qquad 7 + 8 = 18 - 3$$

In this chapter, the class will explore equations with variables, such as $3 \cdot n = 18$.

### Key Concept—Backtracking

The class will learn two methods for solving equations: backtracking and guess-check-and-improve. To find the output $t$ for the equation $4 \cdot n + 5 = t$, start with the input $n$. Multiply the input by 4 and then add 5.

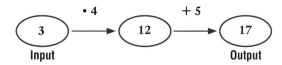

Here is the flowchart for an input of 3.

If given an output of 21, the flowchart could be used to work backward and determine that the input was 4.

### Chapter Vocabulary

| | |
|---|---|
| **backtracking** | **inequality** |
| **equation** | **open sentence** |
| **flowchart** | **solution** |
| **guess-check-and-improve** | |

### Home Activities

- Have your student share newly learned strategies for solving equations.
- Take turns writing equations on slips of paper. Then solve one another's equations.

# Understand Equations

You have been working with equations for many years. An **equation** is a mathematical sentence stating that two quantities have the same value. An equals sign, =, is used to separate the two quantities. Here are three examples of equations.

$$9 + 6 = 15 \qquad 9 + 6 = 5 \cdot 3 \qquad 7 + 5 = 15 - 3$$

## Vocabulary

equation

### Explore

In the game *Equation Challenge,* you will see how many equations you can make using a given set of numbers.

### *Math Link*

The equals sign was first used by Robert Recorde in his book, *Whetstone of Witte,* published in 1557. Recorde states that he chose parallel line segments of equal length because "no two things can be more equal."

### *Equation Challenge* Rules

- Your teacher will call out seven single-digit numbers. Write the numbers on a sheet of paper. (Note: Your teacher may call the same number more than once. For example, the numbers might be 1, 2, 5, 3, 4, 9, and 3.)

- You have five minutes to write down as many correct equations as you can. Use only the seven numbers, an equals sign, operation symbols, decimal points, and parentheses. Follow these guidelines when writing your equations.

  – Use each of the seven numbers only once in each equation.

  – You do not need to use all the numbers in each equation.

  – You can combine the numbers to make numbers with more than one digit. For example, you could combine 1 and 2 to make 12, 21, 1.2, or $\frac{1}{2}$.

- Check your equations to make sure they are correct.

Here are some sample equations for the numbers 1, 2, 5, 3, 4, 9, and 3.

$$9 + 5 = 14 \qquad \frac{35 + 1}{4} = 9 \qquad 9 \cdot 5 = 3 \cdot 3 \cdot (4 + 1)$$

Play *Equation Challenge.* Be creative when writing your equations.

# Investigation  Equations and Inequalities

## Vocabulary

**inequality**

An *equation* is a mathematical sentence stating that two quantities have the same value. A mathematical sentence stating that two quantities have *different* values is called an **inequality**. Inequalities use the symbols $\neq$, $<$, and $>$. The table below explains these symbols.

| Symbol | What It Means | Example |
|:---:|:---:|:---:|
| $\neq$ | is not equal to | $7 + 2 \neq 5 + 1$ |
| $<$ | is less than | $4 + 5 < 20$ |
| $>$ | is greater than | $6 \cdot 9 > 6 + 9$ |

Just as sentences with words can be true or false, so can equations and inequalities. Consider these six sentences.

Stop signs are yellow.      $4 = 32 \div 8$

The sun is hot.      $5 \cdot 6 > 6 \cdot 5$

Alaska is south of Texas.      $5 \cdot 4 = 27 - 3$

### Think & Discuss

Which of the sentences above are true? Which are false?

Find a way to make each false sentence true by changing or adding just one word, symbol, or number.

### ✓ Develop & Understand: A

**The sentences below are false. Make each sentence true by changing one symbol or number.**

1. $17 + 5 < 3^2 + 12$
2. $14 + 5 = 12 + 11$
3. $23 - 11 = 22 \div 2$
4. $6 \cdot 5 > (4 \cdot 7) + 8$

**Tell whether each sentence is true or false. If it is false, make it true by replacing the equals sign with < or >.**

5. $5 + 13 = 2 + 4^2 + 3$
6. $7 + (2 \cdot 3) = (6 \cdot 2) + 1$
7. $24 \div 5 = 2 + 3$
8. $\frac{2}{5} = \frac{1}{2}$
9. $0.25 = \frac{1}{4}$
10. $8 + 12 \div 4 = (8 + 12) \div 4$

## Investigation 2 Equations with Variables

### Vocabulary

open sentence

solution

You can determine whether the equations in Investigation 1 are true by finding the value of each side. But what if an equation contains a variable? For example, consider this equation.

$$3 \cdot n = 18$$

You cannot tell whether this equation is true or false unless you know the value of $n$. An equation or inequality that can be true or false depending on the value of the variable is called an **open sentence**.

> ### Think & Discuss
>
> For each open sentence, find a value of $n$ that makes it true. Then find a value of $n$ that makes it false.
>
> $$3 \cdot n = 18 \qquad n \div 2 = 2.5 \qquad n + 5 = 25$$

Finding the values of the variable or variables that make an equation true is called *solving* the equation. A value that makes an equation true is a **solution** of the equation. Consider this equation.

$$6 \cdot n - 1 = 29$$

Finding a solution of $6 \cdot n - 1 = 29$ is the same as answering the following question.

For what value of $n$ is $6 \cdot n - 1$ equal to 29?

---

**Math Link**

These are three ways to write "6 times."

$6 \times n \quad 6 \cdot n \quad 6n$

---

Jing tried several values for $n$. Here is what she found:

| $n$ | $6 \cdot n - 1$ | Test | Solution? |
|---|---|---|---|
| 3 | 17 | $6 \cdot n - 1 < 29$ | no |
| 4 | 23 | $6 \cdot n - 1 < 29$ | no |
| 5 | 29 | $6 \cdot n - 1 = 29$ | yes |
| 6 | 35 | $6 \cdot n - 1 > 29$ | no |
| 7 | 41 | $6 \cdot n - 1 > 29$ | no |
| 8 | 47 | $6 \cdot n - 1 > 29$ | no |

Jing found that 5 is a solution of $6 \cdot n - 1 = 29$ since $6 \cdot 5 - 1 = 29$. The results kept increasing as $n$ increased, so she concluded that 5 must be the only solution.

## ✅ Develop & Understand: A

**Each of these equations has one solution. Solve each equation.**

1. $6 \cdot n - 1 = 41$          2. $6 \cdot n - 1 = 11$

3. $2p + 7 = 19$          4. $4 + 4 \cdot b = 20$

5. $\frac{5}{4} = \frac{25}{d}$          6. $m + 3 = 2m$

7. Try the numbers 1, 2, 3, 4 in the equation $2 \cdot d + 3 = 15 - 2 \cdot d$ to test whether any of them is a solution.

8. Write three equations with a solution of 13. Check that your equations are correct by substituting 13 for the variable.

All of the equations you have seen so far have one solution. It is possible for an equation to have more than one solution or to have no solution at all.

For some equations, every number is a solution. Such equations are always true, no matter what the values of the variables are.

**Math Link**

Equations with squared variables, for example, $m^2 - 3m = 0$, are called *quadratic equations.* They can be used to describe the path in which a projectile travels.

## Think & Discuss

Equations that include a squared variable often have two solutions. Find two solutions for the equation $m^2 - 3m = 0$.

Explain why each equation below is always true.

$$n + 3 = 3 + n \qquad a \cdot 5 = a + a + a + a + a$$

9. Try the values 1, 2, 3, and 4 to test whether any are solutions of this equation.

$$t^2 + 8 = 6 \cdot t$$

**Tell whether each equation below is always true, sometimes true, or never true. Explain how you know.**

10. $m - m = 0$

11. $\dfrac{r}{3} = r$

12. $q + 7 = q - 7$

13. $p \div 7 = \dfrac{1}{7} \cdot p$

14. $n \cdot 2 = n + 1$

15. $(a + 3) \cdot 2 = 2a + 6$

16. Which number is a solution to the equation below?

$$5n + 7 = 27$$

  a. 1  b. 2  c. 3  d. 4

17. Which number is a solution to the equation below?

$$t^2 + 3 = 12$$

  a. 1  b. 2  c. 3  d. 4

18. Which equation does **not** have at least one solution?

  a. $4x + 3 = 16$  b. $5m - 2 = 28$
  c. $3y - 1 = 3y + 1$  d. $n^2 - 2 = 23$

19. Which equation is true for all values of $d$?

  a. $5d - 2 = 8$  b. $5d + 2 = 20$
  c. $5d \div 4 = 5d \cdot 2$  d. $5d + 1 = 5d + 1$

20. **Challenge** Tell whether each equation has a whole-number solution. Explain how you know.

  a. $2 \cdot n - 1 = 37$  b. $2n + 1 = 18$
  c. $3 \cdot n + 5 = n + 7$  d. $n^2 + 2 = 1$

**Real-World Link**

In *meteorology*, the science of weather, an equals sign represents fog.

· · · · · · · · · · · · · · · · · · · · ·

## Share & Summarize

1. Solve the equation $3p + 5 = 11$. Explain how you found the solution.

2. Give an example of an equation that is always true and an example of an equation that is never true.

# Investigation ③ Instructions and Directions

## Materials

- 1 red block and 1 blue block
- 1 yellow counter and 1 green counter
- Lompoc, California, street map

In this investigation, you will practice "undoing" sets of instructions. In Lesson 9.2, you will see how the strategies used in this investigation can help you solve equations.

## Undoing Instructions

Starting with a blank sheet of paper, follow these instructions.

- Draw a small X (in pencil) in the center of the paper.
- Put the red block on the X.
- Put the yellow counter on the red block.
- Put the blue block on the yellow counter.
- Put the green counter on the blue block.

**1.** Write a list of steps that you think would undo these instructions, leaving you with a blank sheet of paper. Do not touch any of the items on the paper until you have finished writing your instructions.

**2.** Follow your steps from Question 1. Did your steps undo the above instructions? If not, rewrite them until they do.

**3.** How do your steps compare with the original set of steps?

## Reversing Directions

Madeline lives in Lompoc, California, on the corner of Nectarine Avenue and R Street. Today, she met her friend T.J. at the town pool. T.J. had given her these directions to get to the pool.

- Start at the corner of Nectarine Avenue and R Street.
- Walk 2 blocks east along Nectarine Avenue.
- Turn right from Nectarine Avenue onto O Street.
- Walk 4 blocks south on O Street.
- Turn left from O Street onto Maple Avenue.
- Walk 5 blocks east on Maple Avenue.
- Turn right from Maple Avenue onto J Street.
- Walk 4 blocks south on J Street.
- Turn left from J Street onto Ocean Avenue.
- Walk 7 blocks east on Ocean Avenue.
- The pool is at the corner of Ocean Avenue and C Street.

*Go on*

Now Madeline must reverse the directions to get home.

**4.** Without looking at the map, write a set of directions Madeline could follow to get home from the pool.

**5.** On the street map, carefully follow the directions that you wrote in Question 4. Do you end up at the corner of Nectarine Avenue and R Street? If not, make changes to your directions until they work.

**6.** Write a set of directions to get from one place on the map to another. Word your steps like those on page 539.

- When you describe a turn, mention the street where you start, whether you turn left or right, and the street where you end.
- When you describe a walk along a street, mention the number of blocks, the direction, and the street name in which you walked.

When you are finished, try your directions to make sure they are accurate.

**7.** Exchange directions with your partner. Without looking at the map, write the steps that reverse your partner's directions. Then use the street map to test your directions.

**8.** Describe some general strategies that you find useful when reversing a set of directions.

## What Did You Learn?

**9.** In this investigation, you undid two types of instructions.
- Steps for stacking blocks and counters
- Directions for getting from one place to another

Describe how the methods that you used to undo the instructions in each case were similar.

**Real-World Link**
It is believed that humans have been making maps since prehistoric times. Archaeologists have discovered systems of lines drawn on cave walls and bone tablets that may be maps of hunting trails made by prehistoric peoples.

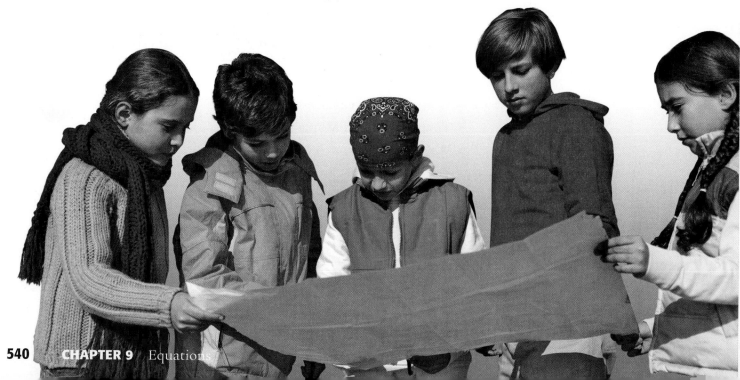

**Practice & Apply**

**Explore** In a round of the game *Equation Challenge,* the numbers 1, 2, 2, 4, 5, 5, and 9 were called.

**1.** Make at least four equations using these numbers.

In Exercises 2 and 3, the sentence is not true. Change or add one number or symbol to make it true.

**2.** $5 + 16 = 3 \cdot 8$

**3.** $8 \cdot 5 \neq 17 + 16 + 7$

In Exercises 4–9, tell whether the sentence is true or false. If it is false, make it true by replacing the equals sign with < or >.

**4.** $3 \cdot 11 = 42 - 9$

**5.** $(3 \cdot 5) + 4 = 4 + 5 + 1$

**6.** $\frac{1}{3} + \frac{4}{6} = \frac{3}{7} + \frac{16}{28}$

**7.** $0.95 = \frac{9}{10}$

**8.** $3 \cdot 13 = 54 - 16$

**9.** $16 - 8 \div 4 = (16 - 8) \div 4$

**Solve each equation.**

**10.** $x \cdot 12 = 48$

**11.** $56 + m = 100$

**12.** $6p + 10 = 28$

**13.** $50 - 4 \cdot z = 30$

**14.** Consider the equations $s + 13 = 20$ and $p + 13 = 20$.

   **a.** Solve each equation.

   **b.** How do the solutions to the two equations compare? Explain why this makes sense.

**15.** Test the values 0, 1, 3, 4, and 6 to see whether any are solutions of $7m - m^2 + 10 = 16$.

**Tell whether each equation is always true, sometimes true, or never true. Explain how you know.**

**16.** $5 \cdot m = \frac{25 \cdot m}{5}$

**17.** $5s = 25$

**18.** $t - 1 = t + 1$

**19.** $p^2 = p \cdot p$

**20.** $n + 6 = 7 \div n$

**21.** $7p = p \div 7$

**22.** Write three equations with a solution of 3.5.

**Real-World Link**

The solution to the equation in Exercise 10, $x \cdot 12 = 48$, is the number of 12-egg cartons needed to hold four dozen eggs.

**23.** Of these three equations, one has no solution, one has one solution, and one has two solutions. Decide which is which, and find the solutions.

**a.** $p^2 + 6 = 5 \cdot p$

**b.** $3p + 5 = 3p - 5$

**c.** $4 \cdot p + 5 = 7$

**Connect & Extend**

**24. What's My Rule?** Isabela and Jada were playing a game of *What's My Rule.* Here is Isabela's secret rule.

$o = 37 - 4 \cdot i$, where $o$ is the output and $i$ is the input

**a.** What input value gives an output of 17? Check your answer by substituting it into the rule.

**b.** What input value gives an output of 5? Check your answer.

**c.** What input value gives an output of 0? Check your answer.

**25.** Pretend you are playing a game of *What's My Rule.* Make up a secret rule for calculating an output value from an input value. Your rule should use one or two operations.

**a.** Write your rule in symbols.

**b.** Find the input value for which your rule gives an output of 25.

**c.** Find the input value for which your rule gives an output of 11.

**26.** Paul and Katarina were playing a game of *What's My Rule.* Here is Katarina's secret rule.

$m = \dfrac{n}{10}$, where $n$ is the input and $m$ is the output

**a.** Write an equation to find the input value that gives an output of 1.5.

**b.** Solve your equation from Part a to find the input value $n$.

**c.** Write an equation to find the input value that gives an output of 20.

**d.** Solve your equation from Part c to find the input value $n$.

**27. In Your Own Words** Explain the meaning of equation, inequality, open sentence, and solution.

**Balancing Equations** You can think of an equation as a balanced scale. The scale at right represents the equation $3 \cdot 5 = 9 + 6$. The scale is balanced because both sides have the same value.

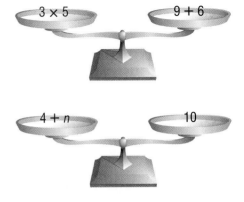

The second scale represents the equation $4 + n = 10$. To solve this equation, you need to find the value of $n$ that will make the scale balance.

In Exercises 28–32, you will solve puzzles involving scales. Thinking about these puzzles may give you some ideas for solving equations.

**28.** This scale is balancing bags of peanuts and boxes of popcorn.

**a.** How many bags of peanuts will balance one box of popcorn?

**b.** If a bag of peanuts weighs 5 ounces, how much does a box of popcorn weigh?

**29.** Consider this scale.

**a.** Write the equation this scale represents.

**b.** What number will balance one $n$?

In Exercises 30–32, refer to the information on page 543.

**30.** Consider this scale.

**a.** Write the equation this scale represents.

**b.** What number will balance one *b*?

**31.** These two scales hold jacks, marbles, and a block.

**a.** How many jacks will balance the block?

**b.** How many jacks will balance one marble?

**c.** If the block weighs 15 grams, how much does a jack weigh? How much does a marble weigh?

**32. Challenge** These scales hold blocks, springs, and marbles.

**a.** How many marbles will balance one spring?

**b.** If a marble weighs 1 ounce, how much does a spring weigh? How much does a block weigh?

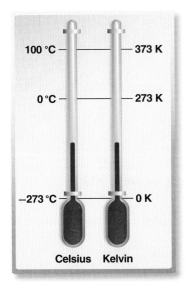

**Mixed Review**

**33. Science** You are familiar with the Fahrenheit and Celsius temperature scales. The Kelvin scale is a temperature scale that is used frequently in science. This equation shows how Kelvin temperatures $K$ are related to Celsius temperatures $C$.

$$K = C + 273$$

**a.** The mean surface temperature on Mercury is about 180°C. Express this temperature in Kelvins.

**b.** The maximum surface temperature on Mars is about 290 Kelvin. Express this temperature in degrees Celsius.

**c.** Which planet is hotter, Mercury or Mars? Explain your answer.

**34. Economics** The unit of currency in South Africa is the rand. In November 2007, one U.S. dollar was worth about 6.48 rand. This relationship can be expressed as an equation, where $R$ stands for the number of rand and $D$ stands for the number of dollars.

$$6.48 \cdot D = R$$

**a.** On that day, Jacob exchanged $75 for rand. How many rand did he receive?

**b.** Jacob wanted to buy a small statue that cost 20 rand. He thought this was about $3.08. His sister thought it was about $129.60. Who was correct? Explain your answer.

**35. Statistics** Jaleesa received the following scores on her 20-point spelling quizzes.

> 20    15    18    19    20    0    14    17    16

**a.** Find the mean, median, and mode of her scores.

**b.** Which measure of center do you think best represents Jaleesa's typical quiz score? Give reasons for your choice.

**c.** Jaleesa's teacher has agreed to drop each student's lowest score. Drop the lowest score. Compute the new mean, median, and mode of Jaleesa's scores.

**Find the indicated percentage.**

**36.** 50% of 12

**37.** 25% of 12

**38.** 75% of 12

# Backtracking

Jay was playing *What's My Rule* with his friend Marla. He figured out that this was the rule.

*To find the output, multiply the input by 3 and then add 7.*

He wrote the rule in symbols, using $n$ to represent the input and $t$ to represent the output.

$$t = 3n + 7$$

Jay wanted to find the input that would give an output of 43. This is the same as solving the equation $3n + 7 = 43$.

> 3 times the number plus 7 is equal to 43.

> So there must have been 36 before the 7 was added.

> 3 times a number gives 36—so the number must have been 12.

> If I let *n* be 12, then $3n + 7$ is $3 \times 12 + 7$, which is 43. So I'm right—the input number must be 12.

### Think & Discuss

Using Jay's rule, what input gives an output of 40?

Explain the reasoning you used to find the input. Check your answer by substituting it into the equation $3n + 7 = 40$.

Jay's method of solving $3n + 7 = 43$ involves working backward from the output value to find the input value. In this lesson, you will learn a technique for working backward called *backtracking*.

# Investigation  Learn to Backtrack

## Vocabulary

backtracking

flowchart

---

### Real-World Link

Computer programmers and engineers use complex flowcharts to represent the steps in computer programs, manufacturing operations, construction projects, and other procedures.

---

Hannah's class was playing *What's My Rule*. Hannah found that the rule was $t = 4n + 5$, where $n$ is the input and $t$ is the output.

To find an output with this rule, you do the following.

> *Start with an input.*
> *Multiply it by 4.*
> *Add 5.*

Hannah drew a diagram called a **flowchart** to show these steps.

The oval at the left side of the flowchart represents the input. Each arrow represents a *mathematical operation*. The oval to the right of an arrow shows the result of a mathematical operation. The oval at the far right represents the output.

Here is Hannah's flowchart for the input value 3.

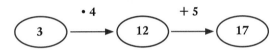

In Exercises 1–4, you will practice working with flowcharts.

## ✔ Develop & Understand: A

**Copy each flowchart. Fill in the ovals.**

1.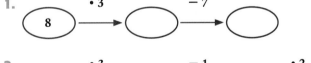

2. 

3. In the rule $j = 7m - 2$, you can think of $m$ as the input and $j$ as the output.

   a. Create a flowchart for the rule but do not fill in the ovals.

   b. Use your flowchart to find the value of $j$ when the value of $m$ is $\frac{5}{3}$.

4. In the rule $d = 3.2 + a \div 10$, you can think of $a$ as the input and $d$ as the output.

   a. Create a flowchart for this rule but do not fill in the ovals. Be sure to think about order of operations.

   b. Use your flowchart to find the value of $d$ when the value of $a$ is 111.

In Exercises 1–4, you used flowcharts to *work forward*, starting with the input and applying each operation to find the output. You can also use flowcharts to *work backward*, starting with the output and undoing each operation to find the input. This process is called **backtracking**. It is useful for solving equations.

## Example

When playing the *What's My Rule* game, Hannah figured out that the secret rule was

$$t = \frac{n + 1}{2}.$$

Hannah wanted to find the input that gives an output of 33. That is, she wanted to solve this equation.

$$\frac{n + 1}{2} = 33$$

To find the solution, she first made a flowchart.

Then she found the input by backtracking.

"Since 33 is the output, I'll put it in the last oval."

"Since the number in the second oval was divided by 2 to get 33, it must be 66."

"One was added to the input to get 66, so the input must be 65."

Hannah checked her solution, 65, by substituting it into the original equation.

$$\frac{65 + 1}{1} = \frac{66}{2} = 33$$

**Math Link**

Operations that undo each other are called *inverse operations*. To undo the division, Hannah used the inverse operation, multiplication. To undo the addition, she used the inverse operation, subtraction.

## ✅ *Develop & Understand: B*

**5.** Marcus used Hannah's rule, $t = \dfrac{n + 1}{2}$, and got the output 53.

Use backtracking to find Marcus' input. Explain each step in your solution.

**Backtracking** Terry solved three equations by backtracking. Below are the flowcharts with which he started. For each flowchart, write the equation he was trying to solve. Use any letter you would like to represent the input variable. Then backtrack to find the solution. Check your solutions.

**6.**

**7.**

**8.**

**9.** Tyrone drew this flowchart.

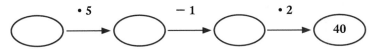

**a.** Copy Tyrone's flowchart. Fill in the ovals.

**b.** Which of these equations can be represented by Tyrone's flowchart? Explain how you know.

$$5 \cdot k - 1 \cdot 2 = 40 \qquad\qquad (k \cdot 5 - 1) \cdot 2 = 40$$

$$2 \cdot (5k - 1) = 40 \qquad\qquad 5k - 1 \cdot 2 = 40$$

## *Share* & *Summarize*

1. Explain what a flowchart is. Demonstrate by making a flowchart for the rule $t = 5n - 3$.

2. Use this flowchart to find the output for input $\dfrac{7}{10}$.

3. Explain what backtracking is. Demonstrate how backtracking can be used to solve $5n - 3 = 45$. Check your solution.

# Investigation 2 Practice Backtracking

In this investigation, you will practice backtracking so you can use it to solve equations quickly and easily.

## ✅ Develop & Understand: A

1. A group of students was playing *What's My Rule* Miguel and Althea thought they knew what the rule was. Both students used $K$ to represent the input and $P$ to represent the output.

   Miguel's rule: $P = 14 \cdot (K + 7)$

   Althea's rule: $P = 14 \cdot K + 7$

   **a.** Make a flowchart for each rule. Do not fill in the ovals.

   **b.** For each rule, use backtracking to find the input that gives the output 105.

   **c.** Are these two rules equivalent? Explain why or why not.

2. Gabriela and Erin were playing a game called *Think of a Number*.

Think of a number. Triple it. Subtract 6. Multiply your result by 5. What do you get?

60.

   Gabriela must figure out Erin's starting number.

   **a.** Draw a flowchart to represent this game.

   **b.** What equation does your flowchart represent?

   **c.** Use backtracking to solve your equation. Check your solution by following Gabriela's steps.

## Think & Discuss

Luke wanted to solve this equation by backtracking.

$$\frac{2 \cdot (n + 1)}{3} - 1 = 5$$

He made this flowchart.

Does Luke's flowchart correctly represent the equation? Why or why not?

Solve the equation. Explain how you found the solution.

The equation in Think & Discuss involves several operations. You can often solve equations like this by backtracking, but you need to pay close attention to order of operations as you draw the flowchart.

## Develop & Understand: B

3. Conor, Althea, and Miguel each made a flowchart to represent this equation.

$$\frac{1 + n \cdot 3}{4} - 11 = 10$$

Tell whose flowchart is correct. Explain the mistakes that the other students made.

**Conor's flowchart**

**Althea's flowchart**

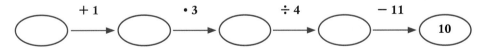

**Miguel's flowchart**

In Exercises 4 and 5, draw a flowchart to represent the equation. Then use backtracking to find the solution. Be sure to check your solution.

4. $\dfrac{n - 13}{2} + 6 = 15$

5. $7\left(\dfrac{n + 4}{7} + 1\right) = 84$

**Solve each equation. Be sure to check your solutions.**

6. $\dfrac{7z + 2}{15} = 2$

7. $(n \cdot 12 + 8) \cdot 100 = 2{,}100$

8. $\dfrac{q - 36}{6} + 16 = 83$

9. $4 \cdot \left(\dfrac{b}{2} - 3\right) + 1 = 97$

## Share & Summarize

Give an example to demonstrate why it is important to pay close attention to order of operations when you make a flowchart.

# Investigation ③ Use Backtracking to Solve Problems

These "ladders" are made from toothpicks.

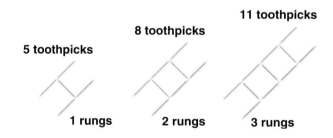

5 toothpicks — 1 rungs

8 toothpicks — 2 rungs

11 toothpicks — 3 rungs

The rule for the number of toothpicks $n$ in a ladder with $r$ rungs is $n = 3r + 2$.

## Think & Discuss

Can you explain why the rule for the number of toothpicks $n$ in a ladder with $r$ rungs is $n = 3r + 2$?

Suppose you have 110 toothpicks. What size ladder can you make? How can you use backtracking to help you find the answer?

### ✅ Develop & Understand: A

1. Write and solve an equation to find the number of rungs on a toothpick ladder made with 53 toothpicks.

2. Look at this pattern of toothpick shapes.

**1 trapezoid**     **2 trapezoids**     **3 trapezoids**

a. Write a rule for finding the number of toothpicks *n* you would need to make a shape with *t* trapezoids.

b. Write and solve an equation to find the number of trapezoids in a shape with 125 toothpicks.

3. Look at the pattern in this table.

| *n* | 0 | 3 | 6 | 9 | 12 |
|---|---|---|---|---|---|
| *y* | 19 | 20 | 21 | 22 | 23 |

a. Write a rule that relates *n* and *y*.

b. Write and solve an equation to find the value of *n* when *y* is 55.

In Exercises 4–7, you will see that backtracking can also be used to solve everyday situations.

### ✅ Develop & Understand: B

4. Leong makes candy apples to sell at the farmer's market on Saturday. He makes a profit of 35¢ per apple.

a. Write a rule Leong could use to calculate his profit if he knows how many candy apples he sold. Tell what each letter in your rule represents.

b. Leong wants to earn $8 so he can see a movie Saturday night. Write an equation Leong could solve to find the number of candy apples he must sell to earn $8.

c. Use backtracking to solve your equation. How many candy apples does Leong need to sell?

5. The plumbers at DripStoppers charge $45 for a house call plus $40 for each hour of work.

   a. Write a rule for the cost of having a DripStoppers plumber come to your home and do *n* hours of work.

   b. DripStoppers sent Mr. Valdez a plumbing bill for $105. Write and solve an equation to find the number of hours the plumber worked at Mr. Valdez's home. Check your solution.

6. When you hire a taxi, you are usually charged a fixed amount of money when the ride starts plus an amount that depends on how far you travel. Suppose a taxi charges $2 plus $0.75 for every quarter mile.

   a. Write an equation to find out how far you can travel for $20.

   b. Solve your equation. How far could you travel for $20? Check your answer.

7. Caroline and Althea are making a kite. The materials for the main part of the kite cost $4.50. The string costs 9¢ per yard.

   a. Write an equation to find how long the string could be if the friends have $30 to spend on their project.

   b. Solve your equation to find how long the string could be.

### Share & Summarize

Jing used toothpicks to create this pattern.

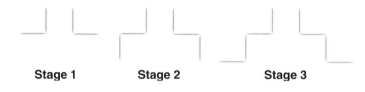

Stage 1          Stage 2          Stage 3

For one of the stages, she needed 112 toothpicks. Which stage was it? Explain how you found your answer.

**Practice & Apply**

Copy each flowchart. Fill in the ovals.

1.
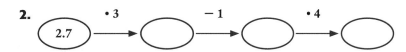

2.

**Backtracking** Hannah solved three equations by backtracking. Below are the flowcharts with which she started. For each flowchart, write the equation she was trying to solve. Then backtrack to find the solution. Check your solutions.

3.

4.
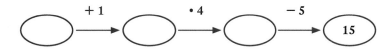

5. **Challenge**

6. In a game of *What's My Rule*, Rosita wrote the rule $b = 3a \div 4$, where $a$ is the input and $b$ is the output.

   **a.** Make a flowchart for Rosita's rule.

   **b.** Use your flowchart to find the output when the input is 18.

   **c.** Backtrack to solve the equation $3a \div 4 = 101$.

**Flowcharts** Make a flowchart to represent each equation. Then use backtracking to solve the equation. Be sure to check your solutions.

7. $4k + 11 = 91$

8. $4 \cdot (m - 2) = 38$

**9.** Neva and Jay were playing *Think of a Number.* Neva said,

*Think of a number. Subtract 1 from your number. Multiply the result by 2. Then add 6.*

Jay said he got 10. Neva must figure out Jay's starting number.

**a.** Draw a flowchart to represent this game.

**b.** What equation does your flowchart represent?

**c.** Use backtracking to find the number with which Jay started. Check your solution by following Neva's steps.

**10.** For a round of the *What's My Rule* game, Mia and Desmond wrote the rule $y = 9(2x + 1) + 1$, where $x$ is the input and $y$ is the output.

**a.** Draw a flowchart for Mia and Desmond's rule.

**b.** Use your flowchart to solve the equation $9(2x + 1) + 1 = 46$.

Draw a flowchart to represent each equation. Then use backtracking to find the solution. Be sure to check your solutions.

**11.** $\dfrac{3 \cdot m \cdot 2}{6} - 2 = 1$    **12.** $\dfrac{8p + 2}{5} - 5 = 19$    **13.** $3\left(\dfrac{5 + n}{3} - 4\right) = 15$

**14.** Look below at the toothpick sequence. The rule for the number of toothpicks $t$ needed to make stage $n$ is $t = 2n + 3$.

|  Stage 1  |  Stage 2  |  Stage 3  |

**a.** Explain why this rule works for every stage.

**b.** Write and solve an equation to find the number of the stage that requires 99 toothpicks.

**15.** Look at the pattern in this table.

| x | 0 | 5 | 10 | 15 | 20 |
|---|---|---|----|----|----|
| y | 100 | 102.5 | 105 | 107.5 | 110 |

**a.** Write a rule that relates $x$ and $y$.

**b.** Write and solve an equation to find the value of $x$ when $y$ is 197.5.

**Real-World Link**

The earliest canoes had frames made from wood or whale bone and were covered with bark or animal skin.

**16. Economics** At Marshall Park, you can rent a canoe for $5 plus $6.50 per hour.

**a.** Write a rule for calculating the cost $C$ of renting a canoe for $h$ hours.

**b.** Conor, Jing, and Miguel paid $27.75 to rent a canoe. Write and solve an equation to find the number of hours the friends used the canoe. Check your solution.

**17. Economics** Avocados cost $1.89 each. Hannah plans to make a large batch of guacamole for a party. She wants to buy as many avocados as possible. She has $14.59 to spend.

   **a.** Write an equation to find how many avocados Hannah can buy for $14.59.

   **b.** Solve your equation. How many avocados can Hannah buy?

*Connect & Extend*   **18.** Evita wants to make a fence out of wooden poles. She drew a diagram to help her figure out how many poles she would need.

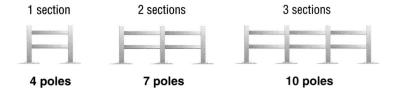

   **a.** Write a rule connecting the number of poles $p$ to the number of sections $s$.

   **b.** The lumber yard has 100 poles in stock. Write an equation to find the number of fence sections Evita can build with 100 poles.

   **c.** Solve your equation. How many fence sections can Evita build?

   **d.** If each pole is 2 yards long, how long will a 100-pole fence be?

**19. Physics** A bus is traveling at an average speed of 65 miles per hour.

   **a.** Copy and complete the table to show the distance the bus would travel in the given numbers of hours.

| Time (hours), $t$ | 1 | 2 | 3 | 4 | 5 |
|---|---|---|---|---|---|
| Distance (miles), $d$ | | | | | |

   **b.** On a grid like the one below, plot the points from your table. If it makes sense to do so, connect the points with line segments.

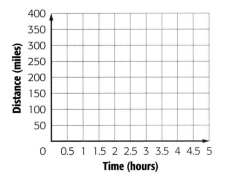

   **c.** Use your graph to estimate how long it would take the bus to travel 220 miles.

**d.** Write a rule that relates the time traveled *t*, in hours, to the distance traveled *d*, in miles.

**e.** Write and solve an equation to find how long it would take the bus to travel 220 miles. How does the solution compare to your estimate from Part c?

**20.** Julie and Noah were playing *Think of a Number*. Noah said,

*Think of a number. Square it. Add 3. Divide your result by 10. What number do you get?*

Julie said she got 8.4. Noah must figure out Julie's starting number. He drew this flowchart to represent the game.

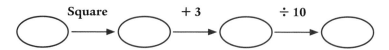

**a.** What equation does Noah need to solve to find Julie's starting number?

**b.** Use backtracking to solve your equation. Check your solution by following Noah's steps.

**21. Economics** Althea wants to buy a jacket at Donovan's department store. The store is having a "20% off" sale. In addition, Althea has a coupon for $5 off any item in the store.

**a.** Write a rule Althea could use to calculate the price *P* she would pay for a jacket with an original price of *d* dollars. (Note: The $5 is subtracted *after* the 20% discount is calculated.)

**b.** Althea pays $37.80 for a jacket. Write and solve an equation to find the original price of the jacket.

**22. Preview** This equation gives the height *h* of a baseball, in feet, *t* seconds after it is thrown straight up from ground level.

$$h = 40 \cdot t - 16 \cdot t^2$$

**a.** How high is the ball after 0.5 second?

**b.** How high is the ball after 1 second?

**c.** Based on what you learned in Parts a and b, estimate how long it would take the ball to reach a height of 20 feet. Explain your estimate.

**d.** Substitute your estimate for *t* in the equation to find out how high the ball would be after that number of seconds. Were you close? Was your estimate too high or too low?

23. **Sports** Lana is making fishing lures. Each lure requires 2¢ worth of fishing line. Lana uses feathers and weights to make the lures. The feathers cost 17¢ apiece. The weights cost 7¢ apiece.

   a. Write a rule for the cost $C$ of a lure made with $f$ feathers and $w$ weights.

   b. Lana does not want to spend more than 65¢ on each lure. Write an equation you could solve to find how many feathers she could use on a lure made with two weights.

   c. Solve your equation to find how many feathers Lana can use on a lure with two weights.

24. **In Your Own Words** In this lesson, you used flowcharts to help solve equations. Explain how to make a flowchart for this equation.

$$\left(\frac{d}{3} - 5\right) \cdot 2 + 4 = 30$$

   Also, show how you would use your flowchart to find the value of $d$.

25. **Economics** Jordan has a part-time job as a telemarketer. He earns $14.00 per hour plus 50¢ for every customer he calls.

   a. Write a rule for computing how much Jordan will earn on a three-hour shift if he calls $c$ customers.

   b. Jordan would like to earn $100 on his 3-hour shift. How many customers must he call?

*Mixed Review*

**Change each fraction or decimal to a percent.**

26. $\frac{2}{5}$          27. 0.78          28. $\frac{1}{3}$

29. 0.7          30. $\frac{112}{70}$          31. 3.06

**Geometry** Find the area of each figure in Exercises 32–35.

32.
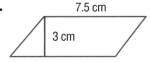

33.

34.
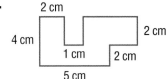

35.
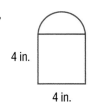

# Guess-Check-and-Improve

Backtracking is useful for solving many types of equations. However, as you will see in this lesson, some equations are difficult or impossible to solve by backtracking.

Johanna and Rosita were playing *What's My Rule*. Johanna made this table to keep track of her guesses.

From her table, Johanna figured out that Rosita's secret rule was

$$n = m \cdot (m + 1)$$

where $m$ is the input and $n$ is the output.

| Input | Output |
|-------|--------|
| 10 | 110 |
| 20 | 420 |
| 30 | 930 |
| 40 | 1,640 |

Now the two friends want to figure out what input gives an output of 552. That is, they want to solve the equation $m \cdot (m + 1) = 552$.

**Think & Discuss**

Rosita suggests they solve the equation by backtracking. Try to solve the equation this way. Are you able to find the solution? Why or why not?

What advice would you give Rosita and Johanna to help them solve the equation?

## Investigation  Use Guess-Check-and-Improve

**Vocabulary**

guess-check-
and-improve

As you have seen, backtracking does not work for every equation. In this lesson, you will learn another solution method. This method is called **guess-check-and-improve** because that is exactly what you do.

The following example shows how Rosita and Johanna used guess-check-and-improve to solve the equation $m \cdot (m + 1) = 552$.

## Example

From the table Johanna had made during the game, she could see that the output for $m = 20$ was too low, and the output for $m = 30$ was too high.

| Input | Output |
|-------|--------|
| 10 | 110 |
| 20 | 420 |
| 30 | 930 |
| 40 | 1,640 |

Using this information, the friends decided to try 25, the number halfway between 20 and 30. They checked their guess by substituting it for $m$ in the expression $m \cdot (m + 1)$.

$$m \cdot (m + 1) = 25 \cdot (25 + 1)$$
$$= 25 \cdot 26$$
$$= 650$$

The output 650 is too high. Johanna recorded the guesses and the results in the table.

| $m$ | $m \cdot (m + 1)$ | Comment |
|-----|-------------------|---------|
| 20 | 420 | too low |
| 30 | 930 | too high |
| 25 | 650 | too high but closer |

The friends now decided the solution must be between 20 and 25. The table below shows their next two guesses.

| $m$ | $m \cdot (m + 1)$ | Comment |
|-----|-------------------|---------|
| 20 | 420 | too low |
| 30 | 930 | too high |
| 25 | 650 | too high but closer |
| 22 | 506 | too low but close |
| 23 | 552 | 23 is the solution |

A solution to $m \cdot (m + 1) = 552$ is 23.

Review the process that Johanna and Rosita used.

- They *guessed* the solution.
- They *checked* their solution by substituting it into the equation.
- They used the result to *improve* their guess.

Now it is your turn to try guess-check-and-improve.

## ✅ *Develop & Understand: A*

1. Conor, Marcus, and Jing are playing *What's My Rule.* Below is Jing's secret rule.

   $d \cdot (d + 3) = J$, where $d$ is the input and $J$ is the output

   a. Conor gave Jing an input. Jing calculated the output as 8,554. Write an equation to find Conor's input.

   b. Find a solution of your equation using guess-check-and-improve.

   c. Marcus gave Jing an input. Jing calculated the output as 32.56. Write an equation to find Marcus' input.

   d. Find a solution of your equation from Part c using guess-check-and-improve.

**For each equation, use guess-check-and-improve to find a solution.**

2. $2n + 6 = 20$
3. $19 = \dfrac{4}{q} + 3$
4. $s^2 + 2s = 19.25$

5. Miguel is trying to solve this equation using guess-check-and-improve.

   $$25 - 3 \cdot d = 17.8$$

   The table below shows his first two guesses. Miguel asked, "Why was the output for 8 lower than the output for 7? Shouldn't a greater input give a greater output?"

   | d | 25 − 3 · d | Comment |
   |---|---|---|
   | 7 | 4 | too low |
   | 8 | 1 | still too low |

   a. Answer Miguel's questions.

   b. What input do you think Miguel should try next? Explain.

6. Hannah and Luke were trying to solve $7.25t - t^2 = 12.75$. They made the table below using guess-check-and-improve.

   | t | 7.25t − t² | Comment |
   |---|---|---|
   | 5 | 11.25 | too low |
   | 6 | 7.5 | too low |
   | 4 | 13 | too high |
   | 2 | 10.5 | too low |

**Math Link**

Imagine that a garage contains two tricycles and several bicycles and that there are 20 bicycle wheels in all. You can solve the equation in Exercise 2 to find the number of bicycles in the garage.

Hannah thinks the solution must be between 2 and 4. Luke thinks it must be between 4 and 5.

**a.** Is there a solution between 2 and 4? If so, find it. If not, explain why not.

**b.** Is there a solution between 4 and 5? If so, find it. If not, explain why not.

## Share & Summarize

Describe any strategies you have discovered for finding a solution efficiently by using guess-check-and-improve.

# Investigation 2 Solve Problems Using Guess-Check-and-Improve

In this investigation, you will solve exercises by writing equations and then using guess-check-and-improve. As you work, you will find that you cannot always give an exact decimal value for a solution. In such cases, you can approximate the solution.

## ✓ Develop & Understand: A

1. The floor of Mr. Cruz's basement is shaped like a rectangle. The length of the floor is 2 meters greater than the width.

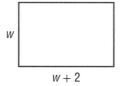

$w$

$w + 2$

**a.** Write a rule to show the connection between the floor's area $A$ and the width $w$.

**b.** The area of the basement floor is 85 square meters. Write an equation to find the floor's width.

**c.** Use guess-check-and-improve to find an approximate solution of your equation. Give the solution to the nearest tenth.

**d.** What are the dimensions of the basement floor?

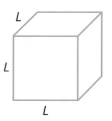

**2.** A *cube* is a three-dimensional shape with six identical square faces.

The *surface area* of a cube is the sum of the areas of its six faces.

This cube has edges of length $L$.

a. What is the area of one face of this cube?

b. Write a rule for finding the surface area $S$ of the cube.

c. Suppose the cube has a surface area of 100 square centimeters. Write an equation to find the cube's edge length.

d. Use guess-check-and-improve to find the edge length of the cube to the nearest 0.1 centimeter.

In all of the equations you have solved so far, the variable appears on only one side of the equation. You can also use guess-check-and-improve to solve equations in which the variable appears on both sides.

## Example

Solve this number puzzle.
*12 more than a number is equal to 3 times the number. What is the number?*

If you let $m$ stand for the number, you can write the puzzle as an equation.

$$m + 12 = 3m$$

To solve the equation, substitute values for $m$ until the two sides of the equation are equal.

| $m$ | $m + 12$ | $3m$ | Comment |
|-----|----------|------|---------|
| 1 | 13 | 3 | not the same |
| 2 | 14 | 6 | a bit closer |
| 5 | 17 | 15 | very close |
| 6 | 18 | 18 | Got it! |

So, 6 is the solution of the equation $m + 12 = 3m$.

Check that 6 is the solution of the original puzzle. Twelve more than 6 is 18, which is equal to 3 times 6.

In Exercises 3–9, you will practice solving equations in which the variable appears on both sides.

# Develop & Understand: B

In Exercises 3–5, write an equation to represent the number puzzle. Then find the solution using guess-check-and-improve.

3. Five added to a number is the same as 1 subtracted from twice the number. What is the number?

4. Four times a number plus 1 is equal to 4 added to twice the number. What is the number?

5. Two added to the square of a number is equal to 5 times the number minus 4. What is the number? There are two solutions. Try to find them both.

6. Consider the equation $2m = 5m - 18$.

   a. Make up a number puzzle that matches the equation.

   b. Solve the equation. Check that the solution is also the answer to your puzzle.

7. Peta and Ali went to the pet store to buy fish for their tanks. Peta bought three black mollies for $b$ dollars each and one peacock eel for $12. Ali bought seven black mollies for $b$ dollars each and an Australian rainbow fish for $2. Peta and Ali spent the same amount of money.

   a. How much money did Peta spend? Your answer should be an expression containing the variable $b$.

   b. How much money did Ali spend? Your answer should be an expression containing the variable $b$.

   c. Using your two expressions, write an equation that states that Peta and Ali spent the same amount of money.

   d. Solve your equation to find the value of $b$. How much did each black molly cost?

   e. How much did each friend spend?

Use guess-check-and-improve to find a solution of each equation.

8. $3n = 9 - 2n$

9. $p + 1 = 5(p - 4)$

## Share & Summarize

Make up a number puzzle like those with which you have worked in this investigation. Exchange puzzles with your partner. Solve your partner's puzzle.

# Investigation 3 Choose a Method

Now you have two methods for solving equations, *backtracking* and *guess-check-and-improve*. The exercises in this investigation will help you decide which solution method is more efficient for a particular type of equation.

In the cartoon, Marcus and Rosita are trying to solve some equations. Marcus uses backtracking. Rosita uses guess-check-and-improve.

For which equation in the cartoon is backtracking a better method than guess-check-and-improve? Why?

For which equation is guess-check-and-improve a better method than backtracking? Why?

For which equation do both methods seem to work well?

## ✅ Develop & Understand: A

In Exercises 1–4, do Parts a and b.

**a.** Find a solution of the equation using one method while your partner finds a solution using the other. Switch methods for each equation so you have a chance to practice both.

**b.** Discuss your work with your partner. Indicate whether both methods work. Tell which method seems more efficient.

**1.** $3p - 8 = 25$

**2.** $1.6r + 3.96 = 11$

**3.** $k + 4 = 6k$

**4.** $(j - 2) \cdot j = 48$

**Solve each equation. Tell which solution method you used. Explain why you chose that method.**

**5.** $4w + 1 = 2w + 8$

**6.** $5 \cdot \dfrac{2k + 4}{6} = 10$

**7.** Luke says, "When the variable appears only once in an equation, I use backtracking. If it occurs more than once, I use guess-check-and-improve." Discuss Luke's strategy with your partner. Would it be effective for the equations in Exercises 1–6? Explain.

### Share **&** Summarize

Tell which solution method you would use to find a solution of each equation. Explain your choice.

**1.** $n^2 + n = 30$

**2.** $4 + (v - 3) = 8$

**3.** $3g = 6g - 7$

**4.** $2e - 7 = 4$

**Practice & Apply**

1. Hannah, Althea, and Jahmal are playing *What's My Rule*. Here is Althea's secret rule.

   $$3 \cdot p + p^2 = q, \text{ where } p \text{ is the input and } q \text{ is the output}$$

   **a.** Hannah gives Althea an input. Althea calculates the output as 24.79. Write an equation you could solve to find Hannah's input.

   **b.** Find a solution of your equation using guess-check-and-improve.

   **c.** Jahmal gives Althea an input. Althea calculates the output as 154. Write an equation you could solve to find Jahmal's input.

   **d.** Find a solution of your equation in Part c using guess-check-and-improve.

   **For each equation, use guess-check-and-improve to find a solution.**

2. $16 - 5k = 2$

3. $h \cdot (5 + h) = 26.24$

4. $y^2 + 72 = 17y$ There are two solutions. Try to find them both.

5. **Geometry** Reina is planning her summer garden.

   **a.** Reina wants to plant strawberries in a circular plot covering an area of 15 square meters. Write an equation to find the radius Reina should use to lay out the plot.

   **b.** Use guess-check-and-improve to find the strawberry plot's radius to the nearest tenth of a meter.

   **c.** Reina also wants to plant tomatoes in five identical circular plots with a total area of 25 square meters. Write an equation to find the radius of one of the tomato plots.

   **d.** What should be the radius of each tomato plot, to the nearest tenth of a meter?

**Math Link**

The area of a circle is given by the formula $A = \pi \cdot r^2$, where $r$ is the radius of the circle.

**Real-World Link**

Tomatoes are the world's most popular fruit, in terms of tons produced each year.

In Exercises 6–8, write an equation to represent the number puzzle. Then find the solution using guess-check-and-improve.

**6.** Three times a number plus 5 is equal to 5 times the number. What is the number?

**7.** Ten plus the square of a number is equal to 6 times the number plus 2. What is the number? (This puzzle has two solutions. Find them both.)

**8.** Three times a number plus 1 is equal to 9 plus the number. What is the number?

**9.** Marjorie said, "If you double my macaw's age and then subtract 21.75, your answer will be half my macaw's age."

    **a.** Write an equation to find how old Marjorie's macaw is.

    **b.** Use guess-check-and-improve to find the macaw's age. Check your answer in Marjorie's original statement.

**10.** Consider the equation $3m - 11 = m + 3$.

    **a.** Make up a number puzzle that matches the equation.

    **b.** Solve the equation. Check that the solution is also the answer to your puzzle.

**Find a solution of each equation.**

**11.** $3.3h - 7 = 2.801$           **12.** $2l = 4l - 20$

**13.** $j \cdot (3 - 2j) = 1$           **14.** $2 = m^2 - m$

**15.** $9 \cdot \dfrac{g \div 2 + 1}{5} = 12$       **16.** $143 = (q + 1) \cdot (q - 1)$

In Exercises 17–19, tell whether you would use backtracking or guess-check-and-improve to find a solution. Explain your choice.

**17.** $n^2 + n = 30$

**18.** $47(2v - 3.3) = 85$

**19.** $s = 17.5s - 0.5$

**Real-World Link**

The macaw is a long-tailed parrot. In the rain forest, macaws have an average life span of 25 to 30 years. In captivity, they can reach ages of 70 years or more.

**Connect & Extend**

**20. In Your Own Words** You have learned how to solve equations using two methods: backtracking and guess-check-and-improve. Compare these two methods, and explain why it is important to be able to solve equations in more than one way.

**21.** Aisha and Terrell were playing *Think of a Number*. Terrell said,

> *Think of a number. Multiply the number by 1 less than itself. What number do you get?*

Aisha got 272. Terrell must figure out Aisha's starting number.

**a.** What equation does Terrell need to solve to find Aisha's number?

**b.** Use guess-check-and-improve to find Aisha's number. Check your answer by following the steps to verify that you get 272.

**22. Geometry** The elevator in Rafael's apartment building has a square floor with an area of 6 square meters.

**a.** Write an equation to find the dimensions of the elevator floor.

**b.** Use guess-check-and-improve to find the dimensions of the elevator floor to the nearest tenth of a meter.

**23. Physical Science** When an object is dropped, the relationship between the distance it has fallen and the amount of time it has been falling is given by the rule

$$d = 4.9 \cdot t^2$$

where $d$ is the distance in meters and $t$ is the time in seconds.

**a.** Jing dropped a ball from a pier. It took the ball 1.1 seconds to hit the water. How many meters did the ball travel?

**b.** A bolt falls 300 meters down a mine shaft. Write an equation to find how long it took the bolt to fall.

**c.** Find how long it took the bolt to fall, to the nearest tenth of a second.

24. **Nutrition** Three blueberry muffins and a plain bagel have the same number of calories as two blueberry muffins and one bagel with cream cheese. A bagel has 150 calories, and cream cheese adds 170 calories. How many calories are in a blueberry muffin?

25. **Number Sense** This formula can be used to find the sum $S$ of the whole numbers from 1 to $n$.

$$S = \frac{n \cdot (n + 1)}{2}$$

For example, you can use the formula to find the sum of the whole numbers from 1 to 100.

$$S = \frac{n \cdot (n + 1)}{2}$$
$$= \frac{100 \cdot (100 + 1)}{2}$$
$$= \frac{100 \cdot 101}{2}$$
$$= \frac{10,100}{2}$$
$$= 5,050$$

a. Use the formula to find the sum of the whole numbers from 1 to 9. Calculate $1 + 2 + 3 + 4 + 5 + 6 + 7 + 8 + 9$ to check your answer.

b. If the sum of the numbers from 1 to $n$ is 6,670, what must $n$ be?

c. If the sum of the numbers from 1 to $n$ is 3,003, what must $n$ be?

**For each equation, use guess-check-and-improve to find a solution.**

26. $x^3 + x = 130$

27. $4n^4 = 48$

28. $2a^4 - 4a^2 = 16$

29. $c^4 = \frac{1}{100}$

**30. Geometry** The *volume* of a three-dimensional shape is the amount of space inside of it. Volume is measured in cubic units, such as cubic centimeters and cubic inches. You can calculate the volume *V* of a cylinder with radius *r* and height *h* using this formula: $V = \pi \cdot r^2 \cdot h$.

**a.** Calculate the volume of a cylinder with radius 2 centimeters and height 5 centimeters.

**b.** A can of fruit has a volume of 350 cubic centimeters and a height of 15 centimeters. Write an equation you could solve to find its radius.

**c.** Solve your equation to find the can's radius to the nearest tenth of a centimeter.

**Mixed Review**

**31. Geometry** A rectangle has an area of 48 square feet and a perimeter of 32 feet. What are the dimensions of the rectangle?

**32.** Which has the greater area, a circle with diameter 11 meters or a square with side length 10 meters?

**33. Sports** In a track-and-field competition, ten women participated in the 100-meter run. Here are their times in seconds.

12.2   11.3   13.5   11.5   11.7   12.6   15.5   11.8   13.4   11.5

**a.** Find the mean and the median time.

**Fill in the blanks.**

**34.** 25% of _____ = 14

**35.** _____ % of 75 = 25

**36.** 80% of 200 = _____

**37.** 125% of _____ = 300

**38.** _____ % of 280 = 238

**39.** 1% of 30 = _____

9

# Review & Self-Assessment

## Chapter Summary

You started this chapter by looking at numeric equations and inequalities. Then you turned your attention to equations containing variables. You learned that a value of a variable that makes an equation true is called a *solution* of the equation. You saw that while many equations have only one or two solutions, some have every number as a solution. Others have no solution at all.

You then created flowcharts to represent rules. You also saw how *backtracking*, working backward from an output using a flowchart, can be used to solve some equations. You were then faced with equations that could not be solved by backtracking, and you learned how to use *guess-check-and-improve* to solve them. Finally, you learned some strategies for determining which solution method to use for a given equation.

## Strategies and Applications

The questions in this section will help you review and apply the important ideas and strategies developed in this chapter.

### Understanding equations and inequalities

1. Explain what the symbols $=$, $>$, $<$, and $\neq$ mean. For each symbol, write a true mathematical sentence using that symbol.

2. Give an example of an open sentence. Provide a value of the variable that makes your sentence true.

3. Explain why the sentence $P + 5 = P$ is never true.

4. Explain why the sentence $2 \cdot (x + 3) = 2 \cdot x + 2 \cdot 3$ is always true.

### Solving equations by backtracking

5. Solve this equation by creating a flowchart and backtracking.

$$10 \cdot \frac{4 + 9x}{7} - 25 = 45$$

Explain each step in your solution.

6. Admission to the town carnival is $4.50 per person. Tickets for rides cost 75¢ each. Russ has $10 to spend at the carnival, and he wants to go on as many rides as possible.

    a. Write an equation you could solve to find the number of tickets $t$ Russ can buy.

    b. Solve your equation by backtracking. How many tickets can Russ buy?

### Solving equations using guess-check-and-improve

**7.** The Smallville community garden is made up of 15 identical square plots with a total area of 264.6 m$^2$.

   **a.** Write an equation to find the side length of each plot.

   **b.** Solve your equation using guess-check-and-improve. Make a table to record your guesses and the results. What is the side length of each plot?

**8.** A number multiplied by 2 more than the number is 9 times the number, plus 8.

   **a.** Write an equation to represent the number puzzle.

   **b.** Solve your equation using guess-check-and-improve.

### Choosing a solution method for an equation

**9.** Explain why you could not use backtracking to solve the equation $(n + 3.5) \cdot n = 92$.

**10.** Tell which solution method you would use to solve the equation $6.34 + 10.97 \cdot y = 208.188$. Give reasons for your choice.

## Demonstrating Skills

**Tell whether each sentence is *true* or *false*. If it is false, make it true by changing one number or symbol.**

**11.** $(4 + 5) \cdot 6 = 24 + 30$

**12.** $4^2 = 4 \cdot 2$

**13.** $30 \div (3 + 2) = 30 \div 3 + 30 \div 2$

**14.** $7 + 4 - 1 > 5 + 6$

**Tell whether each equation is *always true*, *sometimes true*, or *never true*. Explain how you know.**

**15.** $c^3 = c \cdot c \cdot c$           **16.** $n - 5 = n + 1$

**17.** $x - 9 = 0$              **18.** $2m - 2m = 0$

**Find a solution of each equation.**

**19.** $4.7x + 12.3 = 42.85$       **20.** $m \cdot (m + 5) = 336$

**21.** $5n + 7 = 14$           **22.** $\dfrac{5(x - 1)}{3} = 7.5$

**23.** Sarah lives near the library. In the summer, she likes to walk to the library. The following directions describe how to get from Sarah's house to the library.

- Start at the corner of Winfree Drive and Dempsey Road.
- Walk two blocks west on Dempsey.
- Turn right from Dempsey to Hempstead Road.
- Walk 3 blocks north on Hempstead Road.
- Turn left on Schrock Road.
- Walk 2 blocks west on Schrock Road.
- Turn right on Otterbein Road.
- Walk 4 blocks north on Otterbein Road.
- Turn left on Walnut Street.
- Walk 5 blocks west on Walnut Street.
- The library is on the corner of Walnut Street and State Street.

Write a set of directions that Sarah could follow to get home from the library.

## Test-Taking Practice

**SHORT RESPONSE**

**1** Tell whether the sentence is true or false. If it is false, make it true by changing one number or symbol.

$$13 + 26 = (3 + 4) \cdot 7$$

*Show your work.*

*Answer* _____

**MULTIPLE CHOICE**

**2** Which is a solution to the equation $7x = 63$?

**A** 9

**B** 56

**C** 70

**D** 441

**3** Which operation can be used to solve the equation $\frac{h}{3} = 6$?

**F** add 3 to both sides of the equation

**G** subtract 3 from both sides of the equation

**H** divide both sides of the equation by 3

**J** multiply both sides of the equation by 3

**4** Solve the following equation.
$$6m + 11 = 35$$

**A** 4

**B** 6

**C** 7

**D** 8

**5** You buy 3 pens for $4.50. The equation that can be used to find the cost of each pen is $3p = 4.5$. How much does each pen cost?

**F** $1.00

**G** $1.25

**H** $1.50

**J** $1.75

# Data and Probability

## Real-Life Math

**Fat Chance!** A probability is a number between 0 and 1 that tells you the chance that an event will occur. The closer a probability is to 0, the less likely the event is to happen.

Just how likely is it that some everyday events happen?

- If you toss a coin to determine the answers of a 10-question true-false test, the probability you will get all the answers correct is $\frac{1}{1,024}$, or about 0.001.

- If you randomly dial a three-number combination on a dial lock with numbers from 0 to 29, the probability you will open the lock is $\frac{1}{27,000}$, or about 0.00004.

**Think About It** To better understand how unlikely these events are, compare these probabilities to the probability that it will rain tomorrow in your community.

**Math Online**

Take the **Chapter Readiness Quiz** at glencoe.com.

# Dear Family,

The class will close its exciting year of mathematics by exploring probability and analyzing data with graphs. Probability tells you that it is very unlikely for you to win the grand prize in a state lottery. Suppose you have to pick six different numbers from 1 to 54. To win the grand prize, all six numbers must match those selected in a random drawing. The probability that you will win is only 1 in 25,827,165, or 0.00000004.

Along with analyzing data, numbers, facts, or other measurable information, it is important to display your findings. Graphs can make it easier to see patterns and draw conclusions.

## Key Concept–Probability

The probability that some event will happen can be described by a number between 0 and 1.

- A probability of 0 means that the event has no chance of happening. So, the probability of winning a state lottery is very close to 0.
- A probability of 1 means that the event is certain to happen.
- A probability of $\frac{1}{2}$, or 50%, means that the event is just as likely to happen as not to happen.

For example, if a weather forecaster states that the probability of rain tomorrow is 90%, it's probably a good idea to take your umbrella, although it might not rain after all. If the probability of rain is 10%, it is unlikely that it will rain.

In this chapter, the class will use mathematical reasoning to calculate probabilities in simple situations like tossing coins or drawing names from a hat. Students will also do some experiments in which they actually toss coins or draw names and then compare the results with the calculated probabilities.

## Chapter Vocabulary

| | | |
|---|---|---|
| distribution | Fundamental Counting Principle | simulation |
| equally likely | | theoretical probability |
| experimental probability | histogram | |
| | probability | Venn diagram |

## Home Activities

- Look for situations in everyday life that involve probability, such as the chance of rain or the odds in sports games.
- Encourage your student's exploration by playing games of chance together.
- Have your student teach you the games we play in class. Ask him or her to describe what part probability plays in each game.

# Data Displays

People in many professions use data to help make decisions. When data are first collected, they are often just lists of numbers and other information. Before they can be understood, data must be organized and analyzed. In this chapter, you will investigate several tools for understanding data.

In many activities in this chapter, your class will play the role of a company called Data, Inc., a consulting group that specializes in organizing and analyzing data. Various people and organizations will come to Data, Inc. for advice and suggestions.

In Chapter 8, you saw how graphs can help you discover patterns and trends. In this lesson, you will look at several types of graphs and compare the kinds of information each tells you about a set of data.

## Math Link

The word *data* is plural and means "bits of countable or measurable information." The singular form of data is *datum*. When you talk about data, you should use the plural forms of verbs, for example *are*, *were*, or *show*.

### Explore

These graphs have no labels or scale values. Tell whether one of these graphs could describe each situation below. If it could, tell what the axes would represent and what the graph would reveal about the situation. If neither graph could describe the situation, describe or sketch a graph that could.

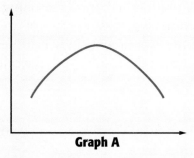

**Graph A**

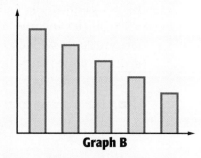

**Graph B**

- the number of visitors to a zoo over the past year
- the weight of a young hippo from birth to age one
- the distance from a ball to the ground after the ball is dropped
- the number of minutes of daylight each day during a year
- the number of school days remaining on the first of each month, from February through June
- the number of children born each month in one year in Canada

# Investigation ① Line Graphs

## Materials

- copies of the "clues"

The Smallville police department is investigating the disappearance of a man named Gerald Orkney. Here is what they know so far.

- Mr. Orkney lives alone with his pet iguana, Agnes.
- Mr. Orkney did not show up for work on December 15. When his friends came to check on him, he and Agnes were gone.
- An atlas and the graphs below, which have no scales or labels, were found in Mr. Orkney's apartment.

The police department has asked Data, Inc. to help the investigators figure out what happened to Gerald Orkney.

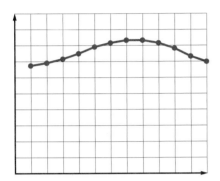 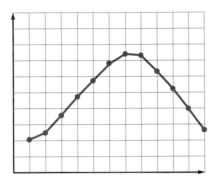 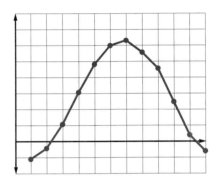

## Think & Discuss

Look carefully at the three graphs found by police. Think about the shape of each graph and about the axes. How are the graphs alike? How are they different?

Since the graphs do not have labels, it is impossible to know exactly what they represent or how they are related to Gerald Orkney's disappearance. However, you may be able to make a *hypothesis,* or an educated guess, based on the information you do have.

Try to think of some ideas about what the graphs might show. Consider both what the graphs look like and how they might be related to the other information found by police.

Using the graphs and the other information given, try to create at least one hypothesis about what might have happened to Gerald Orkney.

### ✅ *Develop & Understand: A*

1. During a search of Gerald Orkney's office, the police found more detailed versions of the graphs. Your teacher will give you copies of the new graphs. Look at them closely.

   a. What new information do the graphs reveal? Does this information fit any of the ideas your class had in Think & Discuss? Now what do you think the graphs might show?

   b. Does the new information support any of the hypotheses your class made about what happened to Gerald Orkney? Explain.

   c. Make a new hypothesis, or make changes to an earlier hypothesis, to fit all the information you have so far.

2. The police department has discovered more clues, consisting of a list and a note. Your teacher will give you a copy of this new information.

   a. What might the list have to do with the graphs? Do you have new ideas about what each graph might show? If so, add appropriate titles and axis labels to the graphs.

   b. Does this new information support your hypothesis about what happened to Gerald Orkney? Explain.

   c. Now what do you think happened to Gerald Orkney? Make a new hypothesis, or make changes to an earlier hypothesis, to fit this new information.

   ### Share & Summarize

   Write a letter to the Smallville police department summarizing your group's investigation and presenting your hypothesis about what happened to Gerald Orkney.

# Investigation ② Bar Graphs

The environmental group Citizens for Safe Air has asked Data, Inc. to analyze some data about *hydrocarbons.* These compounds are part of the emissions from cars and other vehicles that pollute the air. The group wants to know how the total amount of hydrocarbon emitted by vehicles has changed over the past several decades and how it might change in the future.

The table shows estimates of the typical amount of hydrocarbon emitted per vehicle for each mile driven in the United States for years from 1960 to 2015. The values from 2000 to 2015 are predictions.

**Average Per-Vehicle Emissions**

| Year | Grams of Hydrocarbon per Mile |
|------|-------------------------------|
| 1960 | 17 |
| 1965 | 15.5 |
| 1970 | 13 |
| 1975 | 10.5 |
| 1980 | 7.5 |
| 1985 | 5.5 |
| 1990 | 3 |
| 1995 | 1.5 |
| 2000 | 1 |
| 2005 | 0.75 |
| 2010 | 0.5 |
| 2015 | 0.5 |

**Source:** "Automobiles and Ozone," Fact Sheet OMS-4 of the Office of Mobile Sources, the U.S. Environmental Protection Agency.

## ✅ Develop & Understand: A

1. For which five-year period is the decrease in per-vehicle emissions greatest? For which five-year period is it least?

2. On a set of axes like this one, draw a bar graph showing the typical per-vehicle emissions for each year given in the table.

**Average Per-Vehicle Emissions**

**3.** Describe what your graph indicates about the change in per-vehicle emissions over the years. Discuss high and low points, periods of greatest and least change, and any other patterns you see.

**4.** To determine when the greatest decrease in per-vehicle emissions occurred, is it easier to use the table or the graph? Explain.

You have seen that the amount of hydrocarbon *each vehicle* emits *per mile* has decreased over the years. But this is not enough information to conclude that the *total amount* of hydrocarbon emitted by *all vehicles* is decreasing. You also need to consider the total number of miles driven by all vehicles.

This table shows estimates of the number of miles driven, or expected to be driven, by all vehicles in the United States for various years between 1960 and 2015.

**Miles Driven in U.S.**

| Year | Miles in Billions |
|------|-------------------|
| 1960 | 750 |
| 1965 | 950 |
| 1970 | 1,150 |
| 1975 | 1,250 |
| 1980 | 1,500 |
| 1985 | 1,500 |
| 1990 | 2,000 |
| 1995 | 2,300 |
| 2000 | 2,600 |
| 2005 | 2,850 |
| 2010 | 3,150 |
| 2015 | 3,400 |

**Source:** "Automobiles and Ozone," Fact Sheet OMS-4 of the Office of Mobile Sources, the U.S. Environmental Protection Agency, Jan. 1993.

5. Look at the table on page 582. During which five-year period does the number of miles driven increase most? During which five-year period does it increase least?

6. On a set of axes like the one below, draw a bar graph showing the billions of vehicle miles traveled for each year given.

**Estimated Vehicle Miles Traveled**

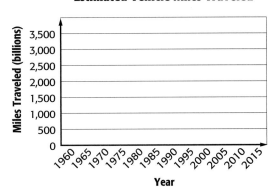

7. Describe what your graph indicates about the change in the total number of miles driven over the years. Discuss high and low points, periods of greatest and least change, and any other patterns you see.

You now know that, over time, the typical amount of hydrocarbon emitted per vehicle has decreased. You also know that more miles are driven each year. In Exercises 8–10, you will combine this information to answer this question.

Is the *total amount* of hydrocarbon emitted from vehicles increasing or decreasing?

**Think & Discuss**

How could you use the data in the two previous tables to calculate estimates of the total amount of hydrocarbon emitted by all vehicles each year?

8. Copy and complete the table to show the total amount of hydrocarbon emitted by U.S. vehicles each year.

**Estimated Total Emissions**

| Year | Billions of Grams of Hydrocarbon |
|------|----------------------------------|
| 1960 | |
| 1965 | |
| 1970 | |
| 1975 | |
| 1980 | |
| 1985 | |
| 1990 | |
| 1995 | |
| 2000 | |
| 2005 | |
| 2010 | |
| 2015 | |

9. Make a bar graph of the data in the table.

10. Describe what your graph indicates about the change in the total hydrocarbon emissions over the years. Discuss high and low points, periods of greatest and least change, and any other patterns you see.

**Share & Summarize**

Write a letter to Citizens for Safe Air. Describe Data, Inc.'s investigation of hydrocarbon emissions, summarizing your findings about how total emissions have changed over the past few decades and how they might change in the future.

# Investigation 3 Histograms

**Vocabulary**

distribution

histogram

There are many types of graphs. The graph that is best for a given situation depends on the data you have and the information you want to convey.

## ✅ Develop & Understand: A

This bar graph shows the times of some of the participants in the women's team sprint cross-country skiing event at the 2006 Winter Olympics.

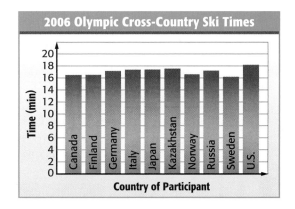

**2006 Olympic Cross-Country Ski Times**

*Real-World Link*

The oldest known skis are 4,000 to 5,000 years old. In the tenth century, Viking soldiers used skis for transportation. From the 15th to the 17th centuries, several northern European armies had companies of ski troops.

1. How many countries are represented?

2. From which country was the gold medalist? What was her time?

3. How many participants completed the race with a time between 17 minutes and 18 minutes?

4. The bars in the graph are arranged alphabetically by country. Think of another way the bars could be ordered. What kinds of questions would be easier to answer if they were ordered that way?

It probably took you some time to figure out the answer to Exercise 3 above. Although it is easy to use the bar graph to find the time for each participant, it is not as easy to find the number of skiers who finished within a particular time interval.

You will now use a *histogram* to display the ski times. In a **histogram**, data are divided into equal intervals with a bar for each interval. The height of each bar shows the number of data values in that interval. There are no gaps between intervals.

## ✅ Develop & Understand: B

In Exercises 5–8, you will make a table of frequencies. *Frequencies* are counts of the number of data values in various intervals. You will use your frequency table to create a histogram.

5. Copy this table. Use the bar graph on page 585 to count the number of participants who finished in each time interval. Record this information in the "Frequency" column.

| Time (minutes:seconds) | Frequency |
|---|---|
| 15:00–15:59 | |
| 16:00–16:59 | |
| 17:00–17:59 | 5 |
| 18:00–18:59 | |
| 19:00–19:59 | |

**Math Link**

*Relative frequency* is the ratio of the number of data in an interval to the total number of data in all intervals. For example, the relative frequency of the data in the 17:00–17:59 interval is $\frac{5}{10}$, or 0.50.

6. Copy the axes below. Create a histogram by drawing bars showing the number of participants who finished in each time interval. The bar for the interval 17:00–17:59 has been drawn for you.

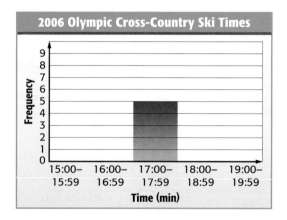

7. In which one-minute interval did the greatest number of skiers finish?

8. The shape of a histogram reveals the **distribution** of the data values. In other words, it shows how the data are spread out, where there are gaps, where there are many values, and where there are only a few values. What can you say about the distribution of times for this event?

# ✅ Develop & Understand: C

Rather than showing the *number* of values in each interval, some histograms show the *percent* of values in each interval. For example, this histogram shows how the test scores for Mr. Wilson's Math exam were distributed. The maximum possible score was 75 points.

9. Describe the shape of the histogram. Tell what the shape indicates about the distribution of test scores.

10. Which interval includes the greatest percent of test scores? About what percent of scores are in this interval?

11. Which interval includes the least percent of test scores? About what percent of scores are in this interval?

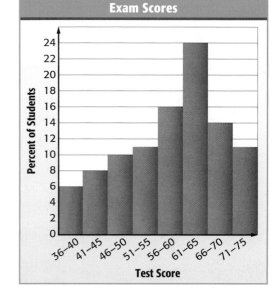

12. Suppose 64 students took Mr. Wilson's test. How many of them received a score from 66 to 70?

13. If you were to add the percents for all the bars, what should be the total? Why?

## Math Link

*Cumulative frequency* is the total number of all data values less than the upper limit of a certain interval. This is found by adding together the frequencies of the interval and all other intervals that come before it. For example, the cumulative frequency of the data less than 50 is $6 + 8 + 10$, or 24.

*Cumulative relative frequency* is the ratio of the cumulative frequency for an interval to the total number of data in all intervals. For example, the cumulative relative frequency of the 46–50 interval is $\frac{24}{100}$, or 0.24.

## Share & Summarize

1. What type of information does a histogram display? Give an example of a situation for which it would make sense to display data in a histogram.

2. In this investigation, you looked at a bar graph and a histogram of Olympic ski data. What are some things the bar graph shows better than the histogram? What are some things the histogram shows better than the bar graph?

# Investigation  4 Venn Diagrams

**Vocabulary**

**Venn diagram**

The principal of Smallville High School has asked Data, Inc. to help him prepare his annual report to the school board.

At Smallville High School, all 100 freshmen are required to take a Science class, either Biology or Computer Science. The principal of Smallville High made this bar graph showing the number of freshmen enrolled in each Science class.

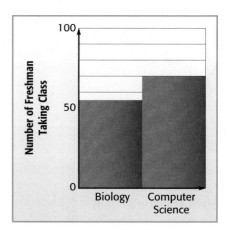

## Think & Discuss

Does the number of students in the graph add up to 100?

Assuming the principal did not make a mistake, what might explain the way the graph looks?

How could the principal change his bar graph to represent the situation more clearly?

When you have data about how things are sorted into groups, one useful way to display it is with a **Venn diagram**. A Venn diagram is especially useful for illustrating how groups overlap. This Venn diagram shows the number of freshmen at Smallville High in the different Science classes. You can see that 25 students are taking two Science classes.

**Freshmen in Science Class**

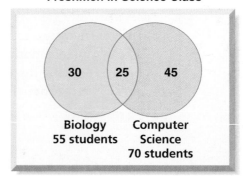

In a Venn diagram, each circle represents a group. The data that is in the circle belongs to that group, and the data that is outside the circle does not belong to the group. If an item belongs to two groups, it goes in the space where the two circles overlap. That is, it is in both circles at the same time.

## Real-World Link

Biologists classify living things according to a set of nested categories. The categories are Kingdom, Phylum, Class, Order, Family, Genus, and Species. For example, two of the Kingdoms are Plants and Animals. The Animal Kingdom is divided into several Phyla, which is the plural of Phylum. Mammals, Birds, and Insects are all classes within the Phylum of Vertebrates, or animals with backbones.

· · · · · · · · · · · · · · · · · · · · · ·

## ✅ Develop & Understand: A

The juniors at Smallville High all have to take at least one History class. They can take American History, Ancient Civilizations, or both. According to the principal's data, 65% of juniors take American History and 50% take Ancient Civilizations.

**1.** Complete this Venn diagram showing which History classes the juniors take.

**2.** What percentage of juniors are enrolled in both History classes?

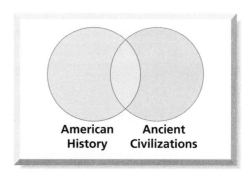

The juniors can also choose between two foreign languages, Chinese and Spanish. Some students take both languages, while some students do not study a foreign language at all. Here is a bar graph showing the enrollment data for foreign language classes.

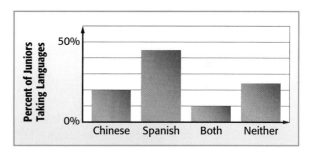

**3.** Draw a Venn diagram to represent this data.

**4.** How does your diagram represent the students who are not in a foreign language class?

**5.** What percentage of juniors study Chinese?

**6.** What percentage of juniors study Spanish?

You can use Venn diagrams to sort individual items, categories of things, or numerical data like the class enrollment data you have seen.

**Shape Blocks**

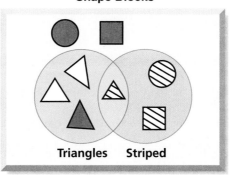

Triangles    Striped

**Animal Traits**

Dog    Cat

Numerical data can be represented in different ways. For example, the student enrollment data could be represented as numbers of students, instead of percentages.

## ✅ Develop & Understand: B

7. There are 200 juniors at Smallville High. Seventy students are in Ancient Civilizations and thirty students take both American History and Ancient Civilizations. Make a new Venn diagram that shows the number of students in each History class.

   The principal reports that all the freshmen study Math; 105 of them take Algebra 1 and 85 take Geometry. Use this information for Exercises 8 and 9.

8. What additional data do you need from the principal in order to display the freshmen Math enrollment numbers in a Venn diagram?

9. What additional data do you need if you want to display the enrollment data as percentages in a Venn diagram?

10. The principal also reports that 75 freshmen study Chinese and 80 study Spanish. What additional data do you need in order to display this information in a Venn diagram?

### Math Link

In theory, a Venn diagram could have four or more circles, but it gets very difficult to draw. When all the categories are circles, you cannot get more than three to overlap in a way that makes an area for each possible combination of categories.

## ✅ Develop & Understand: C

This Venn diagram shows the number of seniors who take Chemistry, Spanish, and Chorus classes. There are 200 seniors attending Smallville High.

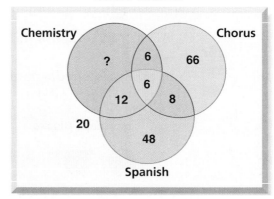

**Senior Courses**

11. What does the gray-shaded area of the diagram represent?

12. What does the green-shaded area represent?

13. How many students in the senior class take Chemistry but not Chorus or Spanish?

14. How many students in the senior class are in Chorus?

**15.** This Venn diagram shows how many seniors are enrolled in Trigonometry, Computer Science, and Physics.

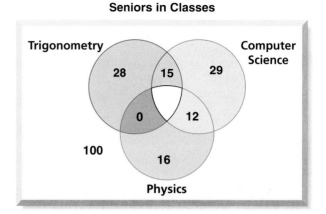

**Seniors in Classes**

**a.** How many students are enrolled in two of these three classes?

**b.** What percentage of the senior class takes at least one of these three classes?

**c.** The principal needs to schedule these three classes so that there are no schedule conflicts. Can any of the three classes be scheduled at the same time? Explain your reasoning.

**16.** The principal gave this table to Data, Inc.

| Class | Latin | World History | Drama | Latin and World History | Latin and Drama | World History and Drama | All Three Classes | None |
|---|---|---|---|---|---|---|---|---|
| Seniors in the Class | 24 | 103 | 23 | 6 | 10 | 4 | 2 | 50 |

Display this data in a Venn diagram. Fill in all the areas of the diagram, including the area outside the circles. Indicate empty areas with zeros.

## Share & Summarize

What do the different areas in a Venn diagram represent?

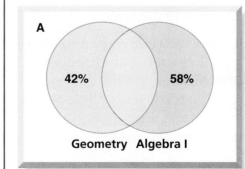

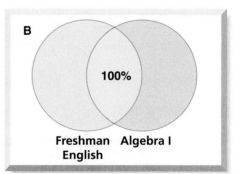

What can you say about the data in Venn diagram A?

What can you say about the data in Venn diagram B?

**Practice & Apply**

**1.** In Parts a–d, tell which graph could represent the situation.

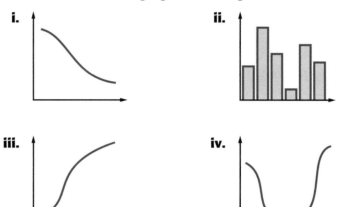

**i.**

**ii.**

**iii.**

**iv.**

**a.** a child's activity level from before a nap until after a nap

**b.** the populations of six cities

**c.** the change in water level from high tide to low tide

**d.** the change in the weight of a cat from birth until age two

**2.** This table shows the number of people who visited Milo's Restaurant in the years from 1996 to 2007.

**a.** The change in the number of customers is greatest from 2006 to 2007. Between which two years is the change in the number of customers least?

**b.** Make a bar graph showing the number of customers during the years shown in the table.

**c.** Describe what your graph indicates about the change in the number of customers during the time period. Discuss high and low points, periods of greatest and least change, and any other patterns you see.

**Customers at Milo's Restaurant**

| Year | Visitors (thousands) |
|------|----------------------|
| 1996 | 39.4 |
| 1997 | 42.7 |
| 1998 | 47.3 |
| 1999 | 45.8 |
| 2000 | 44.8 |
| 2001 | 43.3 |
| 2002 | 46.5 |
| 2003 | 47.8 |
| 2004 | 46.4 |
| 2005 | 48.5 |
| 2006 | 50.9 |
| 2007 | 45.5 |

**3.** Think about the multiplication facts from 0 • 0 to 12 • 12. You can group the products into intervals of 10. For example, a product can be between 0 and 9, between 10 and 19, between 20 and 29, and so on.

**a.** Would you predict that the products are evenly distributed among the intervals of 10, or do you think some intervals contain more products than others?

**b.** Copy and complete this multiplication table.

| × | 0 | 1 | 2 | 3 | 4 | 5 | 6 | 7 | 8 | 9 | 10 | 11 | 12 |
|---|---|---|---|---|---|---|---|---|---|---|----|----|----|
| 0 | | | | | | | | | | | | | |
| 1 | | | | | | | | | | | | | |
| 2 | | | | | | | | | | | | | |
| 3 | | | | | | | | | | | | | |
| 4 | | | | | | | | | | | | | |
| 5 | | | | | | | | | | | | | |
| 6 | | | | | | | | | | | | | |
| 7 | | | | | | | | | | | | | |
| 8 | | | | | | | | | | | | | |
| 9 | | | | | | | | | | | | | |
| 10 | | | | | | | | | | | | | |
| 11 | | | | | | | | | | | | | |
| 12 | | | | | | | | | | | | | |

**c.** Make a table, like that on the right, showing the number of products that fall in each interval of 10.

**d.** Make a histogram that shows the number of products in each interval of 10. Be sure to include axes labels and scale values.

**e.** What does the shape of your histogram reveal about the distribution of the products?

**f.** Now make another histogram showing the number of products that fall into intervals of 20, that is, 0–19, 20–39, 40–59, and so on.

**g.** Describe the similarities and differences in the two histograms.

| Product | Frequency |
|---------|-----------|
| 0–9 | |
| 10–19 | |
| 20–29 | |
| 30–39 | |
| 40–49 | |
| 50–59 | |
| 60–69 | |
| 70–79 | |
| 80–89 | |
| 90–99 | |
| 100–109 | |
| 110–119 | |
| 120–129 | |
| 130–139 | |
| 140–149 | |

**4.** This Venn diagram displays data about after-school clubs at Smallville High.

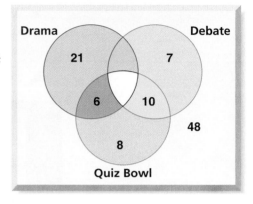

**After-school Clubs**

**a.** What number of students at Smallville High are in the Drama club?

**b.** What number of students are in both the Drama club and the Quiz Bowl club?

**c.** What number of students do not participate in any of these three clubs?

**d.** What number of students are in only one club?

**e.** Based on this diagram, which two clubs might meet at the same time? Why?

**5.** A local radio station sent a survey to its listeners asking what kinds of music they enjoyed. This bar graph summarizes the responses to the survey. The percentages are percentages of all the listeners who answered the survey.

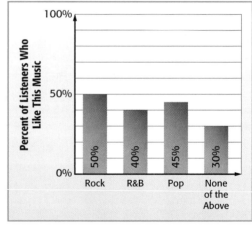

Which of these Venn diagrams might represent the same data as the bar graph? Explain your reasoning.

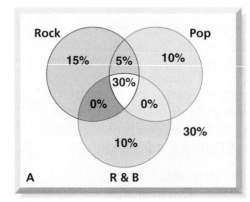

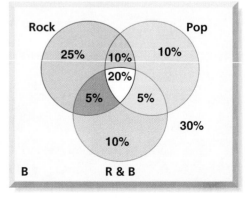

**6.** Use the data in this table to answer the following questions.

| Sport | Soccer (total) | Basketball (total) | Track (total) | Soccer & Basketball | Soccer & Track | Basketball & Track | All Three Sports | None or Other |
|---|---|---|---|---|---|---|---|---|
| **Number of Students who Participate** | 21% | 34% | 30% | 2% | 6% | 9% | ? | 36% |

**a.** Display the data in a Venn diagram.

**b.** What percentage of students play all three sports?

**c.** What percentage of students only play soccer?

**d.** There are 900 students at Smallville High. Make a copy of your Venn diagram, displaying the data as numbers of students.

**7.** Use the Venn diagram to answer Parts a–e.

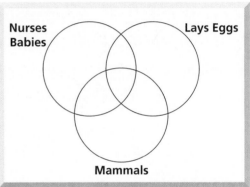

**Mammal Characteristics**

**a.** All mammals nurse their babies, and everything that nurses its babies is a mammal. Shade the areas of the Venn diagram that cannot have any animals in them.

**b.** The duck-billed platypus and echidna are examples of a kind of mammal called monotremes. Monotremes nurse their babies, like all mammals, but their babies hatch out of eggs. Write Monotremes in the appropriate area of the Venn diagram.

**c.** Placentals are mammals that bear live babies, like cats and squirrels. Marsupials, like opossums and kangaroos, are mammals that carry their babies in a pouch until the babies are big enough to be on their own. Place placentals and marsupials in the appropriate areas of the Venn diagram.

**d.** Birds are a separate class from mammals. All birds lay eggs. Place birds in the appropriate area of the Venn diagram.

**e.** Fish are also a separate class. Some fish lay eggs, and others bear live babies. How do fish fit into this Venn diagram? Explain your reasoning.

**Real-World Link**

The U.S. women's soccer team won the 1999 World Cup competition, defeating China 5–4. The game was tied after two overtime periods and had to be decided on penalty kicks.

**8.** This table shows the number of U.S. girls of various ages who played soccer in leagues recognized by the American Youth Soccer Organization in a recent year.

  **a.** Create a histogram showing these data. The first bar, which includes five- and six-year-old girls, has been drawn for you.

  **b.** Describe the shape of the histogram. Tell what the shape indicates about the distribution of ages.

**Soccer Players**

| Ages | Girls |
|---|---|
| 5 and 6 | 23,805 |
| 7 and 8 | 45,181 |
| 9 and 10 | 46,758 |
| 11 and 12 | 39,939 |
| 13 and 14 | 26,147 |
| 15 and 16 | 11,518 |
| 17 and 18 | 4,430 |

**Source:** American Youth Soccer Organization

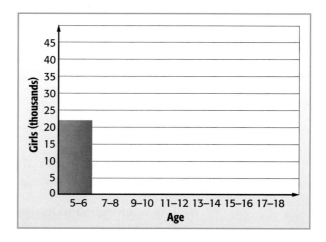

*Connect* **&** *Extend*

**Math Link**

To make a line graph, plot the data points and connect them with line segments.

**9. Economics** Here are data about the number of motor vehicles manufactured in the United States, Europe, and Japan from 1993 to 2001.

  **a.** On the same set of axes, make a line graph of the data for each group. Use a different point shape or line color for each group.

**Motor Vehicles Manufactured (millions)**

| Year | U.S. | Europe | Japan |
|---|---|---|---|
| 1993 | 10.9 | 15.2 | 11.2 |
| 1994 | 12.3 | 16.2 | 10.6 |
| 1995 | 12.0 | 17.0 | 10.2 |
| 1996 | 11.8 | 17.6 | 10.3 |
| 1997 | 12.1 | 17.8 | 11.0 |
| 1998 | 12.0 | 16.3 | 10.1 |
| 1999 | 13.1 | 17.6 | 9.9 |
| 2000 | 12.8 | 17.7 | 10.1 |
| 2001 | 11.5 | 17.7 | 9.8 |

**Source:** *World Almanac and Book of Facts 2003.*
Copyright © 2003 Primedia Reference Inc.

  **b.** Is there one group that consistently produces more motor vehicles than the others? If so, which group is it?

**c.** Write two or three sentences comparing the number of vehicles manufactured in the United States to the number manufactured in Japan for the years from 1993 to 2001.

**d.** Given the trends in these data, which group do you think produced the most motor vehicles in 2002? Which group do you think produced the fewest? Give reasons for your answers.

**Jackson Square, New Orleans**

10. **Earth Science** The *latitude* of a location indicates how far it is from the equator, which has latitude 0°. The further from the equator a place is, the greater its latitude. The latitude measure for a location includes the letter N or S to indicate whether it is north or south of the equator.

This table gives the lowest average monthly temperature and the latitude of nine cities.

| City | Latitude | Lowest Average Monthly Temp. (°F) |
|---|---|---|
| Albuquerque, New Mexico, U.S.A. | 35° N | 34 |
| Georgetown, Guyana | 7° N | 79 |
| New Orleans, Louisiana, U.S.A. | 30° N | 51 |
| Portland, Maine, U.S.A. | 44° N | 22 |
| Porto Alegre, Brazil | 30° S | 58 |
| Recife, Brazil | 8° S | 75 |
| San Juan, Puerto Rico | 18° N | 72 |
| St. John's, Newfoundland, Canada | 48° N | 23 |
| Stanley, Falkland Islands | 52° S | 36 |

**Source:** www.worldclimate.com

**a.** Make a line graph of the latitude and temperature data. When you graph the latitude values, ignore the N and S, graphing only the numbers. This way, you will be graphing each city's distance from the equator.

**b.** Does there appear to be an overall relationship between the latitude of a city and its lowest average monthly temperature? If so, describe the relationship.

**c.** The island of Nassau in the Bahamas has a latitude of about 25° N. Predict Nassau's lowest average monthly temperature. Explain how you made your prediction.

**11. Preview** Middle school students were asked in a survey how much time they spend with their parents or guardians on a typical weekend. Here are the results.

| Time Spent with Parents | Boys (percent) | Girls (percent) |
|---|---|---|
| Almost all | 39.6 | 49.6 |
| One full day | 18.6 | 21.8 |
| Half a day | 17.5 | 17.1 |
| A few hours | 24.3 | 11.5 |

If you wanted to compare the boys' responses with the girls' responses, you could display these data in two circle graphs.

**Time Spent with Parents**

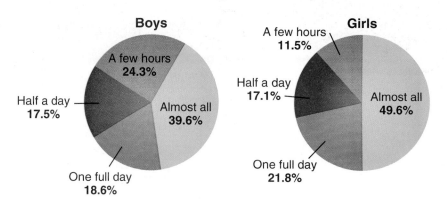

**a.** You could also show the data in a *double bar graph*. For each time category, the graph will have two bars, one showing the percent of boys in that category and the other showing the percent of girls. Copy and complete the graph below.

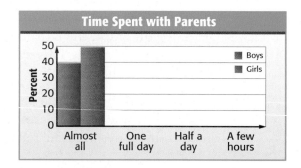

**b.** Which display do you think makes it easier to compare the two categories of data? Give reasons for your choice.

**12.** Drake's mother told him he could not play video games after school until his performance in math class improved significantly. Drake's math teacher gives a 20-point quiz each week. Drake made this bar graph to show his mother how much his scores had improved over the past five weeks.

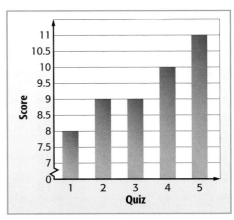

**a.** The bar for Quiz 5 is three times the height of the bar for Quiz 1. Is Drake's score on Quiz 5 three times his score on Quiz 1?

**b.** Drake's mother says his graph is misleading because it makes his improvement look more dramatic than it really is. What features of the graph make it misleading?

**c.** Make a new bar graph that you feel gives a more accurate view of Drake's performance on the weekly quizzes.

**13. In Your Own Words** Describe two types of graphs you have used to display data. For each type of graph, give an example of a set of data you might display with that type of graph.

**14.** Use the following clues to help you complete this Venn diagram.

**a.** There are no numbers outside of the circles.

**b.** The sum of the numbers in the three overlapping areas equals the number in the center area.

**c.** The total of the numbers in circle C is 10.

**d.** The total of the numbers in circle A is twice the total of the numbers in circle B.

**e.** The total of the numbers in all the areas of the diagram is 31.

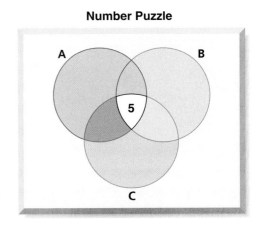

**15.** Draw a different Venn diagram that fits all the clues from Exercise 8.

**Mixed Review**

**Solve each equation.**

**16.** $5n - 3 = 17$

**17.** $x + 5 = 12$

**18.** $2 + 3a = 14$

**19.** $\dfrac{x}{10} = \dfrac{25}{50}$

**20.** $5a - 5 = 15$

**21.** $4d + 3 = 19$

**Shade the given percent of a 100-grid. Then express the part of the area that is shaded as a fraction and as a decimal.**

**22.** 20%                          **23.** 30%

**24.** 2%                           **25.** 25%

**26.** 120%                         **27.** 75%

**28.** A group of girls is selling cookies. Miss Susanne wants the girls to sell the same number of boxes. Last year, 20 girls sold 80 boxes of cookies.

   **a.** Complete this ratio table based on last year's information.

| Girls |  | 10 | 15 | 20 |  |  |
|---|---|---|---|---|---|---|
| Boxes of cookies | 20 |  |  | 80 |  |  |

   **b.** How many girls will sell 20 boxes?

   **c.** There are 25 girls selling cookies. How many cookies will they sell?

# Collect and Analyze Data

## Materials

• survey form

The editors of *All about Kids!* magazine are researching an article about the activities in which middle school students participate. They would like the article to address these questions.

• In what activities do middle school students participate after school and on weekends?

• What percent of students participate in each activity?

• How many hours a week do students typically spend on each activity?

• What are students' favorite activities?

• Do boys and girls like different activities?

• Do students tend to spend the most time on the activities they like best?

The editors have hired Data, Inc. to help with the article. They would like you to answer the above questions for the students in your class.

They have suggested using the form below to collect your class data.

Are you male or female?

Please fill out this table, listing the time you spend each week in each activity, after school and on weekends.

What is your favorite activity?

| Activity | Time Spent Each Week |
|---|---|
| Doing homework | |
| Spending time with friends | |
| Playing sports | |
| Reading books | |
| Reading newspapers | |
| Using a computer | |
| Taking care of pets | |
| Watching movies or videos | |
| Talking on the phone | |
| Watching TV | |
| Listening to music | |
| Playing video or board games | |
| Shopping | |

Look over the list of activities. Decide with your class whether to add or delete any activities.

Consider each question the editors want answered. Think about whether the survey form will collect the information needed to answer it. Decide as a class whether anything should be added to the form.

Each student in your class should fill out a survey form. All the data from the class will be combined later.

## Investigation 1 Plan Your Analysis

In this investigation, you will think about what types of statistics and graphs might be useful for reporting the results of the survey. You will not do your analysis until the next investigation, but carefully planning your strategy now will make your analysis much easier.

### Develop & Understand: A

Exercises 1–6 list the six questions asked by the magazine editors. For each question, complete Parts a and b.

  a. Tell which collected data you will need to answer the question.

  b. Describe a procedure you could use to answer the question. Be sure to indicate any statistical measures (like mean, median, mode, or range) that you will need to find or computations you will need to do.

1. In what activities do middle school students participate after school and on weekends?

2. What percent of students participate in each activity?

3. How many hours a week do students typically spend on each activity?

4. What are students' favorite activities?

5. Do boys and girls like different activities?

6. Do students tend to spend the most time on the activities they like best?

Save your answers from these exercises for Investigation 2.

Magazine articles often use graphs to present data. In Exercises 7–12, you will think about what types of graphs might be useful to include in the magazine article.

You are familiar with several types of graphs, such as line graphs, bar graphs, histograms, pictographs, and circle graphs. You might also consider one of the special types of bar graphs described in the example below.

**Math Link**

A *pictograph* uses symbols or pictures to represent data.

## Example

A *double bar graph* compares data for two groups. For example, this double bar graph compares the favorite primary colors of boys and girls in one middle school class.

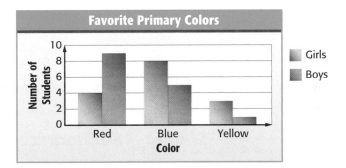

A *stacked bar graph* shows how the data represented by each bar is divided into two or more groups. This graph shows how the number of children who chose each color is divided between boys and girls.

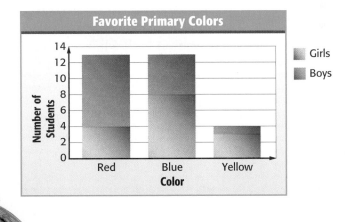

### ✅ *Develop & Understand: B*

For each question, decide whether it would be useful to include a graph with the answer to the question. If so, describe the graph you would use. Include at least one double bar graph or stacked bar graph.

7. In what activities do middle school students participate after school and on weekends?

8. What percent of students participate in each activity?

9. How many hours a week do students typically spend on each activity?

10. What are students' favorite activities?

11. Do boys and girls like different activities?

12. Do students tend to spend the most time on the activities they like best?

Save your answers from these exercises for Investigation 2.

### Share & Summarize

Think of at least one more question you think would be interesting to address in the magazine article, such as training horses. Describe the data you would need to collect to answer the question. Tell what statistics and graphs you would include in your answer.

# Investigation ② Carry Out Your Analysis

## Materials

- results of class survey
- answers for Investigation 1

You have collected data about the activities in which the students in your class participate. Now you will analyze the class data and use your results to answer the questions posed by the magazine editors.

### ✅ Develop & Understand: A

The editors' six questions appear in Exercises 1–6. Work with your group to analyze the data and answer each question. Include the following information for each exercise.

- the results of your computations and the measures you found
- a few sentences answering the question, including statistical measures that support your answers
- a graph, if appropriate, to help illustrate your answer

You can use your answers from Exercises 7–12 in Investigation 1 as a guide, but you may change your mind about what statistics and graphs to include.

As you work, you may want to create tables to organize your data and calculations. Here is an example you might find useful.

| Activity | Number Who Participate | | | Percent Who Participate | | | Mean Time Spent | | | Median Time Spent | | |
|---|---|---|---|---|---|---|---|---|---|---|---|---|
| | Boys | Girls | All | Boys | Girls | All | Boys | Girls | All | Boys | Girls | All |
| Doing homework | | | | | | | | | | | | |
| Spending time with friends | | | | | | | | | | | | |
| Playing sports | | | | | | | | | | | | |
| Reading books | | | | | | | | | | | | |
| Reading newspapers | | | | | | | | | | | | |
| Using a computer | | | | | | | | | | | | |
| Taking care of pets | | | | | | | | | | | | |
| Watching movies or videos | | | | | | | | | | | | |
| Talking on the phone | | | | | | | | | | | | |
| Watching TV | | | | | | | | | | | | |
| Listening to music | | | | | | | | | | | | |
| Playing video or board games | | | | | | | | | | | | |
| Shopping | | | | | | | | | | | | |

1. In what activities do middle school students participate after school and on weekends?

2. What percent of students participate in each activity?

3. How many hours a week do students typically spend on each activity?

4. What are students' favorite activities?

5. Do boys and girls like different activities?

6. Do students tend to spend the most time on the activities they like best?

## Share & Summarize

1. Write a few sentences summarizing your group's work on Exercises 1–6. Discuss how you divided the work among group members and how you organized the data to make it easier to answer the questions.

2. Do you think a nationwide survey of middle school students would give results similar to your class results? Explain why or why not.

# Investigation ③ Make Predictions

## Materials

- bag of beans

People sometimes want to estimate large quantities that are difficult or impossible to count, such as the number of animals or plants of a particular endangered species or the number of people affected by a flu epidemic. If you had to count each individual member of these populations, it would be costly, time-consuming, and probably impossible to get exactly right.

In this investigation, you will learn how proportions can help you to estimate such numbers without actually counting every member.

### ✅ Develop & Understand: A

Each student in your group should take five beans from your bag and mark them. Put all the marked beans back into the bag with the other beans. Mix them carefully.

Without looking in the bag, take out 20 beans. Record the number of marked beans.

1. What is the number of marked beans?

2. Estimate the total number of beans in the bag. How did you make your estimate?

3. What did you need to assume to estimate the total number of beans?

4. Do you think your estimate would be better, worse, or the same if your bag contained thousands of beans? Why? How might you modify the method you used to make the estimate more accurate?

To estimate the numbers of different animals in the wilderness, scientists use a method called *capture-tag-recapture*. This method is similar to the process you used to estimate the number of beans in your bag.

Using this method, scientists capture a certain number of animals and mark them using collars, rings, or other tags. They then release the animals. Some time later, they capture another group of animals and count how many in that group are tagged. They can then solve a proportion to estimate the total number of animals.

### Real-World Link

Bird banding is a helpful way to study wild birds. Scientists attach a small, numbered metal or plastic ring to the birds' legs or wings. Scientists can track the migration, longevity, and feeding habits of a bird with this type of identification.

5. Suppose biologists tagged 45 blue whales in the Antarctic waters. The next year, they caught 75 blue whales and found that 15 of them had tags. Estimate the total number of whales in the area. Explain your method.

6. Red wolves have been classified as extinct in the wild, but some still live in captivity. Some efforts have been made to restore red wolves to forests in North Carolina and Tennessee.

   Suppose biologists caught 20 wolves from those forests, tagged them, and then freed them. Later, they caught 15 wolves and found that five had tags. Estimate the number of red wolves in the forests. Explain your method.

7. Suppose ornithologists tagged and released 240 bald eagles from across the United States. A couple of months later, they caught 100 birds and found that three of them had tags. Estimate the number of bald eagles in the United States.

Surveys do not use the capture-tag-recapture method, but they do use proportions to make estimates of large populations of people.

✅ *Develop & Understand: C*

The population of Massachusetts is about 6.4 million people. A survey questioned 1,000 people across the country. It found that approximately 300 people had two televisions in their homes. About 400 people had three or more televisions. The survey also found that about 700 people had cable television service.

8. Estimate how many households in Massachusetts have two televisions. Assume the proportions for Massachusetts are the same as for the sample.

9. Estimate how many households have three or more televisions.

10. Estimate how many households in Massachusetts receive cable television.

11. Discuss with your partner the methods you used to solve Exercises 8–10. What assumptions did you make and why?

> ### *Share* 🔗 *Summarize*
>
> What method would you use to estimate the number of people in a large crowd, such as the large gathering at Times Square in New York City on New Year's Eve?

# Investigation  Choose a Graph

When you have a choice of graph for organizing and displaying data, what do you do? The "best" graph for a situation is the one that most clearly shows the data for the purpose you need.

These are some of the types of graphs that you have learned and from which you can choose.

- line graphs
- histograms
- bar graphs

For example, the data below, which show the average monthly temperatures for Knoxville, Tennessee, in degrees Fahrenheit, were used to create the three graphs.

| Jan | Feb | Mar | Apr | May | Jun | Jul | Aug | Sep | Oct | Nov | Dec |
|-----|-----|-----|-----|-----|-----|-----|-----|-----|-----|-----|-----|
| 36 | 40 | 49 | 58 | 65 | 73 | 77 | 76 | 70 | 58 | 49 | 40 |

Reprinted with permission from *The World Almanac and Book of Facts*.

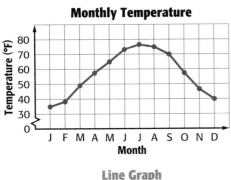

**Line Graph**

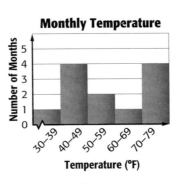

**Histogram**

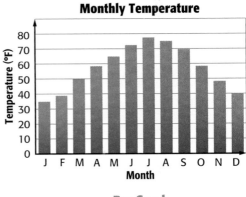

**Bar Graph**

## ✓ Develop & Understand: A

**Math Link**

In a histogram, the height of a bar tells how many data values are in a particular range of values. In a bar graph, the height of a bar gives a particular data value.

1. Which graphs give a sense of the temperature changes over the year?

2. From the graphs you named in Exercise 1, describe the average monthly temperatures in Knoxville.

3. Which graphs seem most useful for displaying the temperature data? Explain.

4. Are the other graphs useless for displaying the data? Explain.

## ☑️ *Develop & Understand: B*

Kyung and Isandro gathered some data on pets for a school project.

**Most Popular Pets**

Kyung surveyed the 27 students in his class about their favorite pets. Here are his findings.

| Dogs | Cats | Birds | Fish | Mice |
|------|------|-------|------|------|
| 8 | 9 | 5 | 4 | 1 |

**Tropical Fish**

Isandro found ten students who kept tropical fish. He asked how many fish they each had in their tanks. The responses were as follows.

$$4 \quad 9 \quad 11 \quad 13 \quad 15 \quad 16 \quad 16 \quad 16 \quad 18 \quad 20$$

**Cost per Week**

Kyung and Isandro gathered information on how much it costs students to care for their pets each week, on average. Here are their results.

- under $5: 23%
- between $5 and $10: 48%
- over $10: 29%

**Life Expectancy of Dogs**

Isandro researched the following data for how long pet dogs live on average.

| Years | 0 to 4 | 5 to 8 | 9 to 12 | 13 to 16 | over 16 |
|-------|--------|--------|---------|----------|---------|
| Percentage | 17% | 8% | 26% | 37% | 12% |

5. The boys wanted to use graphs to effectively display their data. For each of the four types of data they collected, choose one of the following graph types that you think would be useful and one that would be inappropriate. Explain your choices.
   - line graphs
   - histograms
   - bar graphs

6. For each data set, choose an appropriate graph type. Make a graph to display the data.

## *Share & Summarize*

For each type of graph, describe the kind of data or situation that it would be best for displaying.

1. line graph

2. histogram

3. bar graph

## Practice **&** Apply

**Real-World Link**

Americans aged 13 to 18 spend more than 72 hours a week using electronic media, such as the Internet, cell phones, television, music, and video games.

In Exercises 1 and 2, use this information.

The editors of *All about Kids!* would like to publish an article about teenage Internet users. Here are the questions they would like to answer.

**Question 1.** How much time do teen Internet users typically spend on the Internet each week?

**Question 2.** How much time do they typically spend on various Internet activities?

**Question 3.** What are these teenagers' favorite Internet activities?

The magazine editors have collected data from 15 students who are regular Internet users. The green entries indicate the students' favorite activities.

**Weekly Time Spent on the Net (minutes)**

| Student Initials | Chatting in a Chat Room | Playing Games | Doing Homework | Surfing the Web | E-mail |
|---|---|---|---|---|---|
| AB | 90 | 75 | 80 | 90 | 12 |
| BT | 75 | 150 | 0 | 150 | 15 |
| CP | 0 | 75 | 60 | 150 | 5 |
| CT | 0 | 240 | 0 | 0 | 0 |
| GO | 120 | 90 | 90 | 60 | 3 |
| KQ | 135 | 60 | 40 | 80 | 15 |
| LM | 75 | 160 | 30 | 45 | 6 |
| MC | 15 | 0 | 35 | 60 | 15 |
| MH | 80 | 180 | 30 | 90 | 6 |
| NM | 90 | 90 | 45 | 90 | 15 |
| PD | 100 | 150 | 60 | 90 | 0 |
| RL | 100 | 90 | 45 | 90 | 0 |
| SK | 90 | 135 | 60 | 240 | 10 |
| SM | 60 | 135 | 40 | 60 | 22 |
| YS | 120 | 30 | 60 | 45 | 15 |

**1.** Complete Parts a and b for Questions 1, 2, and 3 above.

   **a.** Describe any statistical measures you will need to find or computations you will need to do to answer the question.

   **b.** Describe a graph that would be appropriate to include with the answer to the question.

2. Refer to the information on the previous page. Analyze the given data. Use your results to answer the three questions posed by the magazine editors. Provide the following information for each question.

   **a.** the results of your computations and the measures you found

   **b.** one or more sentences answering the question

   **c.** a graph to help illustrate your answer

3. **Biology** Marine biologists estimated the number of manatees living in the waters off the coast of Florida. They caught and tagged 120 manatees and then let them go. Of the 150 manatees they caught the next year, nine were marked. Estimate the number of manatees living in these waters.

4. A farmer recorded the temperature every two hours during daylight.

| Time | 6 A.M. | 8 A.M. | 10 A.M. | 12 noon | 2 P.M. | 4 P.M. | 6 P.M. | 8 P.M. |
|---|---|---|---|---|---|---|---|---|
| Temp (°F) | 45 | 50 | 60 | 65 | 80 | 75 | 60 | 50 |

   **a.** Draw a graph showing the change in the temperature during the day.

   **b.** Use your graph to estimate the temperature at 11 A.M.

   **c.** During which hours of the day was the temperature above 60°F?

   **d.** It is recommended that crops get irrigated when the temperature is between 50°F and 60°F. From these data, at what times should the farmer irrigate the crops?

5. Alexis lives on a farm. For a Math project, she measured the growth of several chicks in their first week of life. The table shows the averages for her chicks. What type of graph would you use to display these data? Explain.

| Day | 0 | 1 | 2 | 3 | 4 | 5 | 6 | 7 |
|---|---|---|---|---|---|---|---|---|
| Mass (grams) | 25 | 28 | 31 | 34 | 38 | 43 | 48 | 54 |

6. Kirk counted the colors in a package of candy. What type of graph do you suggest he use to display his data? Explain.

| Color | Red | Green | Yellow | Brown | Blue | Orange |
|---|---|---|---|---|---|---|
| Number | 8 | 3 | 10 | 18 | 5 | 13 |

**7.** Ms. Estefan polled her class about their birthday months. What type of graph might she use to display these data? Explain.

| Month | Jan | Feb | Mar | Apr | May | Jun | Jul | Aug | Sep | Oct | Nov | Dec |
|---|---|---|---|---|---|---|---|---|---|---|---|---|
| Students | 5 | 3 | 2 | 3 | 3 | 2 | 0 | 6 | 3 | 2 | 2 | 1 |

**8.** Mr. Malone polled his class on the kinds of exercise they prefer. Of 24 students, 10 said running, 5 said walking, 4 said swimming, and 3 said biking. The other 2 said they never exercise. What kind of graph do you suggest he use to display these data? Explain.

**9. Social Studies** This double bar graph shows the number of licensed drivers per 1,000 people and the number of registered vehicles per 1,000 people in seven states.

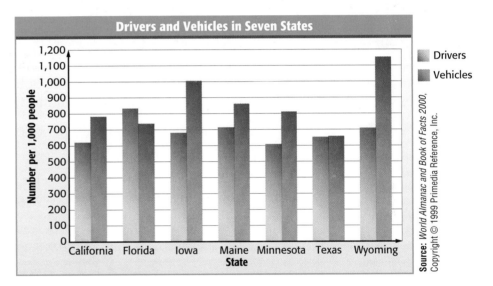

**Drivers and Vehicles in Seven States**

Source: *World Almanac and Book of Facts 2000,* Copyright © 1999 Primedia Reference, Inc.

**Real-World Link**

Automobiles were commercially available in the United States beginning in 1896. In 1903, Missouri and Massachusetts adopted the first driver's license laws. In 1908, Rhode Island became the first state to require a driver's test.

**a.** Which state has the greatest difference between the number of registered vehicles and the number of licensed drivers?

**b.** Which state has about one registered vehicle per licensed driver?

**c.** Which state has less than one registered vehicle per licensed driver?

**d.** Which two states have about the same number of registered vehicles?

**10. In Your Own Words** Think of a question you could answer by collecting data from a particular group of people. Describe the group you would survey and the type of information you would collect. Then tell what statistical measures and graphs might be useful for summarizing the data.

**11. Social Studies** This stacked bar graph shows the number of bachelor's degrees awarded to men and women in the United States in various years.

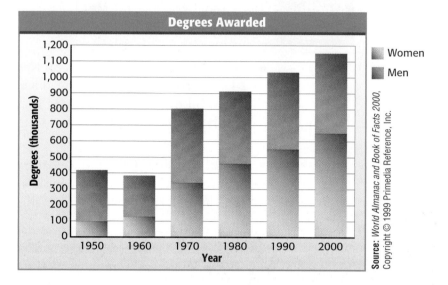

**a.** Describe how the total number of bachelor's degrees awarded changed over the years shown in the graph.

**b.** Describe how the total number of bachelor's degrees awarded to women has changed over the years shown in the graph.

**c.** In 1970, about how many bachelor's degrees were received by men? About how many were received by women?

**d.** In 1950, about what fraction of bachelor's degrees were received by women? In 2000, about what fraction were received by women?

**12. Social Studies** When a political group holds a rally to show support for a cause or to protest, the police will usually estimate the number of people attending. The estimates are often made by counting the number of people in a particular location. That number and the area of the selected location can be used to set up a proportion for estimating the total attendance.

The organizers of such rallies will often make their own estimates. Usually, their estimates are much higher.

Suppose the organizers of an event want their estimate to be as high as possible to show that many people support their cause. How might they choose the location where they will count people attending? Explain.

**Real-World Link**

A healthy adult human has 20 to 30 trillion red blood cells, with an average life span of 120 days. To keep up with this demand, the body replaces about 2 million red blood cells every second.

13. **Life Science** A sample of blood was diluted 100,000 times. The drawing represents a microscopic photograph of the red blood cells in 20 cubic millimeters of the diluted blood. Each dot represents one cell.

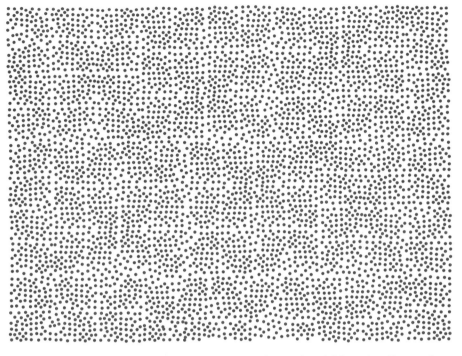

a. Think of a way to estimate the number of red blood cells in the drawing without counting all of them. Explain your method.

b. Use your method to estimate how many red blood cells are in 20 mm$^3$ of the sample of diluted blood.

c. About how many red blood cells were in 1 mm$^3$ of this blood sample before it was diluted?

14. **Geography** The table shows population and land area for four states.

**Population and Area in the U.S.**

| State | 2000 Population | Land Area (mi$^2$) |
|-------|-----------------|--------------------|
| Alaska | 626,932 | 570,374 |
| California | 33,871,648 | 155,973 |
| Illinois | 12,419,293 | 55,593 |
| Vermont | 608,827 | 9,249 |

Reprinted with permission from *The World Almanac and Book of Facts.*

**Real-World Link**

Japan has the world's highest population density, 322 people per square kilometer in 2000. In comparison, the density of the United States was only 31 people per square kilometer.

a. Estimate how many people live in a particular 15 mi$^2$ area in California and in a particular 15 mi$^2$ area in Vermont.

b. Calculate the population density, the number of people per square mile, for each state. Which state is most crowded?

15. **Earth Science** The maximum distances between the planets in the solar system and the Sun are given to the right.

| Planet | Distance from Sun (millions of miles) |
|--------|---------------------------------------|
| Mercury | 43.4 |
| Venus | 67.7 |
| Earth | 94.5 |
| Mars | 154.8 |
| Jupiter | 507.0 |
| Saturn | 936.0 |
| Uranus | 1,867.0 |
| Neptune | 2,818.0 |

a. Display these data using any kind of graph you like. Explain your choice of graph.

b. Write a few sentences describing any observations you can make about the planets from your graph.

Reprinted with permission from *The World Almanac and Book of Facts 1999*. Copyright © 1998 Primedia Reference Inc. All rights reserved.

**Mixed Review**

16. $\frac{2}{3} + \frac{9}{8}$

17. $\frac{6}{10} \div \frac{1}{4}$

18. $\frac{9}{11} \cdot \frac{4}{12}$

19. $\frac{1}{2} - \frac{3}{10}$

**For each square, estimate the percent of the area that is shaded.**

20.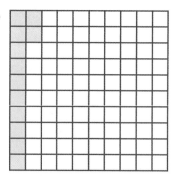

21.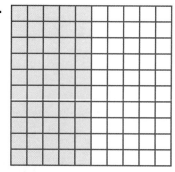

22. Libby is having several friends over to plan their act for the talent show. She wants to serve them bananas for a snack. The farmer's market is selling five bananas for $2.50. The grocery store sells three bananas for $1.80. Where will Libby get the most for her money?

**Tell whether each given rate is a unit rate or a non-unit rate.**

23. 65 mph

24. 500 pages in four hours

25. 65 beats per minute

26. 400 miles in eight hours

# The Language of Chance

People often make comments like the following.

- "It probably won't rain tomorrow."
- "I expect to have a lot of homework this week."
- "Our team has a good chance of winning the game."
- "It's not likely she will eat that entire cake!"
- "The chances are 50/50 that we'll go to the movies tonight."
- "There's a 40% chance of rain tomorrow."

The words *probably, expect, chance,* and *likely* are used when someone is making a prediction.

**Real-World Link**

Some gorillas have been taught to communicate with humans by using sign language. One such gorilla, Koko, has a vocabulary of over 1,000 words.

### Think & Discuss

What are some other words or phrases people use when predicting the chances of something happening?

Listed below are six events. How likely do you think each event is? Talk about them with your class. Come to an agreement about whether each event has the given chance of happening.

- has no chance of happening
- could happen but is unlikely
- is just as likely to happen as not to happen
- is likely to happen
- is certain to happen

*Event 1:* Our class will have homework tonight.

*Event 2:* It will snow tomorrow.

*Event 3:* If I toss a penny in the air, it will land heads up.

*Event 4:* A gorilla will eat lunch with us today.

*Event 5:* You choose a name from a hat containing the names of all the students in your class, and you get a girl's name.

*Event 6:* You draw a number from a hat containing the numbers from 1 to 5, and you get 8.

# Investigation  Probability in Everyday Life

## Vocabulary

experimental
  probability

probability

## Materials

• paper cup

---

### Math Link

The probability *P* of an event *E* is the ratio that compares the number of favorable outcomes *f* to the number of possible outcomes *n*. $P(E) = \frac{f}{n}$

---

The **probability**, or chance, that an event will happen can be described by a number between 0 and 1.

- A probability of 0, or 0%, means the event has no chance of happening.
- A probability of $\frac{1}{2}$, or 50%, means the event is just as likely to happen as not to happen.
- A probability of 1, or 100%, means the event is certain to happen.

For example, the probability of a coin landing heads up is $\frac{1}{2}$, or 50%. This means you would expect a coin to land heads up $\frac{1}{2}$, or 50%, of the time.

The more likely an event is to occur, the greater its probability. If a weather forecaster says the probability of rain is 90%, it is a good idea to take your umbrella when you go outside. Of course, it *might* not rain after all. On the other hand, if the forecaster says the probability of rain is 10%, you might want to leave your umbrella at home. Still, you *might* get wet.

You can represent the probability of an event by marking it on a number line like this one.

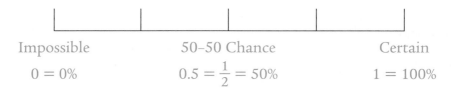

| Impossible | 50–50 Chance | Certain |
| $0 = 0\%$ | $0.5 = \frac{1}{2} = 50\%$ | $1 = 100\%$ |

For example, the next number line shows the probabilities of tossing a coin and getting heads, of a goldfish walking across a room, and of Alaska getting snow this winter.

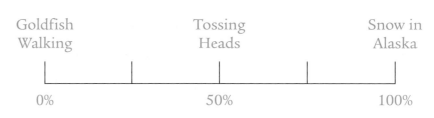

Goldfish Walking     Tossing Heads     Snow in Alaska

0%     50%     100%

## ✓ Develop & Understand: A

1. Describe an event you think has the given chance of happening.

   a. The event has no chance of happening.

   b. The event could happen but is unlikely.

   c. The event is just as likely to happen as not to happen.

   d. The event is likely to happen.

   e. The event is certain to happen.

**2.** Copy this number line. Label the number line with your events from Exercise 1. You can use the letters of the events for the labels.

0%          50%          100%

In Exercises 1 and 2, you used your experience to estimate the chances that certain events would occur. For example, you know from experience that when you toss a coin, it lands tails up about half the time. In some situations, you can use data to help estimate probabilities.

## ✅ Develop & Understand: B

On Saturday, Carolina's baseball team, the Rockets, is playing Marlon's team, the Lions. Carolina decided to look at the scores from the last six times their teams played each other.

| Lions | 3 | 8 | 6 | 4 | 4 | 5 |
|---|---|---|---|---|---|---|
| Rockets | 5 | 2 | 4 | 5 | 7 | 6 |

**3.** How many times did the Rockets win?

**4.** Which team do you think is more likely to win the next game?

**5.** Carolina can estimate the probability that her team will win by dividing the number of times the Rockets won by the number of games played. What probability estimate would she get based on the results of the six games? Give your answer as a fraction and as a percent.

**6.** Suppose Carolina knew the results of only the first three games. What would be her probability estimate? If she knew the results of only the last three games, what would be her probability estimate?

**7.** These two teams have played six games against each other. Suppose they had played eight games, and each team had won one more game. What probability estimate would you give for the Rockets winning the next game?

The probabilities you found in Exercises 3–7 are examples of experimental probabilities. An **experimental probability** is always an estimate. It can vary depending on the particular set of data you use.

If you want to find an experimental probability when you have no data available, you might perform *experiments* to create some data.

### Real-World Link

Little League baseball was started in 1939 in Williamsport, Pennsylvania, with three teams and 45 players. Today, more than 2.5 million children worldwide play Little League baseball.

## ✅ Develop & Understand: C

Much like a science experiment, a probability experiment involves trying something to see what happens. You may have some idea of what will occur, but the actual results can be surprising.

**Real-World Link**

Many sport statistics can be thought of as experimental probabilities. For example, a basketball player's free-throw percentage is the probability he will score on his next free throw.

8. Toss a paper cup so that it spins in the air. Record how it lands. Does it land right side up, upside down, or on its side? This is one *trial* of the experiment.

**Right Side Up**     **Upside Down**     **On Its Side**

Toss the cup 29 more times, for a total of 30 tosses. Record the landing position each time. You may want to use tally marks as shown below.

| Right Side Up | Upside Down | On Its Side |
|:---:|:---:|:---:|
| II | I | IIII |

9. How many trials did you perform in your experiment?

10. How many times did the cup land right side up? Upside down? On its side?

11. Find the portion of the trials for which the cup landed right side up, stating your answer as a fraction or a percent. Your answer is an experimental probability that the cup lands right side up when tossed.

12. Now find an experimental probability that the cup lands upside down and an experimental probability that it lands on its side.

13. Share your results with the class. Consider the results found by your classmates. Suggest at least one way you might use them to find a class experimental probability for the cup landing right side up.

### Share & Summarize

1. What does it mean to say that the probability of an event is 1?

2. Rey is in a basketball league. He was practicing free throws one afternoon, and he made 32 out of 50 shots. Estimate the probability that he makes a free throw, expressing it as a fraction and a percent.

3. Suppose you conducted $y$ trials of an experiment, and a particular event happened $x$ times. Find an experimental probability that the event will occur.

# Investigation ② Theoretical Probability

## Vocabulary

**equally likely**

**theoretical probability**

## Materials

- coin
- die

In some situations, all of the possibilities for a situation, called the *outcomes,* have the same probability of occurring. For example, a coin toss has two possible outcomes, heads or tails. If the coin is fair, about half of the tosses will come up heads and half will come up tails. In situations like these, the outcomes are **equally likely**.

When outcomes are equally likely, you can calculate probabilities by reasoning about the situation. A **theoretical probability** does not depend on an experiment. Therefore, it is always the same for a particular event.

### Example

In a class competition, five students, Althea, Conor, Hannah, Luke, and Rosita, are tied for first place. To break the tie, they will write their names on slips of paper and place them in a bowl. A judge will choose one slip without looking. The student whose name is on that slip will receive first prize. What is the probability that the name chosen has three syllables?

There are five names. The judge chooses a name *at random,* that is, in a way that all five names have the same chance of being selected. *Althea* and *Rosita* are the only names with three syllables.

Since there are five equally likely outcomes and two of them are three-syllable names, you would expect a three-syllable name to be selected $\frac{2}{5}$ of the time. So, the probability of choosing a name with three syllables is $\frac{2}{5}$, or 40%.

### ✅ Develop & Understand: A

Think about the situation described in the example. Decide how likely each of the following events is, and determine its theoretical probability. Explain your answers.

1. The name chosen does not begin with *R*.

2. The name chosen begins with *J*.

3. The name chosen has four letters or more.

4. The name chosen has exactly four letters.

**Real-World Link**

In 2006, the most popular names for girls born in the U.S. were Emily, Emma, Madison, Abigail, Olivia, Isabella, Hannah, Samantha, Ava, and Ashley.

5. The name chosen ends with *A*.

6. The name chosen does *not* end with *A*.

7. Add the probabilities you found in Exercises 5 and 6. Why does this sum make sense?

When people talk about probabilities involved in games, rolling dice, or tossing coins, they usually mean *theoretical* rather than experimental probabilities. In the rest of this investigation, you will consider the relationship between these two types of probability.

## Think & Discuss

Discuss what the students are saying. Who do you think is correct?

**Real-World Link**

In 2006, the most popular names for boys born in the U.S. were Jacob, Michael, Joshua, Matthew, Ethan, Andrew, Daniel, Anthony, Christopher, and Joseph.

### Real-World Link
The United States produced half-cent coins from 1793 to 1857.

# ✓ Develop & Understand: B

8. If you toss a coin 12 times, how many times would you expect to get heads? Explain.

9. Conduct an experiment to find an experimental probability of getting heads. Toss a coin 12 times. Record the results, writing H for each head and T for each tail.

   a. How many heads did you get?

   b. Use your results to find an experimental probability of getting heads when you toss a coin.

10. Compare the theoretical results with your experimental results.

    a. Is your result in Part a of Exercise 9 the same as the answer you computed in Exercise 8?

    b. Is your experimental probability the same as the theoretical probability?

11. Now combine your experimental results with those of other students. Make a table like this one, showing how many times out of 12 each student's coin came up heads.

    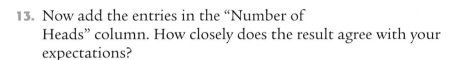

    | Student | Number of Heads |
    |---------|-----------------|
    | James   | 7               |
    | Ali     | 5               |
    |         |                 |
    |         |                 |
    |         |                 |
    |         |                 |

12. Calculate the total number of tosses your class made. How many heads would you expect for that number of tosses?

13. Now add the entries in the "Number of Heads" column. How closely does the result agree with your expectations?

14. What percent of the total number of coin tosses came up heads? In other words, what is the experimental probability of getting heads based on the data for your entire class?

15. Which experimental probability is closer to the theoretical probability, the one you calculated for Part b of Exercise 9 or the one you calculated for Exercise 14?

It is normal for experimental probabilities to be different from theoretical probabilities. In fact, when you repeat an experiment a small number of times, the experimental and theoretical probabilities may not be close at all.

### Real-World Link
Abraham Lincoln is the only U.S. president on a coin who faces to the right.

However, when you repeat an experiment a large number of times, for example, by combining all the coin tosses of everyone in your class or by performing more tosses, the experimental probability will usually grow closer to the theoretical probability.

The theoretical probability tells what is likely to happen *in the long run,* that is, if you try something a large number of times. It does not reveal exactly what will happen each time.

### ✅ Develop & Understand: C

A standard die has six faces, each indicating a different number from 1 to 6. When you roll a die, the *outcome* is the number facing up.

16. Roll a die 12 times. Record each number you roll. How many 3s did you get?

17. Use your results to find an experimental probability of rolling 3.

18. Collect results from others in your class. Make a table with these columns.

| Student | Number of 3s |
|---------|--------------|
|         |              |
|         |              |
|         |              |

19. Find the total number of rolls and the total number of 3s for all the students in your class. Then compute an experimental probability of rolling 3.

20. Now think about the *theoretical* probability of rolling 3.

    a. How many possible outcomes are there for a single roll? Are they all equally likely?

    b. On each roll of a die, what is the probability that you will roll 3?

    c. If you roll a die 12 times, how many 3s would you expect?

21. Compare your theoretical results from Exercise 20 with your own experimental results from Exercise 17 and to your class experimental results from Exercise 19. Which experimental result is closer to the theoretical result?

## Share & Summarize

1. At Jewel's birthday party, her mother assigned each of the ten partygoers, including Jewel and her two sisters, a number from 1 to 10.

   To see who would play *Pin the Tail on the Donkey* first, Jewel's father pulled one of ten balls, numbered 1 to 10, from a box. The person whose number was selected would play first.

   **a.** What is the probability that the number chosen was Jewel's or one of her sisters?

   **b.** What is the probability that the number chosen was *not* Jewel's or one of her sisters? That is, what is the probability that the number belonged to one of the seven guests? Explain how you found your answer.

2. Dalila has a spinner divided into five same-sized sections, numbered from 1 to 5. She spun the arrow 100 times, recording the result each time.

   **a.** How many times would you expect the arrow to land on section 4?

   **b.** If the actual number of fours Dalila recorded was different from the number you answered in Part a, does that mean your calculation was incorrect? Explain.

3. Suppose you perform an experiment 20 times and a friend performs the same experiment 200 times. You both use your results to calculate an experimental probability.

   Whose experimental probability would you expect to be closer to the theoretical probability? Explain.

# Investigation ③ The Spinning Top Game

## Materials

- 4-cornered top with sides marked DN, TA, TH, and P1.
- counters

The *Spinning Top* game is played with a four-cornered top that contains four symbols. There are many variations of this game.

In this investigation, you will look at the probabilities for one of the most common versions of the *Spinning Top* game.

### Play the Game

You will play this game in a group of four. Here are the rules.

- Each player begins with ten counters.
- Each player puts a counter in the center of a table.
- Players take turns spinning the top. One of these four symbols will land face up.

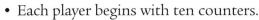

**DN   TA   TH   P1**

The letter facing up tells the player what to do.

**DN**   Do nothing.

**TA**   Take all the counters in the center.

**TH**   Take half of the counters in the center. Round down. For example, if the number of counters is five, take two.

**P1**   Pay one by placing counter in the center.

- Before each turn, each player puts another counter in the center. A player with no counters left is out of the game.
- The game continues until only one player has counters left or your teacher says time is up. The player with the most counters wins.

1. Play the game with your group, keeping a tally of the symbols the players spin.

| DN | TA | TH | P1 |
|----|----|----|----|
| ‖ |    | I  |    |

2. How many times did each symbol land face up in the whole game?

3. Draw a bar graph showing the number of times each symbol landed face up.

## Calculate Probabilities

Now that you have some experience playing the *Spinning Top* game, you can calculate the probabilities of certain outcomes.

4. Begin by calculating an experimental probability that each letter lands face up, based on the results of your game.

Assume each symbol is equally likely to land face up.

5. Calculate the theoretical probability of each symbol landing face up.

6. How do the theoretical probabilities you calculated in Question 5 compare to your experimental probabilities from Question 4?

7. What is the theoretical probability of winning counters in a turn? Explain.

8. What is the probability of losing a counter in a turn? Explain.

9. Jahmal said the first player has a better chance to win all the counters in the center than the other three players do. Do you agree with Jahmal? Explain your answer.

10. Suppose the first player wins all the counters on his or her turn. What is the probability that the second player will also win all the counters? Explain your answer.

## What Have You Learned?

11. If you are equally likely to get each of the four sides of the top, why are you not equally likely to win counters as to lose counters?

12. Change the rules of the *Spinning Top* game so that the probability of winning counters on each turn is $\frac{1}{4}$ and the probability of losing counters is $\frac{1}{2}$.

# Investigation  Geometric Probability

## Materials

- *Rice Drop* game board
- grain of uncooked rice
- inch ruler

To calculate theoretical probabilities for the situations with which you have worked so far, you divided the number of outcomes that mean an event occurred by the total number of outcomes. In this investigation, you will look at a situation where you need to use a different strategy.

### ✅ Develop & Understand: A

*Rice Drop* is a game of chance. To play, a larger version of the game board at right should be placed on a hard, flat surface such as a table or desk. The game board is the area inside the outer border, not the entire page.

Here are the game rules.

- On a player's turn, he or she holds a grain of rice about one foot above the game board, near the center.

- The player drops the grain of rice and watches where it lands. If it bounces outside the outer border, the drop does not count. The player must try again.

- If the grain of rice lands in the square, circle, triangle, or parallelogram, the player scores one point. A grain that lands on the edge of a figure should be counted as inside the figure if half or more of the grain is inside.

- Each player gets ten chances to score. Remember, if the grain of rice bounces off the board, the drop does not count. The player with the greatest score wins.

1. Play the game with your group. On your turn, record the results of your drops by making tally marks in a chart like this one.

| Circle | Square | Triangle | Nonsquare Parallelogram | No Figure |
|---|---|---|---|---|
|  |  |  |  |  |

2. Use the results of your ten drops to estimate the chances that the grain of rice will land in one of the following areas.

   a. the circle

   b. the square

   c. the triangle

   d. the nonsquare parallelogram

   e. no figure

   f. any figure

3. Now combine the results of your group. Calculate new estimates for the probabilities in Exercise 2.

4. Which set of probabilities do you think are more reliable, those from Exercise 2 or Exercise 3? Explain.

**Think & Discuss**

Do you think a grain of rice is as likely to land on the circle as it is to land outside any of the figures?

Can you think of a way you might calculate the theoretical probability of scoring a point on a single drop?

For the rest of this investigation, assume that the rice lands in a completely random spot on the game board. That is, assume the rice is just as likely to land in one spot on the game board as in another.

5. Suppose you use a game board divided into four equal rectangles, like this one. What is the probability that the rice lands on the shaded rectangle?

**6.** The game board below is also divided into four equal sections. What is the probability that the rice lands on the shaded rectangle?

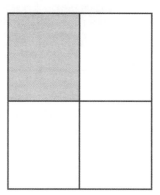

**7.** To create this game board, a square was removed from the shaded rectangle of the game board in Exercise 6. Is the probability that the rice lands in the shaded figure *less than, greater than,* or *the same as* your answer to Exercise 6? Explain your reasoning.

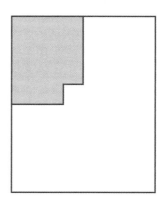

**8.** Now the square that was removed from the rectangle in Exercise 6 has been returned but in a different place. Is the probability that the rice lands in a shaded figure on this board *less than, greater than,* or *the same as* your answer to Exercise 6? Explain your reasoning.

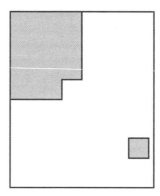

**Real-World Link**

It takes approximately 300 to 600 gallons of water to produce one pound of rice.

### ✅ *Develop & Understand: B*

Rebeca thought she could find the theoretical probabilities for the original *Rice Drop* game board using the areas of the figures and the area of the board.

9. Find the area of the game board and the area of the square.

10. Use your answer to Exercise 9 to find the probability that the rice lands in the square. Express your answer as a percent.

11. Find the probability that the rice lands in each of the remaining figures. Express your answers as percents.

12. What is the probability that a player will score on a single drop? Explain how you found your answer.

13. **Challenge** The theoretical probabilities that you found in Exercise 9–12 assume that the rice lands randomly. Compare the theoretical probabilities that you found in Exercises 9–12 to the experimental probabilities that you found in Exercises 1–4.

   a. Do you think the rice lands in a completely random spot, or are some spots more likely than others? Explain why you think so.

   b. What could you do to test whether your answer to Part a is correct?

## Share & Summarize

1. Suppose this was the game board for a game of *Rice Drop*. The measurements are given in inches.

   Assuming that the rice lands in a completely random place, what is the probability of scoring a point on a single drop?

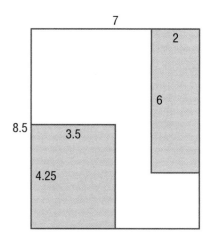

2. Martino and Larisa were playing *Rice Drop* using a checkerboard. They decided that a drop scored a point if the rice landed on a purple square. A drop that does not land on the board does not count.

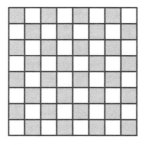

   Assuming that the rice lands in a completely random place, what is the probability that a drop scores a point? Explain how you found your answer.

3. Leon likes playing darts. He throws a dart at the dartboard shown here.

   The points earned for each ring are shown.

   a. Assuming Leon's dart hits the dartboard in a random place, what is the probability that Leon scores at least 3 points? The radius of the inner circle is equal to the width of each ring.

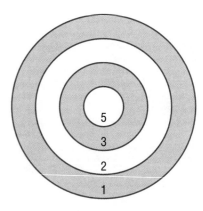

   b. Do you think the assumption in Part a is reasonable? Explain.

**Practice & Apply**

1. Copy this number line. In Parts a–d, add a label to your number line indicating how likely you think the event is.

0%                 50%                 100%

   **a.** I will listen to the radio tonight.

   **b.** I will go to a movie sometime this week.

   **c.** Everyone in my Math class will get a perfect score on the next test.

   **d.** I will wake up before 7:00 A.M. tomorrow.

2. Estimate the probability of each event. Explain your reasoning.

   **a.** The school lunch will taste good tomorrow.

   **b.** Everyone in our class will come to school next Monday.

   **c.** A giraffe will come to school next Monday.

3. LaBron is practicing his archery skills. He hit the bulls-eye with 3 of the first 12 arrows he shot. Use these results to find an experimental probability that LaBron will hit the bulls-eye on his next shot.

4. Get a plastic spoon and conduct this experiment.

   For each trial, drop the spoon and record how it lands, right side up so it would hold water or upside down. Conduct 30 trials for your experiment. Use your results to find an experimental probability that the spoon will land right side up.

   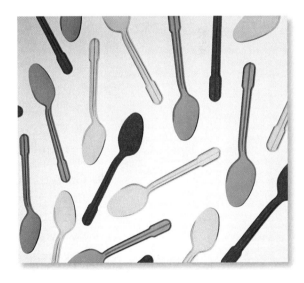

5. A word is chosen at random from *book, paper, pencil,* and *eraser.*

   **a.** What is the probability that the word has only one syllable?

   **b.** What is the probability that the word begins with *P*?

   **c.** What is the probability that the word ends with *P*?

**6.** Lupe tossed a coin ten times and got six heads. Ruby tossed a coin 1,000 times and got 530 heads.

   **a.** Based on Lupe's results, what is an experimental probability of getting heads? Express your answer as a percent.

   **b.** Using theoretical probabilities, how many heads would you expect to get in ten coin tosses? How far was Lupe's result from that number?

   **c.** Based on Ruby's results, what is an experimental probability of getting heads? Express your answer as a percent.

   **d.** Using theoretical probabilities, how many heads would you expect to get in 1,000 coin tosses? How far was Ruby's result from that number?

   **e. Challenge** The difference between the actual number of heads and the expected number of heads is much greater for Ruby than for Lupe. How is it possible that Ruby's experimental probability is closer to the theoretical probability?

**7.** Claudia has a spinner divided into ten equal sections, numbered from 1 to 10. Think about a single spin of the arrow.

   **a.** What is the probability that the arrow will point to section 1?

   **b.** What is the probability that the arrow will point to an odd number?

   **c.** What is the probability that the arrow will point to an even number?

   **d.** What is the probability that the arrow will point to a prime number?

**8.** Garrett and Neela were playing *Rice Drop* using this rectangular game board. The dimensions are in inches. Assume the rice will land on a completely random spot.

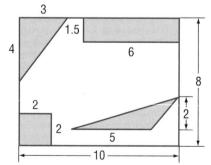

   **a.** Find the probability that the rice lands in a triangle.

   **b.** Find the probability that the rice lands in a shaded quadrilateral.

   **c.** Find the probability that the rice lands in a shaded figure.

   **d.** Find the probability that the rice does not land in a shaded figure.

9. A small area created by buildings built close together is often called a *courtyard*. Suppose a group of six friends are standing in a square courtyard like the one shown here when it starts to rain.

   What is the probability that the first raindrop hits one of the friends? Assume each person occupies a circle about 50 cm in diameter.

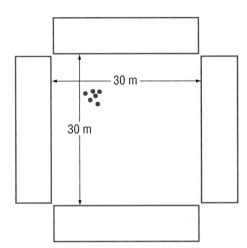

30 m

30 m

10. **Earth Science** Irena and a group of her friends planned a beach outing for a certain day. The local weather service reported a 20% chance of rain on that day. When the day came, it rained. The trip was canceled.

    Irena said the weather service had been wrong when they gave the 20% rain prediction. They said it was not going to rain, but it did.

    a. Do you agree with Irena? Why or why not?

    b. If the weather service prediction did not mean that it would not rain, what do you think it meant?

**Connect & Extend**

11. Elan's radio alarm clock goes off at 6:37 every morning. He complained that, almost every morning, he wakes up to commercials rather than music. Describe an experiment he could conduct to estimate the probability that he will wake up to a commercial. Explain how you would use the result to find an experimental probability.

12. Describe a situation for which the probability that something will occur is $\frac{1}{6}$.

13. A whole number is chosen at random from the numbers 1 to 10.

    a. What is the probability that the number is odd?

    b. What is the probability that the number is prime?

    c. What is the probability that the number is a perfect square?

    d. What is the probability that the number is a factor of 36?

14. A whole number is chosen at random from the numbers 10 to 20.

    a. What is the probability that the number is odd?

    b. What is the probability that the number is prime?

    c. What is the probability that the number is a perfect square?

    d. What is the probability that the number is a factor of 36?

**15.** Two sixth graders and two seventh graders are having a checkers tournament. They decide to randomly choose who the first two players will be and who will use which color. Consider the possible arrangements. For example, the first game might be a seventh grader playing black and a sixth grader playing red.

    **a.** What other possible arrangements are there?

    Assume each of the arrangements is equally likely to occur.

    **b.** What is the probability that the seventh graders will play each other in the first game?

    **c.** What is the probability that at least one seventh grader will play in the first game?

    **d.** What is the probability that exactly one sixth grader will play in the first game?

**16. In Your Own Words** Give an example to illustrate the difference between an experimental probability and a theoretical probability.

**Real-World Link**

The white oak tree is the state tree of Illinois. White oaks can live up to 400 years.

**17.** While visiting a friend, Belinda parked her car under a large oak tree. She left open her moon roof. (A *moon roof* is a small window in the roof.)

The tree drops acorns in a circular area around its trunk, as shown in the diagram. Assume the acorns fall randomly within the circle, which has a radius of 20 feet. No acorns fall where the trunk is. The trunk has a radius of 5 feet.

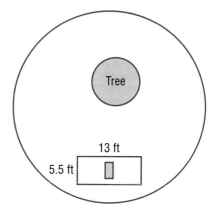

    **a.** Find the area of the region in which an acorn might fall.

    **b.** While Belinda was at her friend's house, an acorn fell from the tree. Find the probability that the acorn hit Belinda's car, including the moon roof.

    **c.** The dimensions of the moon roof are 32 inches by 16 inches. Find the probability that the acorn fell through the moon roof.

    **d. Challenge** Suppose the acorn hit Belinda's car. Find the probability that it fell through the moon roof. (Hint: Should this probability be *more than* or *the same as* the probability you answered in Part c?)

**18.** Owen videotaped all six episodes of his favorite hour-long television show. His friend missed an episode, and Owen said he would loan her the tape. The number line illustrates where on the tape the show had been recorded.

**Hours**

Owen often rewinds the tape and watches different parts of previous episodes. He does not remember where on the tape he last stopped watching the show. Before giving the tape to his friend, Owen put it in his VCR and pushed the play button. What is the probability that the tape started somewhere within the show his friend wanted to see?

*Mixed Review*

**19.** Albert conducted a survey of his class to determine what animals students liked. Write each fraction as a decimal and a percent. Write each given percent as a decimal and as a fraction.

   **a.** He found that 25% liked cats.

   **b.** He found that $\frac{4}{5}$ liked dogs.

   **c.** He found that only $\frac{1}{20}$ liked lizards.

**20.** Which is greater, 0.3 or $\frac{1}{3}$?

**21.** Which is greater, 0.11 or $\frac{1}{9}$?

**22.** List in order from least to greatest: $\frac{3}{8}$, 25%, $\frac{1}{3}$.

**Write each fraction or decimal as a percent. Round to the nearest tenth of a percent.**

**23.** $\frac{3}{5}$       **24.** 0.02       **25.** $\frac{1}{9}$

**26.** $\frac{1}{2}$       **27.** 0.125       **28.** $\frac{1}{3}$

**29.** 2.6       **30.** $\frac{5}{4}$       **31.** 0.002

**Find the unit rate for each non-unit rate given.**

**32.** 600 pages in six hours

**33.** 200 miles in five hours

**34.** 140 beats in two minutes

**35.** 12 pizzas for 24 campers

# Make Matches

In the probability games you have considered so far, the result of one round or trial does not affect the result of another. For example, if you toss a coin and get heads, your chances of getting heads when you flip the coin again are still 50%. These are called *independent events*.

In this lesson, you will work with situations called *dependent events*. That is, what happens in one case *does* affect what can happen in the next.

## Think & Discuss

Suppose you roll a die several times.

• What is the probability of getting 6 on the first roll? What is the probability of getting 4?

• Suppose the first time you roll the die, you get 6. You roll a second time. What is the probability of getting 6 on the second roll? What is the probability of getting 4?

Now imagine that you and some friends are making a poster for a school party. Your teacher gives you six markers in six different colors. Each of you chooses one without looking. Two of the colors are red and green.

• If you choose first, what is the probability that you will get the red marker? What is the probability that you will get the green marker?

• Suppose you chose first and got the red marker. What is the probability that the second person will get the red marker? What is the probability that the second person will get the green marker?

• Why are the probabilities for the dice situation different from those for the marker situation?

# Investigation 1 Match Colors

## Vocabulary

simulation

## Materials

- counters or blocks (2 in each of 2 colors) or slips of paper
- bucket or bag

A **simulation** is an experiment in which you use different items to represent the items in a real situation. For example, to simulate choosing markers and looking at their colors, you can write the color of each marker on a slip of paper and put all the slips into a bag. You can simulate choosing markers by drawing slips from the bag. Mathematically, the situations are identical.

Using a simulation can help with some of the exercises in this investigation.

## ✓ Develop & Understand: A

Ken woke up early and found that a storm had knocked out the power in his neighborhood. He has to dress in the dark. Ken has four socks in his drawer, two black and two brown. Color is the only difference between them. As long as both socks are the same color, Ken does not care which he wears. He takes two socks out of the drawer.

1. Simulate this situation, using counters, blocks, or slips of paper with colors written on them. If you use counters or blocks, you may have to let other colors stand for the sock colors. For example, a red block might represent a brown sock, and a blue block might represent a black sock. Use a bucket or bag to represent Ken's sock drawer.

   a. Without looking, pick two "socks," one at a time, from the "drawer." Record whether the socks match. Then put back the socks, mix them up, and try again. Repeat this process 16 times. Record the results.

   b. Use your results to find an experimental probability that Ken will choose matching socks.

2. If the first sock Ken picks is brown, what is the theoretical probability that the second sock will also be brown? Explain.

3. If the first sock is black, what is the theoretical probability that the second sock will also be black?

4. Ken says that since he has two colors of socks, he has a 50% chance of getting a matching pair. Do you think he is correct? If not, what do you think the actual probability is? Explain.

In Lesson 10.3, you saw how you could use a table to keep track of the possible outcomes for two coin tosses. You can also draw a *tree diagram* to show all the possibilities. The possible results for the first coin can be shown like this.

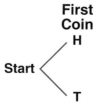

**First Coin**

Start
H
T

The possibilities for the second coin can be shown as branches from each of the first two branches.

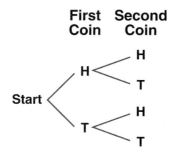

**First Coin   Second Coin**

Start
H — H
H — T
T — H
T — T

You can read off the possible outcomes by following the branches, beginning with Start. For example, following the top set of branches gives the outcome HH.

**First   Second**
**Coin    Coin    Outcome**

Start
H — H    HH
H — T    HT
T — H    TH
T — T    TT

### ✅ *Develop & Understand: B*

5. Suppose you choose one of six markers from a bag in red, orange, yellow, green, blue, or purple. Draw a tree diagram showing the possible colors for the marker.

6. Now consider what happens when you choose a second marker.

   a. Suppose the first marker you chose was red. What are the possible choices for the second marker?

   b. Add a new set of branches to the "red" branch of your tree diagram for Exercise 5, showing the possibilities for your second choice.

   c. Complete your tree diagram by adding branches that show the possibilities when each of the other colors are picked first.

   d. What is the probability that, if you choose the two colors at random, you will get red and green, chosen in either order?

7. Draw a tree diagram to show the possible choices of sock colors for Ken if he chooses two socks from a drawer containing two brown and two black socks. Since there are two socks of each color, label the socks as brown 1, brown 2, black 1, and black 2.

8. How many possible sock pairs are there? How many of them have matching colors?

9. What is the probability that Ken will choose a matching pair?

## ✓ Develop & Understand: C

After choosing socks, Ken has to choose pants and a shirt. His school requires uniforms of blue, tan, or green. He has two pairs of pants, one blue and one tan. He has two shirts, also one blue and one tan. Now he takes one shirt and one pair of pants.

10. Suppose the shirt is blue. What is the probability that the pants will also be blue? Explain how you found your answer.

11. Draw a tree diagram showing the possible choices of shirts and pants.

12. Ken says the probability that he will choose matching shirt and pants is 50%. Is he right? How do you know?

13. Suppose Ken has a third shirt and a third pair of pants, both green. Now what is the probability that he will choose a shirt and pants of the same color? Explain.

14. Suppose Ken has one tan shirt and two blue shirts and one tan pair of pants and two blue pairs of pants. Find the probability that the shirt and pants will match. Show how you found your answer.

# Investigation 2 Match Cards

In an ordinary deck of playing cards, there are four suits.

**Clubs**  **Diamonds**  **Spades**  **Hearts**

There are 13 cards in each suit, one to represent each of the numbers from 1 to 10, and three *face cards*, consisting of jack, queen, and king. Clubs and spades are black, while diamonds and hearts are red.

Many kinds of games, involving various combinations of chance and skill, are played with decks of cards. In this investigation, you will work with some simple games of chance that involve choosing cards from a deck.

**Real-World Link**

The cards described here are English playing cards. In other countries, different suits are used. Traditional German playing cards use the suits hearts, leaves, bells, and acorns. Spanish playing cards use the suits coins, cups, swords, and clubs.

## ✓ Develop & Understand: A

In the first game, a deck of cards is shuffled and placed on a table face down. For one round of the game, players do the following.

- Player 1 chooses a card from the deck without looking and writes down its suit of spades, hearts, diamonds, or clubs.
- Player 1 puts the card back and shuffles the deck.
- Player 2 chooses a card without looking. If it has the same suit as the first card, Player 1 scores a point. Otherwise, Player 2 scores a point.
- Player 2 returns the card and shuffles the deck.

The winner is the player with more points at the end of 20 rounds.

1. What is the probability that Player 1 will choose a heart?

2. If Player 1 chooses a heart, what is the probability that Player 2 will also choose a heart? Explain.

3. What is the probability that Player 2 chooses a card of the same suit as Player 1's card, no matter what that suit was? How do you know?

4. What would you expect the score to be after 20 rounds?

5. Think of a way to change the scoring rules to give both players the same chance of winning.

## ✅ Develop & Understand: B

The second card game is similar to the first. The only difference is that Player 1 does not put the card back before Player 2 chooses. After both players have chosen, the cards are returned to the deck. Player 1 scores a point if the two cards have the same suit. Player 2 scores a point if they have different suits.

6. What is the probability that Player 1 will choose a heart?

7. Suppose Player 1 chooses a heart.

   a. What is the probability that Player 2 also chooses a heart?

   b. Is your answer to Part a different from your answer to Exercise 2? Why or why not?

8. What is the probability that Player 2 chooses a card of the same suit as Player 1, no matter what suit Player 1 chooses?

9. Is this game *more fair, less fair,* or *just as fair* as the game in Exercises 1–5? Explain.

## ✅ Develop & Understand: C

Suppose you want to draw a tree diagram to show the possible choices for the first card game in which Player 1 replaces the card before Player 2 chooses.

10. How many branches would you need to show the possibilities for the first card?

11. How many branches would you have to add to show the possibilities for the second card?

12. How many total branches would your tree diagram have?

Since the game concerns only the suits of the cards, and since the four suits are equally likely for each draw, you can draw a simplified tree diagram showing the four possible suits for each draw.

For example, suppose the first card chosen is a heart. Here is the part of the tree diagram showing the possible suits for the second card.

13. Draw a tree diagram showing all the possible suit combinations for the first game.

14. Hearts and diamonds are red while clubs and spades are black. What is the probability that the two cards have the same color?

First Card    Second Card

Heart — heart, diamond, spade, club

15. Can you use a simplified tree diagram for the second game, in which Player 1 keeps the card instead of returning it to the deck before Player 2 chooses? Explain.

## Share & Summarize

In some probability situations, one event can affect the probability of another.

1. For each pair of events, decide whether the first event affects the probability of the second. If your answer is "yes," explain why.

| First Event | Second Event |
|---|---|
| **a.** getting heads on the flip of a coin | getting heads on the second flip of the coin |
| **b.** getting a king when choosing a card from a deck | getting a king when choosing a second card without returning the first |
| **c.** drawing a certain name from names written on slips of paper and chosen at random from a hat | drawing a second name if the first slip is returned to the hat before the second choice |

2. Create your own sequence of two events for which the first event affects the probability of the second.

3. Now create your own sequence of two events for which the first event *does not* affect the probability of the second.

---

## Investigation ③ The Fundamental Counting Principle

### Vocabulary

Fundamental
   Counting Principle

In the previous lessons, you used tree diagrams to show the possible outcomes in probability situations. In this investigation, you will analyze patterns in the tree diagrams to find a more efficient way to determine the number of possible outcomes for a situation.

### Think & Discuss

Consider the tree diagram you drew for flipping coins on page 640.

How many different outcomes were possible for the first coin?

| First Coin | Second Coin | Outcome |
|---|---|---|
| H | H | HH |
| H | T | HT |
| T | H | TH |
| T | T | TT |

How many outcomes were possible for the second coin?

How many total outcomes, were possible for the situation of flipping two coins, considering HT and TH as different outcomes?

# ✓ Develop & Understand: A

1. Copy the table below, which includes the data from Think & Discuss. Include four additional rows to record information from Exercises 2–4.

| Type of Event | Number of Possibilities for First Event | Number of Possibilities for Second Event | Total Possible Outcomes |
|---|---|---|---|
| Flipping two coins | 2 | 2 | 4 |
| | | | |
| | | | |
| | | | |
| | | | |

2. Look at the tree diagram you drew for choosing among six markers in Exercises 5–9 on pages 640 and 641.

   a. How many choices were there for choosing the first marker?

   b. How many choices were there for choosing the second marker?

   c. How many ways were there to choose the markers with red, green and green, red as different outcomes? In other words, how many branches did the tree diagram have in the last column? Record this information in the table from Exercise 1.

3. Look at the tree diagram you drew for choosing pants and shirts for Ken to wear in Exercises 10–12 on page 641. Fill in the table for the number of options for his shirts, the number of options for pants, and the number of total outfits. Then add a row to the table for the situation in Exercise 13 where he had three shirts and three pants.

4. Look at the tree diagram you drew for the first card game in Exercises 1–5 on page 642. Fill in the number of options for the suit of the first card, the number of options for the suit of the second card, and the number of total ways to draw the suits for two cards.

5. What patterns do you see in the table? Is there any relationship between the possibilities for the first and second events and the total possible outcomes?

The **Fundamental Counting Principle** states that if there are $M$ possibilities for one event, and $N$ possibilities for another event, then there are $M \cdot N$ ways for both events to occur. For example, the total outcomes for flipping two coins is 4 because two possibilities for the first coin times two possibilities for the second coin equals four total possible outcomes.

Besides the tree diagram, you can also organize probability outcomes using a chart. This may make it easier to see why the Fundamental Counting Principle is true.

The information from the tree diagram for flipping two coins would look like this.

|   | H | T |
|---|---|---|
| H | HH | HT |
| T | TH | TT |

However, for a situation like the markers where you found that the events were *dependent*, the chart would be different.

|  | RED | ORN | PUR | BLU | GRE | YEL |
|---|---|---|---|---|---|---|
| RED |  | ORN RED | PUR RED | BLU RED | GRE RED | YEL RED |
| ORN | RED ORN |  | PUR ORN | BLU ORN | GRE ORN | YEL ORN |
| PUR | RED PUR | ORN PUR |  | BLU PUR | GRE PUR | YEL PUR |
| BLU | RED BLU | ORN BLU | PUR BLU |  | GRE BLU | YEL BLU |
| GRE | RED GRE | ORN GRE | PUR GRE | BLU GRE |  | YEL GRE |
| YEL | RED YEL | ORN YEL | PUR YEL | BLU YEL | GRE YEL |  |

## Think & Discuss

Why are some of the squares in the above chart blank? How many squares are blank? How does this chart help explain why the Fundamental Counting Principle works for dependent events?

## ✓ Develop & Understand: B

6. Alyssa and Tucker are testing the Fundamental Counting Principle with a new activity. They have placed five blocks in a paper bag. There are two red, one yellow, one green, and one blue blocks.

   a. Create a chart to show the possibilities if one block is drawn from the bag and replaced, and then another block is drawn.

   b. How many possible outcomes are there for drawing two blocks? How does this relate to the Fundamental Counting Principle?

   c. Recreate the chart to show the possibilities for the blocks if the first block is not returned to the bag.

   d. How many possible outcomes are there for drawing two blocks when the first block is not returned? How does this relate to the Fundamental Counting Principle? How is this shown in the chart?

**7.** In 1982, New York changed its license plate format because it ran out of distinct number/letter combinations.

**a.** The format *before* the change was *number-number-number-letter-letter-letter* where repeated numbers were allowed, repeated letters were allowed, and all digits and letters were allowed. Use the Fundamental Counting Principle to find out how many different license plates were possible with this format. Show your work.

**b.** The format *after* the change was *number-number-number-number-letter-letter-letter*. How many different license plates were possible with this format? Show your work.

**c.** Would a chart or tree diagram be helpful in this situation? Explain your thinking.

**d.** How many different license plates are possible if zero was not used?

## Share & Summarize

The cafeteria serves choices for the main dish, side item, fruit item, and dessert. Each student gets one item from each section of the menu. The menu for today is as follows.

| Main Dish | Side Item | Fruit | Dessert |
|---|---|---|---|
| Enchilada | Salad | Apple | Brownie |
| Hamburger | Carrots | Peaches | Ice Cream |
| Grilled Chicken | | | Popsicle |

**1.** Which of the methods you learned could you use to find the number of possible meals? Explain.

**2.** How many meals are possible?

**3.** How do the total possible outcomes change if there are two drink options?

**Practice & Apply**

1. Suppose there are two cards, numbered 1 and 2. The cards are mixed and placed face down.

   **a.** You arrange the cards in a row and then turn them over. What is the probability that the first card will be a 2?

   **b.** What is the probability that the cards will form the number 21? Explain.

2. Suppose you have three cards, numbered 1, 2, and 3. The cards are shuffled and placed face down in a row.

   **a.** List all the three-digit numbers that can be created from these three cards. Does the counting principle verify the number of outcomes you found? Explain.

   **b.** What is the probability that the cards will form the number 213 when they are turned over?

   **c.** What is the probability that the cards will form a number between 200 and 300?

   **d.** What is the probability that the cards will form an even number?

   **e.** What is the probability that the cards will form a number less than 300?

3. Manuel and Hally are splitting a box of marbles. The box contains three red, two green, and one orange marble. Each friend chooses a marble at random.

   **a.** If Hally chooses first and gets a red marble, what is the probability that Manuel's marble will also be red?

   **b.** Draw a tree diagram showing all the possible combinations when each friend chooses one marble. Label the red marbles R1, R2, and R3. Label the green marbles G1 and G2. Label the orange marble O. Does the counting principle verify the number of outcomes you found? Explain.

   **c.** What is the probability that the two marbles will be the same color?

**4.** Gabriel and Dana are playing a game with an ordinary deck of playing cards. For each turn, Dana chooses a card and returns it to the deck. She shuffles the deck. Then Gabriel chooses a card.

**a.** If Dana picks the five of clubs, what is the probability that Gabriel will pick the five of clubs?

**b.** If Dana picks a black card in either a spade or a club, what is the probability that Gabriel will pick a red card in either a heart or a diamond? How do you know?

**c.** If Dana picks the six of spades, what is the probability that Gabriel will pick a king?

**d.** If Dana picks a red queen, what is the probability that Gabriel will pick a red queen?

**5.** Gabriel and Dana are playing a game with an ordinary deck of playing cards. For each turn, the deck is shuffled, and the cards are spread out face down. At the same time, Dana and Gabriel each choose a card.

**a.** If Dana picks the five of clubs, what is the probability that Gabriel also picks the five of clubs?

**b.** If Dana picks a black card in either a spade or a club, what is the probability that Gabriel picks a red card in either a heart or a diamond? How do you know?

**c.** If Dana picks the six of spades, what is the probability that Gabriel picks a king?

**d.** If Dana picks a red queen, what is the probability that Gabriel picks a red queen?

**For each situation below, use the Fundamental Counting Principle to determine the number of possible outcomes. Verify each using either a chart or a tree diagram.**

**6.** roll one die and flip one coin

**7.** playing *Rock, Paper, Scissors*

**8.** flipping three coins

**9.** pair a digit 1 or 2 with a letter A–Z

**10.** form a two-digit number: select a digit 1 to 4, select a digit from those remaining

**For each situation below, use the Fundamental Counting Principle to determine the number of possible outcomes.**

**11.** selecting a card from a stack of face cards with aces and a card from a stack of number cards (ordinary card deck)

**12.** making outfits with 6 shirts, 3 pants, 2 jackets and 2 shoes

**13.** selecting two students out of a class with 25 students

**14.** Kristen has created a game using these six cards.

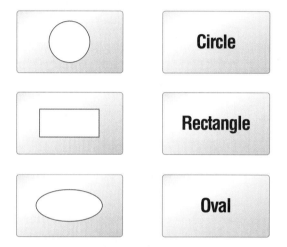

The three shape cards are placed face up. The three word cards are shuffled. One word card is placed face down next to each shape card. A player scores one point for each word card that matches a shape card.

**a.** Write all the possible arrangements of the three word cards. Use C to stand for the circle card, R to stand for the rectangle card, and O to stand for the oval card. How many possibilities are there?

**b.** What is the probability that a player will match all three word cards correctly?

**c.** What is the probability that a player will match at least one word card correctly?

**15.** Cheyenne has written letters to four friends. They are Caroline, Raul, Jing, and Ernest. She has four envelopes, each with the name and address of one of the friends. Cheyenne's little brother wants to help, so he puts one letter in each envelope. Since he cannot read yet, he puts the letters in the envelopes at random.

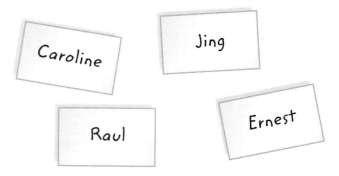

**a.** How many ways can the letters and envelopes be paired?

**b.** What is the probability that everyone receives the correct letter?

**16. Life Science** *Genes* determine many things about a person, including how he or she looks. For example, a person has two genes that determine eye color. The gene for blue eyes is *recessive*. The gene for brown eyes is *dominant*. This means that if a person has one blue-eye gene and one brown-eye gene, he or she has brown eyes. A person with two brown-eye genes also has brown eyes. To have blue eyes, both genes must be blue.

A child gets one eye-color gene from each parent. Assume the chances of passing either gene to a child are equal. For example, a father with one blue-eye gene and one brown-eye gene has a 50% chance of passing the blue-eye gene to his child.

**a.** Suppose two people are having a child. One has a blue-eye gene and a brown-eye gene. The other has two brown-eye genes. What is the probability that the child will have blue eyes? (Hint: You can find the possible gene combinations for the child by making a table or a tree diagram.)

**b.** Suppose the two parents both have one blue-eye gene and one brown-eye gene. What is the probability that the child will have blue eyes?

**c.** Now suppose one of the parents has two blue-eye genes and the other has one blue-eye gene and one brown-eye gene. What is the probability that the child will have blue eyes?

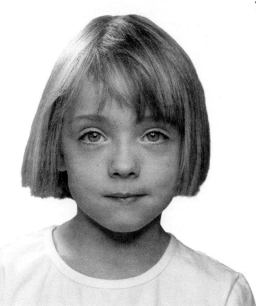

*Real-World Link*

Eye color is a genetic trait and is determined by the amount and type of pigment in the eye's iris.

**Connect & Extend**

**17.** Tavio is going to play a game where he needs to roll two dice. He wants to examine how many different ways the dice can land before he plays the game. He selects one red die and one green die so he can tell them apart as he collects his information.

   **a.** Create a chart to show what the possibilities are as he rolls the two dice.

   **b.** How many possible outcomes are there for rolling two dice? Is this verified by the Fundamental Counting Principle?

   **c.** What is the probability that Tavio will roll a 1 on at least one of the die?

   **d.** In the chart, you found that one possible outcome is rolling a 1 on the red die and a 6 on the green die, or (1, 6). Another outcome is rolling a 6 on the red die and a 1 on the green die, or (6, 1). If your goal is to roll a sum of 7, then both of these outcomes are *favorable* outcomes. List all of the outcomes that result in a sum of 7 on the two dice.

   **e.** What is the probability of rolling a sum of 7 when rolling two dice?

**Mixed Review**

**Find the area of each square, rectangle, parallelogram, and triangle.**

**18.**
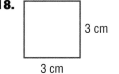
3 cm
3 cm

**19.**
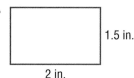
1.5 in.
2 in.

**20.**

1.25 cm    1.6 cm
4 cm

**21.**

5 cm
3 cm

**Find the area of a square with the given side length.**

**22.** 2 in.

**23.** $\frac{1}{2}$ cm

**24.** 12 in.

**25.** 10 in.

**26.** What is the area of a circle with a radius of 3.6 in.? Give your answer to the nearest hundredth square inch.

**Find the volume of each rectangular prism given its dimensions.**

**27.** $l = 2$ in.; $w = 5$ in.; $h = 6$ in.

**28.** $l = 2.5$ cm; $w = 8$ cm; $h = 3$ cm

**Which customary unit is most reasonable to measure the capacity of the following containers?**

**29.** a pond

**30.** a coffee cup

**31.** a child's bucket

**32.** a baby pool

**Find the opposite of each number.**

**33.** 2.4

**34.** −5

**35.** −5.7

**Find the absolute value of each expression.**

**36.** $|-3|$

**37.** $|2.1|$

**38.** $|-2.6|$

**Evaluate each expression.**

**39.** $|30 - 15|$

**40.** $|20 - 30|$

**41.** $|-16| - |16|$

**Give the coordinates of each point plotted on this grid.**

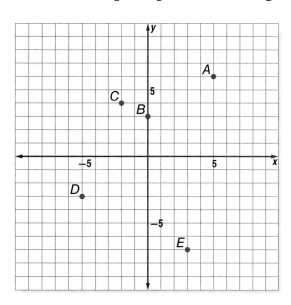

# Review & Self-Assessment

## Chapter Summary

Probability is useful in many areas of life, from playing games to making plans based on weather predictions. In this chapter, you learned how to find *experimental* and *theoretical probabilities* for events in which the possible outcomes are *equally likely*.

You examined probabilities in several types of situations, including some in which the possible outcomes were not easy to determine. For certain games of chance, you came up with strategies for play based on your knowledge of probabilities. You also used *simulation* and tree diagrams to examine situations in which the number of possible outcomes were affected by what had previously happened.

## Strategies and Applications

The questions in this section will help you review and apply the important ideas and strategies developed in this chapter.

### Understanding probability

Aretha took the king of hearts and the king of clubs from a standard deck of cards, leaving only 50 cards. She told Leah that the probability of selecting a queen was now 8%. But she did not tell her how many or what cards she had removed.

**1.** What does it mean that the probability was 8%?

**2.** Leah selected a card from Aretha's deck, looked at it, and then put it back. Aretha shuffled the cards. They repeated this process until Leah had chosen a card 100 times.

**a.** How many times would you expect Leah to have picked a queen?

**b.** Leah chose a queen 7 times. She said that this means Aretha was wrong and that the actual probability is 7%. Aretha and Leah both calculated the probabilities they gave. Is either incorrect in her calculation? Explain.

**c.** Whose is the more accurate probability, Leah's or Aretha's?

**d.** Leah kept selecting cards until she had 1,000 trials. She chose a queen 88 times. Aretha said, "The difference between the 88 queens you selected and the 80 you should have expected was 8. But the difference was only 1 when you drew 100 cards. Your experimental probability will be less accurate for the 1,000 draws than it was for the 100 draws."

Is Aretha correct? Explain.

## Vocabulary

distribution

equally likely

experimental probability

Fundamental Counting Principle

histogram

probability

simulation

theoretical probability

Venn diagram

### Identifying outcomes

**3.** Name two strategies for identifying the outcomes of a probability situation. Illustrate each strategy by using them to find the number of outcomes for turning this spinner twice.

### Finding probabilities of events

**4.** Josh said, "Suppose you roll a standard die. To calculate the probability of getting a prime number, you have to divide 3 by 6, giving 0.5."

**a.** Why did Josh choose 6 for the divisor?

**b.** Why did Josh choose 3 for the dividend?

**c.** Consider this *Rice Drop* game board. Explain why the procedure for calculating the probability that the rice lands in a shaded square is the same as the one Josh used for getting a prime number on a die roll.

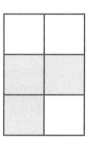

### Using probabilities to analyze games

**5.** A bag contains five slips of paper numbered 1 to 5. In the game *Find the Difference,* each player chooses one of the cards below. Players take turns drawing two numbers from the bag. If the difference of the numbers is on the player's card, the player covers that difference. The numbers are returned to the bag after each turn. The first player to cover all his numbers wins.

| Card A | | Card B | | Card C | |
|---|---|---|---|---|---|
| 1 | 2 | 1 | 2 | 3 | 4 |
| 3 | 4 | 2 | 1 | 4 | 3 |

Which card gives a player the best chance of winning? Explain.

### Working with situations in which the probabilities depend on previous results

**6.** King and Kenna were playing a board game in which they rolled two dice. Rolling the same number on both dice, lets you take an extra turn. Kenna rolled two 3s and then two 5s. As she was getting ready to take another extra turn, King said, "The chances of you getting doubles again are next to nothing!" Is King correct? Explain your answer.

**7.** Describe a probability experiment in which the result of one trial changes the probabilities for the next trial's result. You might want to use dice, cards, spinners, or slips of paper drawn from a bag in your experiment.

## Demonstrating Skills

**8.** At a fundraising carnival, Mario operated a game in which each player spun a wheel. The section on which the wheel stopped would indicate what prize, if any, the player won.

The table shows how many equal-sized spaces listed each type of prize as well as how many people won each prize by the end of the day.

| | Key Chain | Troll Doll | Baseball Cap | Stuffed Animal | Beach Ball | No Prize |
|---|---|---|---|---|---|---|
| **Number of Spaces** | 5 | 4 | 3 | 2 | 1 | 45 |
| **Number of Winners** | 14 | 16 | 13 | 6 | 3 | 148 |

**a.** Find an experimental probability of winning each prize.

**b.** Find the theoretical probability of winning each prize.

Use the information below for Questions 9 and 10.

At the beginning of a computer game called *Geometry Bug,* players take turns choosing circles, squares, and triangles on the screen. After all the shapes have been chosen, a small "bug" appears and flies over the shapes. The bug lands on a random place on the screen. If it lands on one of the players' shapes, that player scores a point. The winner is the player with the most points after 50 landings.

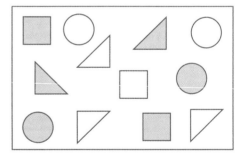

Rosa and Cari were playing with this screen. Rosa's shapes are purple, and Cari's shapes are white. The screen is eight inches wide and five inches high. The circles have a radius of $\frac{1}{2}$ inch. The squares have a side length of one inch. The triangles are right triangles with legs one inch long.

**9.** Consider probabilities for a single bug landing. Write your answers as decimals to the nearest thousandth.

**a.** What is the probability that the bug lands in Cari's square?

**b.** What is the probability that the bug lands in one of Rosa's circles?

**c.** Find each player's probability of scoring on a single bug landing.

**10.** The girls decide to play with an optional rule. When the bug lands on a shape, the shape is removed from the board. For example, if the bug lands on Cari's square, Cari scores a point but the square disappears.

**a.** Find the probability that the bug lands on one of Cari's triangles.

**b.** Suppose the first shape the bug landed on was Cari's triangle in the bottom right. Now what is the probability that the bug lands on any of Cari's shapes?

## Test-Taking Practice

*SHORT RESPONSE*

**1** Amber rolled a number cube 20 times. Her results are recorded in the table below. What was the experimental probability of Amber rolling a 6? What is the theoretical probability of rolling a 6?

| Number on Number Cube | Number of Times Rolled |
|:---:|:---:|
| 1 | 3 |
| 2 | 5 |
| 3 | 1 |
| 4 | 6 |
| 5 | 1 |
| 6 | 4 |

*Show your work.*

*Answer* _____

*MULTIPLE CHOICE*

**2** A bag holds 4 blue marbles, 5 yellow marbles, and 3 green marbles. If a yellow marble is drawn on the first draw and not returned to the bag, what is the probability that the second marble drawn will be yellow?

**A** $\frac{1}{6}$

**B** $\frac{1}{3}$

**C** $\frac{4}{11}$

**D** $\frac{5}{12}$

**3** What is the theoretical probability that a card drawn from a standard deck of 52 cards will be of the clubs suit?

**F** $\frac{1}{52}$

**G** $\frac{1}{13}$

**H** $\frac{1}{4}$

**J** $\frac{1}{2}$

# Glossary/Glosario

**Cómo usar el glosario en español:**

1. Bussca el término en inglés que desees encontrar.

2. El término en español, junto con la definición, se encuentran en la columna de la derecha.

## English

## Español

**A**

**absolute value** (p. 513) The absolute value of a number is its distance from 0 on the number line, and is indicated by drawing a bar on each side of the number. For example, $|-20|$ means "the absolute value of $-20$." Since $-20$ and $20$ are each 20 units from 0 on the number line, $|20| = 20$ and $|-20| = 20$.

**valor absoluto** (pág. 5) El valor absoluto de un número es su distancia desde 0 en la recta numérica, lo cual se indica trazando una barra en cada lado del número. Por ejemplo, $|-20|$ significa "el valor absoluto de $-20$". Como $-20$ y $20$ se encuentran a 20 unidades de 0 en la recta numérica, $|20| = 20$ y $|-20| = 20$.

**acute angle** (p. 27) An angle that measures less than $90°$. Each of the angles shown below is an acute angle.

**ángulo agudo** (pág. 27) Ángulo que mide menos de $90°$. Cada uno de los siguientes es un ángulo agudo.

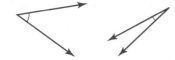

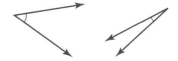

**additive identity** (p. 185) The sum of any number and 0 is the number. For any number $a$, $a + 0 = 0 + a = a$. Example: $6 + 0 = 0 + 6 = 6$

**identidad aditiva** (pág. 185) La suma de cualquier número más 0 es igual al número. Para cualquier número $a$, $a + 0 = 0 + a = a$. Ejemplo: $6 + 0 = 0 + 6 = 6$

**additive inverse** (p. 186) A number that when added to a given number results in a sum of zero. Example: The *additive inverse* of 4 is $-4$ because $4 + (-4) = 0$.

**inverso aditivo** (pág. 186) Número que sumado a otro da como resultado cero. Ejemplo: El *inverso aditivo* de 4 es $-4$ porque $4 + (-4) = 0$.

**algebraic expression** (p. 145) A rule written with numbers and symbols. Examples: $n + n + n + 2$, $3n + 2$.

**expresión algebraica** (pág. 145) Regla escrita con números y símbolos. Ejemplos: $n + n + n + 2$, $3n + 2$.

**angle** (p. 9) Two rays with the same endpoint. For example, the figure below is an angle.

**ángulo** (pág. 9) Dos rayos que parten del mismo punto. Por ejemplo, la siguiente figura muestra un ángulo.

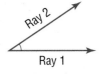

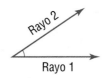

**area** (p. 398) The amount of space inside a two-dimensional shape.

**associative property** (p. 178) The way in which three numbers are grouped when they are added or multiplied does not change their sum or product. For any numbers $a$, $b$, and $c$, $(a + b) + c = a + (b + c)$, and $(ab)c = a(bc)$. Example: $(2 + 3) + 4 = 2 + (3 + 4)$ or $(2 \cdot 3) \cdot 5 = 2 \cdot (3 \cdot 5)$.

**axes** (p. 469) The horizontal line and vertical line that are used to represent the variable quantities on a graph. For example, in the graph below, the horizontal axis represents width and the vertical axis represents length.

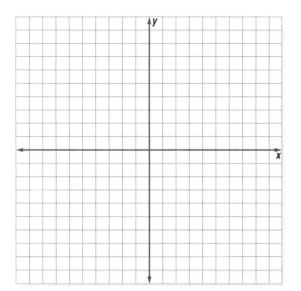

**area** (pág. 398) La cantidad de espacio dentro de una figura bidimensional.

**propiedad asociativa** (pág. 178) La forma en que se agrupen tres números cuando se suman o multiplican no altera el resultado. Sean cuales fueren los números $a$, $b$, y $c$, $(a + b) + c = a + (b + c)$, y $(ab)c = a(bc)$. Ejemplo: $(2 + 3) + 4 = 2 + (3 + 4)$ ó $(2 \cdot 3) \cdot 5 = 2 \cdot (3 \cdot 5)$.

**ejes** (pág. 469) La recta horizontal y la recta vertical que se usan para representar las cantidades variables en una gráfica. Por ejemplo, en la siguiente gráfica el eje horizontal representa longitud.

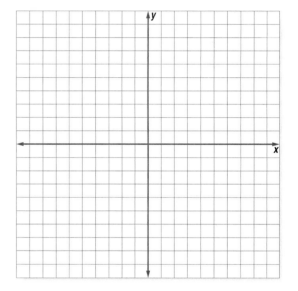

**B**

**backtracking** (p. 548) The process of using a flowchart to work backward, starting with the output and undoing each operation to find the input.

**base of a parallelogram** (p. 410) Any of the sides of a parallelogram. The base of a parallelogram is used in computing its area.

**base of a prism** (p. 437) The parallel faces of a prism.

**base of a triangle** (p. 413) Any of the sides of a triangle. The base of a triangle is used in computing its area.

**vuelta atrás** (pág. 548) El proceso de usar un flujograma para trabajar en sentido inverso, comenzando con la salida y anulando cada operación hasta llegar a la entrada.

**base de un paralelogramo** (pág. 410) Cualquiera de los lados de un triángulo. La base de un paralelogramo se usa para calcular su área.

**base de un prisma** (pág. 437) Las caras paralelas de un prisma.

**base de un triángulo** (pág. 413) Cualquiera de los lados de un triángulo. La base de un triángulo se usa para calcular su área.

C

**chord** (p. 44) A segment connecting two points on a circle. (See the figure in the glossary entry for radius. This figure shows a chord of a circle.)

**cuerda** (pág. 44) Segmento que conecta dos puntos en un círculo. (Ver la figura en el inciso del glosario para radio. Dicha figura muestra una cuerda de un círculo.)

**circumference** (p. 44) The perimeter of a circle (distance around a circle).

**circunferencia** (pág. 44) El perímetro de un círculo (distancia alrededor del círculo).

**commutative property** (p. 176) The order in which two numbers are added or multiplied does not change their sum or product. For any numbers $a$ and $b$, $a + b = b + a$ and $ab = ba$. Example: $2 + 3 = 3 + 2$ or $2 \cdot 3 = 3 \cdot 2$

**propiedad conmutativa** (pág. 176) El orden en que se suman o multiplican dos números no altera el resultado. Sean cuales fueren los números $a$ y $b$, $a + b = b + a$ y $ab = ba$. Ejemplo: $2 + 3 = 3 + 2$ ó $2 \cdot 3 = 3 \cdot 2$

**concave polygon** (p. 12) A polygon that looks like it is "collapsed" or has a "dent" on one or more sides. Any polygon with an angle measuring more than 180° is concave. The figures below are concave polygons.

**polígono cóncavo** (pág. 12) Polígono que parece que se hubiera "hundido" o que tiene una hendidura en uno más de sus lados. Cualquier polígono con un ángulo mayor que 180° es cóncavo. Las siguientes figuras son polígonos cóncavos.

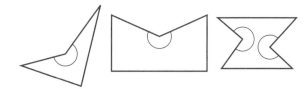

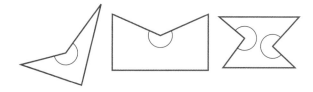

**congruent** (p. 321) Having the same size and the same shape.

**congruente** (pág. 321) Que tiene el mismo tamaño y la misma forma.

**coordinates** (p. 491) Numbers that represent the location of a point on a graph. For example, if a point is 3 units to the right and 7 units up from the origin, its coordinates are 3 and 7.

**coordenadas** (pág. 491) Números que representan la posición de un punto en una gráfica. Por ejemplo, si un punto se encuentra a 3 unidades a la derecha y 7 unidades hacia arriba del origen, sus coordenadas son 3 y 7.

**corresponding angles** (p. 328) Angles of two similar figures that are located in the same place in each figure. For example, in the figure, angle $B$ and angle $E$ are corresponding angles.

**ángulos correspondientes** (pág. 328) Ángulos de dos figuras semejantes ubicadas en el mismo lugar en cada figura. Por ejemplo: En la figura, el ángulo $B$ y el ángulo $E$ son ángulos correspondientes.

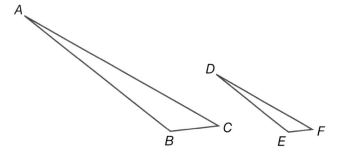

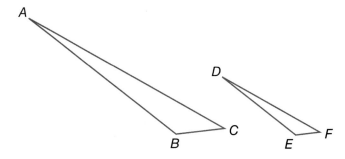

**corresponding sides** (p. 328) Sides of two similar figures that are located in the same place in each figure. For example, in the figure in the glossary entry for corresponding angles, sides *AB* and *DE* are corresponding sides.

**counterexample** (p. 324) In testing a conjecture, an example for which the conjecture is not true.

**lados correspondientes** (pág. 328) Lados de dos figuras semejantes ubicados en el mismo lugar en cada figura. Por ejemplo, en la figura de la definición de ángulos correspondientes del glosario, los lados *AB* y *DE* son lados correspondientes.

**contraejemplo** (pág. 324) Al probar una conjetura, un ejemplo para el cual la conjetura no es verdadera.

**D**

**diameter** (p. 44) A chord that passes through the center of a circle. Diameter also refers to the distance across a circle through its center. (See the figure in the glossary entry for radius. This figure shows a diameter of a circle.)

**distribution** (p. 586) The distribution of a data set shows how the data are spread out, where there are gaps, where there are lots of values, and where there are only a few values.

**distributive property** (p. 181) To multiply a sum by a number, multiply each addend of the sum by the number outside the parentheses. For any numbers $a$, $b$, and $c$, $a(b + c) = ab + ac$ and $a(b - c) = ab - ac$. Example: $2(5 + 3) = (2 \cdot 5) + (2 \cdot 3)$ and $2(5 - 3) = (2 \cdot 5) - (2 \cdot 3)$.

**diámetro** (pág. 44) Cuerda que pasa por el centro de un círculo. El diámetro también se refiere a la distancia a través de un círculo, pasando por su centro. (Ver las figuras en el inciso del glosario para radio. Dicha figura muestra un diámetro de un círculo.)

**distribución** (pág. 586) La distribución de un conjunto de datos muestra la extensión de los datos, las brechas entre los datos, los lugares donde hay muchos valores y donde hay pocos valores.

**propiedad distributiva** (pág. 181) Para multiplicar una suma por un número, multiplíquese cada sumando de la suma por el número que está fuera del paréntesis. Sean cuales fueren los números $a$, $b$, y $c$, $a(b + c) = ab + ac$ y $a(b - c) = ab - ac$. Ejemplo: $2(5 + 3) = (2 \cdot 5) + (2 \cdot 3)$ y $2(5 - 3) = (2 \cdot 5) - (2 \cdot 3)$.

**E**

**equally likely** (p. 621) Outcomes of a situation or experiment that have the same probability of occurring. For example, if one coin is tossed, coming up heads and coming up tails are equally likely outcomes.

**equation** (p. 534) A mathematical sentence stating that two quantities have the same value. An equal sign, =, is used to separate the two quantities. For example, $5 + 8 - 3 = 10$ is an equation.

**equivalent fractions** (p. 62) Fractions that describe the same portion of a whole, or name the same number. For example, $\frac{3}{4}$, $\frac{9}{12}$, and $\frac{30}{40}$ are equivalent fractions.

**equiprobables** (pág. 621) Resultados de una situación o experimento que tienen la misma posibilidad de ocurrir. Por ejemplo, si se lanza una moneda, es equiprobable que la moneda caiga mostrando cara o escudo.

**ecuación** (pág. 534) Enunciado matemático que establece que dos cantidades tienen el mismo valor. Se usa un signo de igualdad, =, para comparar las dos cantidades. Por ejemplo, $5 + 8 - 3 = 10$ es una ecuación.

**fracciones equivalents** (pág. 62) Fracciones que describen la misma parte de un todo o que representan el mismo número. Por ejemplo, $\frac{3}{4}$, $\frac{9}{12}$, y $\frac{30}{40}$ son fracciones equivalentes.

# Glossary/Glosario

**equivalent ratios** (p. 327) Two different ratios that represent the same relationship. For example, 1:3 and 4:12 are equivalent ratios.

**razones equivalents** (pág. 327) Dos razones diferentes que representan la misma relación. Por ejemplo: 1:3 y 4:12 son razones equivalentes.

**experimental probability** (p. 619) A probability based on experimental data. An experimental probability is always an estimate and can vary depending on the particular set of data that is used.

**probabilidad experimental** (pág. 619) Probabilidad que se basa en datos experimentales. Una probabilidad experimental es siempre una estimación y puede variar según el conjunto de datos en particular que se usen.

**exponent** (p. 114) A small, raised number that tells how many times a factor is multiplied. For example, in $10^3$, the exponent 3 tells you to multiply 3 factors of 10: $10 \cdot 10 \cdot 10 = 1,000$.

**exponente** (pág. 114) Número pequeño y elevado que indica cuántas veces se multiplica un factor. Por ejemplo, en $10^3$, el exponente 3 te indica que multipliques 3 factores de 10: $10 \cdot 10 \cdot 10 = 1,000$

**F**

**factor** (p. 114) A factor of a whole number is another whole number that divides into it without a remainder. For example, 1, 2, 3, 4, 6, 8, 12, and 24 are factors of 24.

**factor** (pág. 114) Un factor de un número entero es otro número entero que lo divide sin que quede un residuo. Por ejemplo, 1, 2, 3, 4, 6, 8, 12 y 24 son factores de 24.

**flowchart** (p. 547) A diagram, using ovals and arrows, that shows the steps for going from an input to an output. For example, the diagram below is a flowchart.

**flujograma** (pág. 547) Diagrama que usa óvalos y flechas para mostrar los pasos a seguir desde un dato de entrada hasta uno de salida. Por ejemplo, el siguiente diagrama es un flujograma.

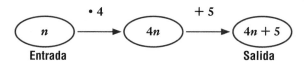

**formula** (p. 42) An algebraic "recipe" that shows how to calculate a particular quantity. For example, $F = \frac{9}{5}C + 32$ is the formula for converting Celsius temperatures to Fahrenheit temperatures.

**formula** (pág. 42) Una "receta" algebraica que muestra cómo calcular una cantidad dada. Por ejemplo: $F = \frac{9}{5}C + 32$ es la fórmula para convertir temperaturas Celsius en temperaturas Fahrenheit.

**fundamental counting principle** (p. 645) If an event $M$ can occur in $m$ ways and is followed by an event $N$ that can occur in $n$ ways, then event $M$ followed by event $N$ can occur in $m \times n$ ways.

**principio fundamental de conteo** (pág. 645) Si un suceso $M$ puede darse en $m$ formas y es seguido por un suceso $N$ que puede darse en $n$ formas, entonces el suceso $M$ seguido por el suceso $N$ puede darse en $m \times n$ formas.

**G**

**guess-check-and-improve** **(p. 560)** A method for solving an equation that involves first guessing the solution, then checking the guess by substituting into the original equation, and then using the result to improve the guess until the correct solution is found.

**conjetura, verifica y mejora** **(pág. 560)** Método para resolver una ecuación que implica primero hacer una conjetura, verificar la conjetura y sustituirla en la ecuación original y luego usar el resultado para mejorar la conjetura hasta hallar la solución correcta.

**H**

**height of a parallelogram** **(p. 410)** The distance from the side opposite the base of a parallelogram to the base. The height of a parallelogram is always measured along a segment perpendicular to the base (or to the line containing the base). The figures below show a base and the corresponding height for three parallelograms.

**altura de un paralelogramo** **(pág. 410)** La distancia desde el lado opuesto a la base de un paralelogramo, hasta la base. La altura de un paralelogramo se mide siempre a lo largo de un segmento perpendicular a la base (o a la recta que contiene la base). Las siguientes figuras muestran una base y la altura correspondiente de tres paralelogramos.

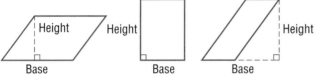

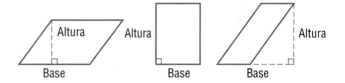

**height of a triangle** **(p. 413)** The distance from the base of a triangle to the vertex opposite the base. The height of a triangle is always measured along a segment perpendicular to the base (or the line containing the base). The figures below show a base and the corresponding height for three triangles.

**altura de un triángulo** **(pág. 413)** La distancia desde la base de un triángulo hasta el vértice opuesto a la base. La altura de un triángulo se mide siempre a lo largo de un segmento perpendicular a la base (o de la recta que contiene la base). Las siguientes figuras muestran una base y la altura correspondiente de tres triángulos.

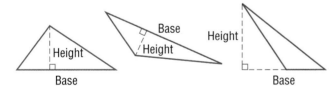

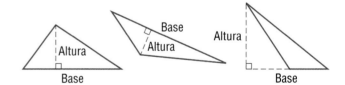

**histogram** **(p. 585)** A bar graph in which data are divided into equal intervals, with a bar for each interval. The height of each bar shows the number of data values in that interval.

**histograma** **(pág. 585)** Gráfica de barras en la cual los datos se dividen en intervalos iguales, con una barra para cada intervalo. La altura de cada barra muestra el número de valores de los datos en ese intervalo.

# Glossary/Glosario

**I**

**improper fraction** (p. 60) A fraction that has a numerator that is greater than or equal to the denominator. Examples: $\frac{21}{4}, \frac{2}{1}$.

**fraccion impropia** (pág. 60) Fraccion que tiene un numerador que es mayor o igual al denominador. Ejemplos: $\frac{21}{4}, \frac{2}{1}$.

**identity element** (p. 176). An addend of zero or a factor of one in the real number system. When a number is added to zero or multiplied by one, the result is the original number.

**elemento identidad** (pág. 176) Sumando de cero o un factor de uno en un sistema de números reales. Cuando a cero se le suma un número o cuando cero se multiplica por uno, el resultado es el número original.

**identity property of multiplication** (p. 185) The product of a factor and one is the factor. Example: $5 \cdot 1 = 5$

**propiedad de identidad de la multiplicación** (pág. 185) El producto de un factor y uno es igual al factor. Ejemplo: $5 \cdot 1 = 5$

**inequality** (p. 535) A mathematical sentence stating that two quantities have different values. For example, $5 + 9 > 12$ is an inequality. The symbols $<, >,$ and $\neq$ are used in writing inequalities.

**desigualdad** (pág. 535) Enunciado matemático que establece que dos cantidades tienen distintos valores. Por ejemplo, $5 + 9 > 12$ es una desigualdad. Los símbolos $<, >$ y $\neq$ se usan para escribir desigualdades.

**input** (p. 113) The value that is substituted into an expression. In a flowchart, it is represented by the oval on the left side.

**entrada** (pág. 113) Valor que se reemplaza en una expresión. En un flujograma, se representa por el óvalo en el lado izquierdo.

**intersecting lines** (p. 30) Lines that are coplanar and have exactly one point in common.

**rectas secantes** (pág. 30) Rectas que son coplanares y tienen exactamente un punto en común.

**inverse element** (p. 176) The opposite of a number for addition and the reciprocal of the number for multiplication. For any nonzero number $n$, $-n$ is the additive inverse and $\frac{1}{n}$ is the multiplicative inverse.

**elemento inverso** (pág. 176) El opuesto de un número para la adición y el recíproco del número para la multiplicación. Para cualquier número no nulo $n$, $-n$ es el inverso aditivo y $\frac{1}{n}$ es el inverso multiplicativo.

**L**

**line graph** (p. 479) A graph in which points are connected with line segments.

**gráfica lineal** (pág. 479) Gráfica en la cual los puntos se conectan con segmentos de recta.

**line plot** (p. 266) A number line with X's indicating the number of times each data value occurs.

**esquema lineal** (pág. 266) Recta numérica que contiene equis que indican el número de veces que ocurre cada valor de los datos.

**line symmetry** (p. 12) A polygon has line symmetry (or reflection symmetry) if you can fold it in half along a line so that the two halves match exactly. The polygons below have line symmetry. The lines of symmetry are shown as dashed lines.

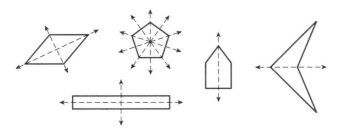

**lowest terms** (p. 62) A fraction is in lowest terms if its numerator and denominator are relatively prime. For example, $\frac{5}{6}$ is in lowest terms because the only common factor of 5 and 6 is 1.

**simetría lineal** (pág. 12) Un polígono tiene simetría lineal (o simetría de reflexión) si se puede doblar por la mitad a lo largo de una línea de modo que las dos mitades coincidan exactamente. Los siguientes polígonos tienen simetría lineal. Los ejes de simetría se muestran como líneas punteadas.

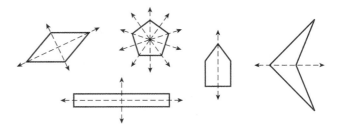

**en términos reducidos o reducida** (pág. 62) Una fracción está en términos reducidos o reducida si su numerador y denominador son primos relativos. Por ejemplo, $\frac{5}{6}$ está reducida dado que el único factor común de 5 y 6 es 1.

**M**

**mean** (p. 270) The number you get by distributing the total of the values in a data set among the members of the data set. You can compute the mean by adding the values and dividing the total by the number of values. For example, for the data 5, 6, 6, 8, 8, 8, 9, 10, 12, the total of the values is 72 and there are 9 values, so the mean is $72 \div 9 = 8$.

**median** (p. 266) The middle value when all the values in a data set are ordered from least to greatest. For example, for the data set 4.5, 6, 7, 7, 8.5, 10.5, 12, 12, 14.5, the median is 8.5.

**mixed number** (p. 60) A whole number and a fraction. For example, $12\frac{3}{4}$ is a mixed number.

**mode** (p. 266) The value in a data set that occurs most often. For example, for the data set 4.5, 6, 7, 7, 7, 8.5, 10.5, 12, 12, the mode is 7.

**media** (pág. 270) El número que se obtiene al distribuir el total de los valores en un conjunto de datos entre los miembros del conjunto de datos. Se puede calcular la media sumando los valores y luego dividiendo el total entre el número de valores. Por ejemplo, para los datos 5, 6, 6, 8, 8, 8, 9, 10, 12, el total de los valores es 72 y hay 9 valores, de modo que la media es $72 \div 9 = 8$.

**mediana** (pág. 266) El valor central cuando todos los valores en un conjunto de datos se ordenan de menor a mayor. Por ejemplo, para el conjunto de datos 4.5, 6, 7, 7, 8.5, 10.5, 12, 12, 14.5, la mediana es 8.5.

**número mixto** (pág. 60) Un número entero y una fracción. Por ejemplo, $12\frac{3}{4}$ es un número mixto.

**moda** (pág. 266) El valor en un conjunto de datos que ocurre con más frecuencia. Por ejemplo, para el conjunto de datos 4.5, 6, 7, 7, 7, 8.5, 10.5, 12, 12, la moda es 7.

# Glossary/Glosario

N

**negative number** (p. 509) A number that is less than 0. For example, −18 (read "negative eighteen") is a negative number.

**número negativo** (pág. 509) Un número menor que 0. Por ejemplo, −18 (que se lee "dieciocho negativo") es un número negativo.

O

**obtuse angle** (p. 27) An angle that measures more than 90° and less than 180°. Each of the angles shown below is an obtuse angle.

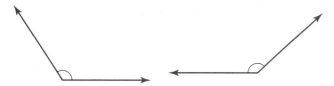

**ángulo obtuso** (pág. 27) Ángulo que mide más de 90° y menos de 180°. Cada uno de los siguientes ángulos es un ángulo obtuso.

**open sentence** (p. 536) An equation or inequality that can be true or false depending on the value of the variable. For example, $5 + n = 20$ is an open sentence.

**enunciado abierto** (pág. 536) Ecuación de desigualdad que puede ser verdadera o falsa dependiendo del valor de la variable. Por ejemplo, $5 + n = 20$ es un enunciado abierto.

**opposites** (p. 510) Two numbers that are the same distance from 0 on the number line, but on different sides of 0. For example, 35 and −35 are opposites.

**opuestos** (pág. 510) Dos números equidistantes de 0 en la recta numérica, pero en lados opuestos de 0. Por ejemplo, 35 y −35 son opuestos.

**order of operations** (p. 126) A convention for reading and evaluating expressions. The order of operations says that expressions should be evaluated in this order:
- Evaluate any expressions inside parentheses and above and below fraction bars.
- Evaluate all exponents, including squares.
- Do multiplications and divisions from left to right.
- Do additions and subtractions from left to right.

For example, to evaluate $5 + 3 \times 7$, you multiply first and then add: $5 + 3 \times 7 = 5 + 21 = 26$. To evaluate $10^2 - 6 \div 3$, you evaluate the exponent first, then divide, then subtract: $10^2 - 6 \div 3 = 100 - 2 = 98$.

**orden de las operaciones** (pág. 126) Una convención para leer y evaluar expresiones. El orden de las operaciones indica que las expresiones se deben evaluar en el siguiente orden:
- Evalúa cualquier expresión entre paréntesis y sobre y debajo de barras de fracciones.
- Evalúa todos los exponentes, incluyendo los cuadrados.
- Efectúa las multiplicaciones y las divisiones de izquierda a derecha.
- Efectúa las sumas y las restas de izquierda a derecha.

Por ejemplo, para evaluar $5 + 3 \times 7$, multiplica primero y luego suma. $5 + 3 \times 7 = 5 + 21 = 26$. Para evaluar $10^2 - 6 \div 3$, evalúa primero el exponente, luego divide y por último resta. $10^2 - 6 \div 3 = 100 - 2 = 98$.

**ordered pair** (p. 491) A pair of numbers that represent the coordinates of a point, with the horizontal coordinate of the point written first. For example, the point with horizontal coordinate 3 and vertical coordinate 7 is represented by the ordered pair (3, 7).

**par ordenado** (pág. 491) Un par de números que representa las coordenadas de un punto, en el cual la coordenada horizontal del punto se escribe primero. Por ejemplo, el punto con la coordenada horizontal 3 y coordenada vertical 7 se representa con el par ordenado (3, 7).

**origin** (p. 469) The point where the axes of a graph meet. The origin of a graph is usually the 0 point for each axis.

**origen** (pág. 469) El lugar donde se encuentran los ejes de una gráfica. El origen de una gráfica es por lo general el punto 0 de cada eje.

**outlier** (p. 274) A value that is much greater than or much less than most of the other values in a data set. For example, for the data set 6, 8.2, 9.5, 11.6, 14, 30, the value 30 is an outlier.

**valor atípico** (pág. 274) Un valor que es mucho mayor o mucho menor que la mayoría de los otros valores en un conjunto de datos. Por ejemplo, para el conjunto de datos 6, 8.2, 9.5, 11.6, 14, 30, el valor 30 es un valor atípico.

**output** (p. 113) The result of a mathematical action or series of actions. In a flowchart, it is represented by the oval on the right side.

**salida** (pág. 113) Resultado de una acción o serie de acciones matemáticas. En un flujograma, se representa por el óvalo en el lado derecho.

**P**

**parallelogram** (p. 410) A quadrilateral with opposite sides that are the same length. The opposite sides of a parallelogram are parallel. Each of the figures shown below is a parallelogram.

**paralelogramo** (pág. 410) Cuadrilátero cuyos lados opuestos tienen la misma longitud. Los lados opuestos de un paralelogramo son paralelos. Cada una de las siguientes figuras es un paralelogramo.

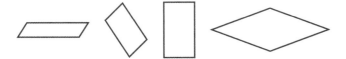

**percent** (p. 349) Percent means "out of 100." A percent represents a number as a part out of 100 and is written with a percent sign. For example, 39% means 39 out of 100, or $\frac{39}{100}$ or 0.39.

**por ciento, porcentaje** (pág. 349) Por ciento significa "de cada 100". Un por ciento representa un número como una parte de 100 y se escribe con un signo de porcentaje. Por ejemplo, 39% significa 39 de cada 100 ó $\frac{39}{100}$ ó 0.39.

**perfect square** (p. 404) A number that is equal to a whole number multiplied by itself. In other words, a perfect square is the result of squaring a whole number. For example, 1, 4, 9, 16, and 25 are perfect squares since these are the results of squaring 1, 2, 3, 4, and 5, in that order.

**cuadrado perfecto** (pág. 404) Número que es igual a un número entero multiplicado por sí mismo. Es decir, un cuadrado perfecto es el resultado de elevar al cuadrado un número entero. Por ejemplo, 1, 4, 9, 16 y 25 son cuadrados perfectos, dado que son el resultado de elevar al cuadrado 1, 2, 3, 4 y 5, en ese orden.

**perimeter** (p. 40) The distance around a two-dimensional shape.

**perímetro** (pág. 40) La distancia alrededor de una figura bidimensional.

# Glossary/Glosario

**perpendicular** (p. 27) Two lines or segments that form a right angle area are said to be perpendicular. For example, see the figures below.

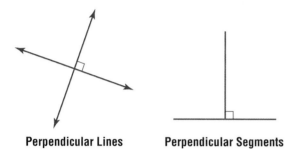

**Perpendicular Lines**      **Perpendicular Segments**

**perpendicular** (pág. 27) Se dice que dos rectas o segmentos que forman un ángulo recto son perpendiculares. Por ejemplo, observa las siguientes figuras.

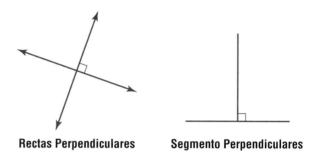

**Rectas Perpendiculares**      **Segmento Perpendiculares**

**polygon** (p. 4) A flat (two-dimensional) geometric figure that has these characteristics:
- It is made of straight line segments.
- Each segment touches exactly two other segments, one at each of its endpoints.

The shapes below are polygons.

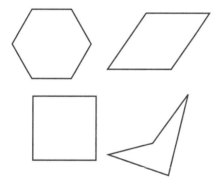

**polígono** (pág. 4) Figura geométrica plana (bidimensional) que posee las siguientes tres características:
- Está compuesta de segmentos de recta.
- Cada segmento interseca exactamente otros dos segmentos, uno en cada uno de sus extremos.

Las siguientes figuras son polígonos.

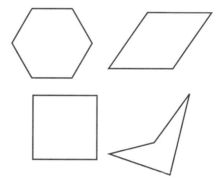

**positive number** (p. 509) A number that is greater than 0. For example, 28 is a positive number.

**número positivo** (pág. 509) Número mayor que 0. Por ejemplo, 28 es un número positivo.

**prism** (p. 435) A figure that has two identical, parallel faces that are polygons, and other faces that are parallelograms.

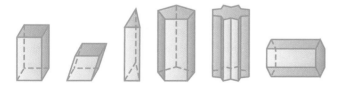

**prisma** (pág. 435) Figura con dos caras paralelas idénticas, las cuales son polígonos, y otras dos caras que son paralelogramos.

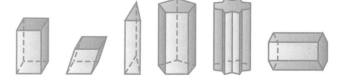

**probability** **(p. 618)** The chance that an event will happen, described as a number between 0 and 1. For example, the probability of tossing a coin and getting heads is $\frac{1}{2}$ or 50%. A probability of 0 or 0% means the event has no chance of happening, and a probability of 1 or 100% means the event is certain to happen.

**property** **(p. 174)** A statement that is true for any number or variables.

**protractor** **(p. 25)** An instrument used to measure angles.

**probabilidad** **(pág. 618)** La oportunidad de que un evento ocurra, descrita como un número entre 0 y 1. Por ejemplo, la probabilidad de lanzar una moneda y que ésta caiga mostrando cara es de $\frac{1}{2}$ ó 50%. Una probabilidad de 0 ó 0% significa que el evento no tiene oportunidad de ocurrir y una probabilidad de 1 ó 100% significa que el evento ocurrirá con seguridad.

**propiedad** **(pág. 174)** Enunciado verdadero para números o variables.

**transportador** **(pág. 25)** Instrumento que se usa para medir ángulos.

**Q**

**quadrants** **(p. 519)** One of the four sections created by the axes on the coordinate plane.

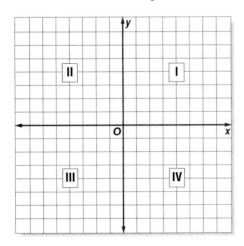

**cuadrante** **(pag. 519)** Una de las cuatro secciones creadas por los-ejes en el plano de coordenadas.

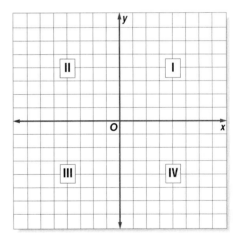

**R**

**radius (plural: radii)** **(p. 44)** A segment from the center of a circle to a point on a circle. Radius also refers to the distance from the center to a point on a circle. The figure below shows a chord, a diameter, and a radius of a circle.

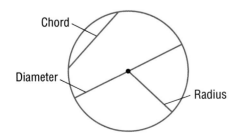

**radio** **(pág. 44)** Un segmento desde el centro del círculo hasta un punto del mismo. Radio también se refiere a la distancia desde el centro hasta un punto del círculo. La siguiente figura muestra una cuerda, un diámetro y un radio de un círculo.

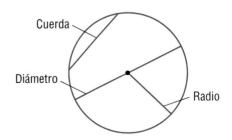

**range** (p. 266) The difference between the minimum and maximum values of a data set. For example, for the data set 4.5, 6, 7, 7, 7, 8.5, 10.5, 12, 12, the range is $12 - 4.5 = 7.5$.

**rango** (pág. 266) La diferencia entre los valores mínimo y máximo en un conjunto de datos. Por ejemplo, para el conjunto de datos 4.5, 6, 7, 7, 7, 8.5, 10.5, 12, 12, el rango es $12 - 4.5 = 7.5$.

**ratio** (p. 291) A way to compare two numbers. For example, when one segment is twice as long as another, the ratio of the length of the longer segment to the length of the shorter segment is 2 to 1, or 2 : 1.

**razón** (pág. 291) Una manera de comparar dos números. Por ejemplo: cuando un segmento es el doble de largo que otro segmento, la razón de la longitud del segmento más largo al segmento más corto es 2 a 1 ó 2 : 1.

**rational numbers** (p. 358) Numbers that can be written as ratios of two integers. In decimal form, *rational numbers* are terminating or repeating. For example, 5, $-0.274$, and $0.\overline{3}$ are rational numbers.

**números racionales** (pág. 358) Números que se pueden escribir como razones de dos enteros. En forma decimal, *los números racionales* son números terminales o periódicos. Por ejemplo: 5, $-0.274$ y $0.\overline{3}$ son números racionales.

**reciprocal** (p. 231) Two numbers are reciprocals if their product is 1. For example, the reciprocal of $\frac{5}{7}$ is $\frac{7}{5}$.

**recíproco** (pág. 231) Dos números son recíprocos si su producto es 1. Por ejemplo, el recíproco de $\frac{5}{7}$ es $\frac{7}{5}$.

**rectangular prism** (p. 435) A solid figure that has two pairs of parallel opposite faces and congruent bases that are all rectangles.

**prisma rectangular** (pág. 435) Figura sólida que tiene dos pares de caras opuestas paralelas y bases congruentes que son rectángulos.

**reference line** (p. 25) The line that goes through 0° on a protractor.

**línea de referencia** (pág. 25) Línea que pasa por 0° en un transportador.

**regular polygon** (p. 12) A polygon with sides that are all the same length and angles that are all the same size. The shapes below are regular polygons.

**polígono regular** (pág. 12) Polígono cuyos lados son todos de la misma longitud y cuyos ángulos tienen la misma medida. Las siguientes figuras son polígonos regulares.

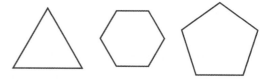

**repeating decimal** (p. 95) A decimal with a pattern of digits that repeat without stopping. For example, 0.232323 ... is a repeating decimal. Repeating decimals are usually written with a bar over the repeating digits, so 0.232323 ... can be written as $0.\overline{23}$.

**decimal periódico** (pág. 95) Decimal con un patrón de dígitos que se repiten indefinidamente. Por ejemplo, 0.232323 ... es un decimal periódico. Los decimales periódicos por lo general se escriben con una barra sobre los dígitos que se repiten, de este modo, 0.232323 ... se puede escribir como $0.\overline{23}$.

**right angle** (p. 27) An angle that measures exactly 90°. Right angles are often marked with a small square at the vertex. Each angle shown below is a right angle.

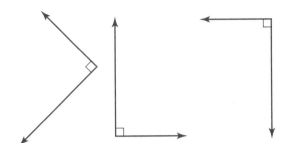

**ángulo recto** (pág. 27) Ángulo que mide exactamente 90°. Los ángulos rectos por lo regular se marcan con un cuadrado en el vértice. Cada ángulo siguiente es un ángulo recto.

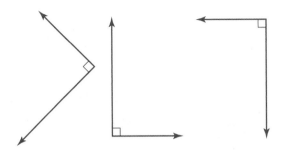

**S**

**sector** (p. 422) A region of a circle bounded by a central angle and its corresponding arc.

**sequence** (p. 122) An ordered list. For example, 2, 5, 8, 11, 14, 17 is a sequence.

**similar** (p. 321) Having the same shape but possibly different sizes.

**simulation** (p. 639) An experiment in which you use different items to represent the items in a real situation. For example, to simulate choosing markers and looking at their colors, you can write the color of each marker on a slip of paper and put all the slips into a bag. You can simulate choosing markers by drawing slips from the bag. Mathematically, the situations are identical.

**solution** (p. 536) A value of a variable that makes an equation true. For example, 4 is the solution of the equation $3n + 7 = 19$.

**surface area** (p. 434) The area of the exterior surface of an object, measured in square units.

**sector** (pág. 27) Región de un círculo limitada por un ángulo central y su arco correspondiente.

**sucesión** (pág. 122) Lista ordenada. Por ejemplo, 2, 5, 8, 11, 14, 17 es una sucesión.

**semejante** (pág. 321) Que tiene la misma forma, pero posiblemente tamaños diferentes.

**simulacro** (pág. 639) Un experimento que usa diferentes artículos para representar una situación real. Por ejemplo, en un simulacro para elegir marcadores y verificar sus colores, puedes escribir el color de cada marcador en tiras de papel y colocar las tiras en una bolsa. Saca tiras de la bolsa para llevar a cabo el experimento de escoger marcadores. Matemáticamente, las situaciones son idénticas.

**solución** (pág. 536) Un valor de una variable que hace verdadera una ecuación. Por ejemplo, 4 es la solución de la ecuación $3n + 7 = 19$.

**área de superficie** (pág. 434) El área de las superficies exteriores de un cuerpo, medida en unidades cuadradas.

# Glossary/Glosario

**T**

**terminating decimals** (p. 92) A decimal whose digits end. Example: 0.2, −1.345

**decimales finitos** (pág. 434) Decimal que tiene fin, es decir, cuyos dígitos no se repiten. Ejemplo: 0.2, −1.345

**term** (p. 122) An item in a sequence. For example, 8 is a term in the sequence 2, 5, 8, 11, 14, 17.

**término** (pág. 122) Un artículo en una sucesión. Por ejemplo, 8 es un término de la sucesión 2, 5, 8, 11, 14, 17.

**theoretical probability** (p. 621) Probability calculated by reasoning about the situation. Since theoretical probabilities do not depend on experiments, they are always the same for a particular event.

**probabilidad teórica** (pág. 621) Probabilidad que se calcula mediante el razonamiento de la situación. Dado que las probabilidades teóricas no dependen de experimentos, son siempre idénticas para un evento en particular.

**trapezoid** (p. 416) A quadrilateral with exactly one pair of parallel opposite sides.

**trapezoide** (pág. 416) Cuadrilátero con exactamente un par de lados paralelos opuestos.

**U**

**unit rate** (p. 300) Term used when one of two quantities being compared is given in terms of one unit. Example: 65 miles per hour or $1.99 per pound.

**tasa unitary** (pág. 300) Término que se usa cuando una de dos cantidades bajo comparación se da en términos de una unidad. Ejemplo: 65 millas por hora o $1.99 por libra.

**V**

**variables** (p. 144) A quantity that varies, or changes. For example, in a problem about the sizes of buildings, the height and width of the buildings would be variables.

**variable** (pág. 144) Cantidad que varía o cambia. Por ejemplo, en un problema sobre el tamaño de edificios, la altura y el ancho de los edificios serían variables.

**vertex (plural: vertices)** (p. 6) A corner of a polygon, where two sides meet. Vertices are usually labeled with capital letters, such as A, B, and C for the vertices of a triangle.

**vértice** (pág. 6) La esquina de un polígono, donde se encuentran dos de sus lados. Por lo general, los vértices se designan con letras mayúsculas, como por ejemplo, A, B y C para los vértices de un triángulo.

**vertical angles** (p. 30) Opposite angles formed by the intersection of two lines. In the figure, the *vertical angles* are ∠1 and ∠3, and ∠2 and ∠4.

**ángulos verticales** (pág. 30) Ángulos opuestos formados por la intersección de dos rectas. En la figura, los *ángulos verticales* son ∠1 y ∠3, y ∠2 y ∠4.

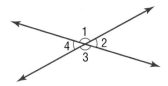

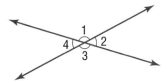

**volume** (p. 434) The space inside a three-dimensional object, measured in cubic units.

**volumen** (pág. 434) El espacio dentro de un cuerpo tridimensional, medido en unidades cúbicas.

# Index

# Index

**Coordinate plane,** 466

  absolute value, 512–514

  graphs, 468–472, 490

  integers, 510–512

  lines and curves, 476–481

  negative coordinates, 515–518

  parts of, 519–521

  points, 472–476, 491–494

  scale values, 495–499

**Coordinates,** 491

**Corresponding angles,** 328

**Corresponding sides,** 328

**Counterexample,** 324

**Counting principle,** fundamental, 644–647

**Curricular Connections**

  biology, 71, 612

  earth science, 51, 507, 525, 597 616, 635

  ecology, 316, 433

  economics, 85, 86, 118, 169, 236, 237, 257, 258, 259, 260, 262, 366, 376, 377, 378, 390, 392, 488, 504, 508, 545, 556–557, 558, 559, 596

  geography, 615

  geometry, 87, 104, 142, 172, 215, 264, 392, 505, 559, 568, 570, 572

  life science, 236, 391, 615, 651

  physical science, 262, 570

  physics, 557

  science, 86, 365, 485, 545

  social studies, 363, 364, 613, 614

**Data**

  analysis of, 602–606

  making predictions and, 607–608

**Data displays,** 578–591. *See also* Graphs

**Decagon,** 5

**Decimals**

  change to fractions, 89–93

divide, 251–254

  measuring with, 79–81

  multiply, 242–250

  understanding, 75–79

**Descartes, Rene,** 131

**Diameter,** 44

**Distribution,** 586

**Distributive Property,** 179–183

**Divide**

  decimals, 251–254

  fractions, 229–232

  multiply or, 254–256

  whole numbers, 226–229

**Divide symbol,** 251

**Double bar graph,** 603

**Equally likely,** 621

**Equation**

  backtracking, 546–559

  balancing, 543–544

  defined, 534

  guess-check-and-improve, 560–572

  and inequalities, 535–536

  with variables, 536–538

**Equivalent fractions,** 62

**Equivalent ratios,** 327

**Equivalent rules,** 152–156

**Experimental probabilities,** 619

**Exponents,** 113–116

**Factor,** 114

**Family Letter,** 3, 57, 109, 197, 289, 347, 397, 467, 533, 577

**Fibonacci sequence,** 108

**Flowchart,** 547

**Formulas**

  area of circle, 420, 568

  area of circle sector, 422

  area of parallelogram, 412

  area of rectangle, 384, 401

  area of trapezoid, 417

  area of triangle, 414

  circumference of circle, 46

  defined, 42

  perimeter of rectangle, 42

  volume of rectangular prism, 437

**Fractions**

  adding, 197, 205–207

  change to decimals, 94–97

  common denominators and, 201–204

  compare, 64–65

  divide by, 229–232

  divide by whole numbers, 226–229

  equivalent, 62

  estimate with, 66–67

  fraction models, 198–201

  model fraction multiplication, 219–221

  multiply with, 222–225

  patterns in, 58–63

  patterns in, decimals and, 97–99

  understand, 59–60

  and whole numbers, multiply, 216–219

**Fundamental counting principle,** 644–647

**Games**

  decimals and fractions, 92–93

  Equation Challenge, 534

  Find the Difference, 655

  Fraction Match, 209

  Guess My Number, 82, 360

  multiplying and dividing decimals, 263

  Percent Ball, 388–389

  polygon classification, 14–15

  Rice Drop, 628–632

  Rolling Fractions, 212

  Spare Change, 74

# Index

# Index

# Photo Credits

While every effort has been made to secure permission for reproduction of copyright material, there may have been cases where we are unable to trace the copyright holder. Upon written notification, the publisher will gladly correct this in future printings.

**viii** Adrian Sherratt/Alamy; **ix** RubberBall/Alamy; **x** Thomas Fricke/Corbis; **xi** Klaus Tiedge; **xii** Hughes Martin/Corbis; **xiii** George Doyle & Ciaran Griffin; **xiv** Jose-Manuel Colomo/Alamy; **xv** Digital Vision/PunchStock; **xvi** Corbis; **xvii** Jack Hollingsworth; **2** Adrian Sherratt/Alamy; **3** Paul Bricknell/DRK/Getty Images; **5** Digital Vision/Alamy; **7** Veer Jim Barber/Photonica/Getty Images; **8** (tl)Photo and Co/Getty Images, (tr)Nikreates/Alamy, (cl)artpartner-images.com/Alamy, (cr)Robert Harding Picture Library Ltd/Alamy, (bl)Spencer Jones/PictureArts/Corbis, (br)Jupiter Images/Brand X/Alamy; **9** Robert Michael/Corbis; **11** Image Source Pink/Alamy; **13** Peter Arnold, Inc./Alamy; **17** Mazer Creative Services; **19** Photographer's Choice RF/Getty Images; **22** Image100/Corbis; **25** Photodisc/Getty Images; **27** ArkReligion.com/Alamy; **29** Steven May/Alamy; **33** Tatsuhiko Sawada/Getty Images; **41** Dag Sundberg/Getty Images; **42** Alex Cao/Digital Vision/Getty Images; **47** JSC Digital Image Collection; **49** Jurgen Vogt/Stone/Getty Images; **56** Image Source/Corbis; **57** Photodisc/Alamy; **60** Corbis Premium RF/Alamy; **61** Foodfolio/Alamy; **63** C Squared Studios/Getty Images; **67** Stockbyte/Alamy; **68** Ingram Publishing/Alamy; **69** D. Hurst/Alamy; **71** Francesco Ruggeri; **72** Asia Images Group/Alamy; **75** D. Hurst/Alamy; **76** ClassicStock/Alamy; **80** Image Source Pink/Alamy; **81** (l to r, t to b 8)Ingram Publishing/Alamy, (2)Mode Images Limited/Alamy, (3)Siede Preis/Getty Images, (4)Alex Cao, (5)Getty Images, (6)Photostock-Dieter Heinemann/Alamy, (7)D. Hurst/Alamy, (9)Tony Latham/Corbis, (10)Photodisc/Getty Images; **83** Mazer Creative Services; **84** Royalty-Free/Corbis; **85** RubberBall/Alamy; **86** (t)courtesy of D. Carr & H. Craighead/Cornell University, (b)John Foxx; **93** C Squared Studios/Getty Images; **98** Harvey Lloyd; **99** EPA/Corbis; **101** Glen Allison/Getty Images; **108** Thomas Fricke/Corbis; **109** ImageState/Alamy; **112** Radlund & Associates/Getty Images; **118** Royalty-Free/Corbis; **119** Brand X; **123** The Garden Picture Library/Alamy; **125,128** Siede Preis/Getty Images; **131** Pictorial Press Ltd/Alamy; **135** The McGraw-Hill Companies, Inc./Jacques Cornell photographer; **136** North Wind Picture Archives/Alamy; **138** IT Stock Free/Alamy; **154** Goodshoot/Corbis; **155** Comstock/PunchStock; **157** Mazer Creative Services; **160** Laurence Mouton; **164** Digital Archive Japan/Alamy; **165** Design Pics Inc./Alamy; **170** Richard Levine/Alamy; **172** Powered by Light/Alan Spencer/Alamy; **173** Brian North/Alamy; **174** Interfoto Pressebildagentur/Alamy; **176** David Young-Wolff; **183** Robert Harding Picture Library Ltd/Alamy; **184** Helene Rogers/Alamy; **186** Ingram Publishing/Alamy; **187** Digital Vision/Getty Images; **188** Jupiter Images/Brand X/Alamy; **198** Mira/Alamy; **204** D. Hurst/Alamy; **207** Studiohio; **208** Dorling Kindersley; **211** Mazer Creative Services/Texas Instruments; **214** Chuck Choi/Arcaid/Corbis; **216** Ryan McVay; **218** The Garden Picture Library/Alamy; **219** Michael & Patricia Fogden; **220** WR Publishing/Alamy; **223** DK Limited/Corbis; **225** Klaus Tiedge; **227** Luna; **228** D. Hurst/Alamy; **232** Wilfried Krecichwost; **235** graficart.net/Alamy; **238** Ingram Publishing (Superstock Limited)/Alamy; **239** Image Source Black/Alamy; **241** D. Hurst/Alamy; **245** GK Hart/Vikki Hart; **246** Mazer Creative Services/Texas Instruments; **248** Paul Springett/Alamy; **250** George Shelley/Corbis; **251** Hal Lott/Corbis; **252** Judith Collins/Alamy; **254** Robert Clare; **255** Rob Walls/Alamy; **257** ImageState/Alamy; **258** Transtock/Corbis; **260** Lawrence Manning/Corbis; **261** Norman Pogson/Alamy; **262** PhotoLink; **264** WidStock/Alamy; **265** Peter Adams/Corbis; **267** TongRo Image Stock/Alamy; **268** Stockbyte; **288** Hughes Martin/Corbis; **290** Tony Cordoza/Alamy; **298** David Buffington/Getty Images; **299** Judith Collins/Alamy; **301** Photodisc/Getty Images; **302** Robert Glusic/Getty Images; **305** Detail Photography/Alamy, (b)Brand X Pictures/PunchStock; **313** (t)ImageSource/Age Fotostock, (b)Corbis; **314** Image Source/Corbis; **315** Mazer Creative Services; **320** Corbis; **325** Comstock/Jupiter Images; **329** Corbis; **333** Nikreates/Alamy; **337** Phil Degginger/Alamy; **339** Travelshots.com/Alamy; **346** George Doyle & Ciaran Griffin; **347** David Cook/www.blueshiftstudios.co.uk/Alamy; **348** Ingram Publishing/Alamy; **351** Corbis Premium RF/Alamy; **353** Moodboard/Corbis; **354** Image Source Black; **355** Corbis; **360** Huw Jones/Alamy; **361** Mazer Creative Services; **363** (t)David Arky/Corbis, (b)Martin Moos; **365** Antonio M. Rosario; **366** GoGo Images Corporation/Alamy; **368** Don Farrall; **371** (t)Profimedia International S.R.O./Alamy, (b)Lew Robertson; **373** Comstock/Corbis; **375** Thomas Northcut; **377** Jeff Harris Photography; **378** Stockbyte/Getty Images; **379 through 381** Photodisc; **383** Archive Holdings Inc.; **385** Amy Neunsinger; **387** Foodfolio/Alamy; **389** Mazer Creative Services; **391** Peter Christopher/Masterfile; **395** (Siede Preis/Getty Images; **396** Jose-Manuel Colomo/Alamy; **397** Alex MacLean/JupiterImages; **398** magebroker/Alamy; **401** Old Paper Studios/Alamy; **403** Photodisc/Getty Images; **406** MM Productions/Corbis; **407** PhotoLink/Getty Images; **409** Thomas Northcut; **414** Image100/Corbis; **421** (tc)Ingram Publishing/Alamy, (tr) Burke/Triolo Productions/Jupiter Images, (bl)Jose Luis Pelaez Inc; **425** Stockdisc/Getty Images; **430** Comstock Images/Alamy; **432** Ron Niebrugge/Alamy; **440** Jeff Greenberg/Alamy; **444** Studiohio; **445** Mazer Creative Services; **455** Dave Lidwell/Alamy; **457** Ingram Publishing/AGE Fotostock; **460** Envision/Corbis; **466** Steve Smith; **469** Photodisc/Gettty Images; **470** Elvele Images/Alamy; **471** Grand Tour/Corbis; **473** Digital Vision/PunchStock; **475** David W. Hamilton/Alamy; **477** Stockdisc/PunchStock; **478** Photodisc/PunchStock; **479** Dana Menussi; **481** Mylife Photos/Alamy; **482** Robert Harding Picture Library Ltd/Alamy; **483** Tim Dolan Photography; **486** Sean Justice/Corbis; **489** Stockdisc/PunchStock; **492** Ilene MacDonald/Alamy; **494** Floresco Productions/Corbis; **495** D. Hurst/Alamy; **498** Brand X Pictures/Alamy; **499** Heide Benser/zefa/Corbis; **502** Eyewire (Photodisc)/PunchStock; **504** Photodisc Collection/Getty Images; **514** PhotoLink/Getty Images; **517** Stockbyte/PunchStock; **521** Corbis; **523** Mazer Creative Services; **538** Corbis; **539** Cyril Laubscher/Getty Images; **541** Photodisc/Getty Images; **543** C Squared Studios/Getty Images; **544** Ed Pritchard/Getty Images; **545** Dorling Kindersley; **546** Walter Meayers Edwards/NGS/Getty images; **546** Mazer Creative Services; **547** Lars Klove/Getty Images; **555** Mark Lewis/Getty Images; **559** Burke/Triolo Productions/Brand X/Corbis; **560** Nick Dolding/Getty Images; **562** AlaskaStock; **568** Stockdisc/PunchStock; **570** The Art Archive/Thyssen-Bornemisza Collection Madrid/Gianni Dagli Orti; **571** Blickwinkel/Alamy; **574** Marielle/Photocuisine/Corbis; **575** Wolfgang Kaehler/Alamy;

# Symbols

## Number and Operations

| | |
|---|---|
| $+$ | plus or positive |
| $-$ | minus or negative |
| $a \cdot b$ $a \times b$ $ab$ or $a(b)$ | $a$ times $b$ |
| $\div$ | divided by |
| $=$ | is equal to |
| $\neq$ | is not equal to |
| $>$ | is greater than |
| $<$ | is less than |
| $\geq$ | is greater than or equal to |
| $\leq$ | is less than or equal to |
| $\approx$ | is approximately equal to |
| $\%$ | percent |
| $a:b$ | the ratio of $a$ to $b$, or $\frac{a}{b}$ |
| $0.7\overline{5}$ | repeating decimal 0.75555… |

## Algebra and Functions

| | |
|---|---|
| $-a$ | opposite or additive inverse of $a$ |
| $a^n$ | $a$ to the $n$th power |
| $|x|$ | absolute value of $x$ |
| $\sqrt{x}$ | principal (positive) square root of $x$ |
| $f(n)$ | function, $f$ of $n$ |

## Geometry and Measurement

| | |
|---|---|
| $\cong$ | is congruent to |
| $\sim$ | is similar to |
| $\circ$ | degree(s) |
| $\overleftrightarrow{AB}$ | line $AB$ |
| $\overrightarrow{AB}$ | ray $AB$ |
| $\overline{AB}$ | line segment $AB$ |
| $AB$ | length of $\overline{AB}$ |
| $\llcorner$ | right angle |
| $\perp$ | is perpendicular to |
| $\|$ | is parallel to |
| $\angle A$ | angle $A$ |
| $m\angle A$ | measure of angle $A$ |
| $\triangle ABC$ | triangle $ABC$ |
| $(a, b)$ | ordered pair with $x$-coordinate $a$ and $y$-coordinate $b$ |
| $O$ | origin |
| $\pi$ | pi $\left(\text{approximately 3.14 or } \frac{22}{7}\right)$ |

## Probability and Statistics

| | |
|---|---|
| $P(A)$ | probability of event $A$ |

# Formulas

| Perimeter | square | $P = 4s$ |
|---|---|---|
| | rectangle | $P = 2\ell + 2w$ or $P = 2(\ell + w)$ |
| Circumference | circle | $C = 2\pi r$ or $C = \pi d$ |
| Area | square | $A = s^2$ |
| | rectangle | $A = \ell w$ |
| | parallelogram | $A = bh$ |
| | triangle | $A = \frac{1}{2}bh$ |
| | trapezoid | $A = \frac{1}{2}h(b_1 + b_2)$ |
| | circle | $A = \pi r^2$ |
| Surface Area | cube | $S = 6s^2$ |
| | rectangular prism | $S = 2\ell w + 2\ell h + 2wh$ |
| | cylinder | $S = 2\pi rh + 2\pi r^2$ |
| Volume | cube | $V = s^3$ |
| | rectangular prism | $V = \ell wh$ or $Bh$ |
| | cylinder | $V = \pi r^2 h$ or $Bh$ |
| | pyramid | $V = \frac{1}{3}Bh$ |

# Measurement Conversions

| Length | 1 kilometer (km) = 1,000 meters (m)<br>1 meter = 100 centimeters (cm)<br>1 centimeter = 10 millimeters (mm) | 1 foot (ft) = 12 inches (in.)<br>1 yard (yd) = 3 feet or 36 inches<br>1 mile (mi) = 1,760 yards or 5,280 feet |
|---|---|---|
| Volume and Capacity | 1 liter (L) = 1,000 milliliters (mL)<br>1 kiloliter (kL) = 1,000 liters | 1 cup (c) = 8 fluid ounces (fl oz)<br>1 pint (pt) = 2 cups<br>1 quart (qt) = 2 pints<br>1 gallon (gal) = 4 quarts |
| Weight and Mass | 1 kilogram (kg) = 1,000 grams (g)<br>1 gram = 1,000 milligrams (mg)<br>1 metric ton = 1,000 kilograms | 1 pound (lb) = 16 ounces (oz)<br>1 ton (T) = 2,000 pounds |
| Time | 1 minute (min) = 60 seconds (s)<br>1 hour (h) = 60 minutes<br>1 day (d) = 24 hours | 1 week (wk) = 7 days<br>1 year (yr) = 12 months (mo) or 52 weeks or 365 days<br>1 leap year = 366 days |
| Metric to Customary | 1 meter ≈ 39.37 inches<br>1 kilometer ≈ 0.62 mile<br>1 centimeter ≈ 0.39 inch | 1 kilogram ≈ 2.2 pounds<br>1 gram ≈ 0.035 ounce<br>1 liter ≈ 1.057 quarts |